Library of
Davidson College

SOILBORNE
PLANT PATHOGENS

SOILBORNE PLANT PATHOGENS

GEORGE W. BRUEHL

Professor Emeritus
Washington State University

MACMILLAN PUBLISHING COMPANY
NEW YORK
Collier Macmillan Publishers
LONDON

Copyright © 1987 by Macmillan Publishing Company
A division of Macmillan, Inc.
All rights reserved. No part of this book may be reproduced
or transmitted in any form or by any means, electronic or
mechanical, including photocopying, recording, or by any
information storage and retrieval system, without permission
in writing from the Publisher.

Macmillan Publishing Company
866 Third Avenue, New York, NY 10022

Collier Macmillan Canada, Inc.

Printed in the United States of America

Printing: 1 2 3 4 5 6 7 8 9 10   Year: 7 8 9 0 1 2 3 4 5 6

Library of Congress Cataloging-in-Publication Data

Bruehl, G. W. (George W.), 1919-
   Soilborne plant pathogens.

   Includes bibliographies and index.
   1. Soilborne plant diseases. 2. Soil microbiology.
3. Micro-organisms, Phytopathogenic. I. Title.
SB732.87.B78 1987   632'.3   85-19901
ISBN 0-02-949130-4

Contents

PREFACE	xi
Acknowledgments	xiv
General References	xv
1 SOIL	**1**
Zonal Soils	1
Topography	4
Soil Profile	5
Soil Texture and Structure	8
Soil pH	11
Comments	14
Addendum	16
2 WATER	**18**
Wetness	18
Classes of Water in Soil	22
Classes of Organisms	25
Interactions	30
Comments	32

3 TEMPERATURE 37

Fusarium Wilts	39
Verticillium Wilts	41
Cortical Rots	45
Temperature Affects Geographic Distribution	46
Planting Date	49
Sequence of Pathogens	49
Diurnal Fluctuations	50
Some Effects of Temperature upon Nematodes	51
Soil Solarization	52
Fire	53
Comments	54

4 SOIL AERATION 58

Quantity of Oxygen Required	58
Oxygen in Relation to Bean Root Rot	62
Influence of Oxygen on Soft Rot of Potato Tubers	63
Oxygen in Rice Fields	64
Tolerance of Fungi to Carbon Dioxide	65
Comments	68

5 HOST DEVELOPMENT AND STRESS 72

Effect of Host Development on Susceptibility	72
Effect of Environmental Stress	75
Complex Stress	81
Nematode Reproduction	82
Trapping Microconidia	84
Trapping Bacteria	85
Translocation Patterns	85
Comments	86

6 FUNGISTASIS 90

Concept of Soilborne	90
Concept of Fungistasis	92
Food as a Timing Factor	94
Induction of Fungistasis	94
Strength of Fungistasis	97
Release from Fungistasis	98
Comments	101

7 THE RHIZOSPHERE   105
Concept: Rhizosphere   105
Exudates   107
Toxic or Repellent Substances   113
Comments   113

8 SPERMOSPHERE   117
Seed Factors   118
Quantity of Exudates   120
Effect of Soil Oxygen and Temperature   121
Comments   122

9 VOLATILES   125
Toxic Volatiles   125
Inhibitory Volatiles   126
Stimulatory Volatiles   127
Effect of Carbon: Nitrogen Ratios of Residue on Volatiles   130
Complex Experiments with Volatiles   131
Comments   132

10 ANTIBIOTICS   135
Conditions Favoring Antibiotic Production   136
Antibiotics in Bulk Soil   138
Antibiotics in the Rhizoplane   139
Staling, Antibiotics, and Fungistasis   139
Tolerance in Plant Pathogenic Bacteria   140
Bacteriocins   140
Comments   141

11 SAPROPHYTIC SURVIVAL   144
Prior Colonization of Substrate   145
Survival Aided by Antibiotic Production   147
Effect of Nitrogen   152
Survival in Wood   154
Comments   156

12 SURVIVAL AS SPORES, HYPHAE, AND MICROSCLEROTIA   159
Germinability   160
Toughness and Longevity of Structures   162

Liberation of Propagules from Infested Debris	164
Rhizoctonia solani in Host Debris	164
Seeds	165
Comments	165

13 SCLEROTIA — 168

Size and Function	170
Formation	170
Effect of Environment on Longevity and Germination	176
Comments	178

14 SURVIVAL OF BACTERIA IN SOIL — 184

Examples	186
Exudate	191
Comments	192

15 COMPETITION — 196

General Soil Microflora	197
Radiate Evolution	200
Intraspecies Selection	201
Competitive Saprophytic Ability	201
Comments	205

16 MICROBIALIZATION — 209

Natural Microbialization	211
Some Early Experiments	212
Mechanisms	213
Bacteria versus Bacteria	215
Supression of Take-all	215
Broad Spectrum Microbialization	216
Comments	218

17 HYPERPARASITES — 221

Fungal Parasites of Fungi	221
Nematodes That Parasitize Fungi	226
Amoeba That Parasitize Fungi and Nematodes	227
Viruses in Fungi	230
Fungal Parasites of Nematodes	231

A Bacterial Parasite of Nematodes — 233
Comments — 234

18 DISEASE SUPPRESSION — **238**

Potato Scab Decline — 239
Take-all Decline — 240
Oat Cyst–Nematode Decline — 241
Rhizoctonia solani Decline — 243
Proper Use of Suppressive — 244
Comments — 245

19 MYCORRHIZAE — **248**

Ectotrophic Mycorrhizae — 248
Vesicular-Arbuscular Mycorrhizae (VAM) — 250
Comments — 252

20 SOILBORNE VIRUSES — **254**

Virus Free in the Soil — 255
Virus in Roots — 255
Is the Vector a Nematode or a Fungus? — 256
Nematode-Transmitted Viruses — 257
Fungal Vectors — 263
Comments — 270

21 HISTORICAL EPIDEMIOLOGY — **277**

Fusarium Wilt — 277
Black Shank of Tobacco — 285
Phytophthora Root and Stem Rot of Soybeans — 289
Comments — 293

22 ACTIVITIES OF MAN — **298**

Tillage — 299
Rotation — 300
Planting — 305
Irrigation — 308
Fertilizers — 311
Chemicals — 313
Comments — 316

23 NUMBERS — 321

Cautions — 322
Counting — 323
Numbers versus Disease — 325
Effect of Cropping on Populations — 330
Effect of Germination Stimulators — 330
Comments — 330

24 EVOLUTION AND TAXONOMY — 335

Antiquity of Plant Pathogens — 335
Polymorphism and Phenotypic Plasticity — 337
Colonists of Cultivated Soils — 338
Nematodes — 340
Radiate Evolution — 342
Variability versus Fitness — 343
Examples of Evolution — 344
Biologic Species, Cryptic Species — 347
Mating Populations Not Considered Biologic Species — 351
Splitting versus Lumping — 352
Comments — 353

INDEX — 359

Preface

In recent years many books treating soilborne pathogens and diseases have appeared. Why the recent spurt of activity? In early plant pathology the most obvious diseases were studied. Who could fail to see ergot, maize smuts, common bunt balls in grain (stinking smut), rotten peaches, rust, or powdery mildew? In the "primitive" state of plant pathology the most obvious diseases were studied first. As sophistication increases, less obvious diseases receive increased attention. Soil fumigants established the importance of many nematodes. New chemicals effective against only a certain group of organisms made establishment of etiology much easier. The early pathologists lacked many of the tools available today.

Why are many soilborne plant pathogens poorly understood? The atmosphere, in spite of pollutants, is less complex chemically, biologically, and physically than the soil. Entire chapters record this complexity (fungistasis, rhizosphere, spermosphere, volatiles, antibiotics, hyperparasites, competition). Complexities in aerial relationships exist, but to a lesser degree. S. D. Garrett devoted a lifetime to synthesizing, interpreting, and developing concepts; in spite of his genius, we still have far to go. But we are not lost in confusion, and we are not overwhelmed. New techniques, tools, and knowledge have accelerated progress. Teachers love generalities that permit presentation of much material in a short time, but it is important to appreciate the limitation of generalities. Each pathogen-host-environment interaction is a distinct relationship. We should restrain our instincts to categorize to excess.

Are diseases caused by soilborne pathogens more difficult to control than those caused by pathogens that are not soilborne? I am not sure they are. Those who take pride in plant pathology were humbled as they watched Dutch elm disease spread among American elms. Those who observe the students of the cereal rusts are not overwhelmed by their success. The pathogens of the above examples are not soilborne, evidence that we should not imbue soilborne plant pathogens with a mystique of great invincibility. The presence of long-lived, dormant structures that resist adversity and remain quiescent but capable of resuming activity rapidly when favorable conditions return is a major advantage of many soilborne plant pathogens. Relatively wide host ranges contribute to the maintenance of inoculum of many. On the other hand, soilborne plant pathogens lack the explosive spread that is characteristic of many aerial pathogens. It is a psychological error to approach soilborne plant pathogens with the idea that their study is particularly difficult.

How important are diseases caused by soilborne plant pathogens? What would the price of peas, watermelons, tomatoes, or bananas be without fusarium wilt resistance? How successful would mechanical harvesters in maize be without the steady, incremental improvements in corn stalks made over the past 60 or so years? Take-all of wheat, in spite of much study, is still a major problem in Australia. In recent years *Phytophthora cinnamomi* killed large numbers of eucalyptus trees in Australia. What would the effect of loss of seed treatments be upon the production of many vegetable and some particular field crops? The above examples are obvious examples of the importance of these diseases. Less obvious are the insidious, nonlethal but debilitating losses from root diseases, particularly those caused by several *Pythium* spp. Soilborne plant pathogens have been, are, and will continue to be a hazard in forests, orchards, gardens, fields, and lawns. I present no figures, but the facts are documented throughout the history of modern plant pathology.

Land and climate well adapted for a particular plant is a world resource we cannot afford to lose. Plants tend to be produced commercially where they are best adapted, and they are planted repeatedly in such sites. If a soilborne pathogen is present and is favored by that environment, the soil infestation increases rapidly. Consequently, without control, much of the land best suited to a crop is soon lost. This happened in bananas with Panama wilt and in tobacco with Granville wilt (Southern bacterial wilt). Plant pathology has triumphed in most cases.

What are some of the gaps in our knowledge or understanding? Where do we need greater effort? Total knowledge or understanding will never be achieved, but some subjects need greater emphasis. What does soil pH really do? We have clues as to how pH affects *Fusarium oxysporum, Ver-*

ticillium spp., *Phymatotrichum omnivorum*, common potato scab, cephalosporium stripe, club root, but we know far too little of what really happens. Plant pathologists and soil chemists should collaborate more. The application of modern soil physics to water relations has resulted in great advances in plant pathology since 1970, and we should start an era of accelerated progress by applying soil chemistry to plant pathology. In recent years biochemistry has risen to the stars, especially molecular biology. But, after reviewing the past, greater advances will probably come from macrophysiology, the study of the interrelationships of roots, stems, leaves, and fruits, especially in respect to reducing losses from cortical rots and root rots. Macrophysiology should also emphasize the effect of heat, water, and nutrient and oxygen stress upon disease. New developments in seed microbialization and in the use of hyperparasites are so exciting that they will reduce the study of general background soil microflora. And particularly dangerous are the great efforts to make monoculture successful. For crops capable of being grown in rotation, rotations should receive greater study.

Some fields of research have been harmed by over enthusiasm. When antibiotics first became prominent in human medicine, many thought they would revolutionize plant pathology. When bacteriophages were discovered, hopes were high that they would control many diseases of humans and animals. Some early students of nematology made exuberant claims to the extent that many pathologists not working with nematodes reacted negatively. Grand claims were made for seed "bacterization" before reliable or practical application was possible. Nematode-trapping fungi were promoted zealously before real evidence of their significance in nature was available. In time, however, all fields advance to the point where some useful applications of research in each area are achieved.

The progress of science is like that of an amoeba—advancing here, dragging there. The advancing parts lead and eventually become part of the body; the dragging parts are pulled into the main body; while the other areas advance and retreat in between; but the main body moves relentlessly forward. *Agrobacterium tumefaciens* K84, the lethal antagonist of the crown gall bacterium, is now part of the body of science. *Nematophthora gynophila*, the fungal parasite of *Heterodera avenae*, has been established as an important ecological factor. It must be fun to be in the advancing leading edge of science, but that part is no more important and may be much less important than the main body, or even those parts that drag at a particular point in time. Oxygen, water, food, temperature, heredity may be old parts of the body of knowledge, no longer "fun" or the leading edge, but they are important.

Once upon a time, people in an area spoke the same language. They worked diligently together to build a tower. Then a deity became angry with this project and gave each group a different language. Unable to communicate with each other, they became frustrated. They abandoned the tower and went their separate ways. It distresses me that each group of scientists develops its own jargon. Every effort should be made to maintain a common language and to keep jargon to a minimum. This would facilitate communication among scientists and reduce the number of terms students must learn.

This book attempts to bring together sufficient information to impart an understanding of soilborne plant pathogens; it requires an introductory knowledge of plant pathology. It has been impossible to be detailed and remain within the desired length, forcing considerable selectivity as to what is discussed and as to the examples used. Greater detail can be obtained from the literature cited, especially from "Key References" listed at the end of each chapter. No chapter is devoted to methods, but some general methods are included throughout.

Soilborne Plant Pathogens is organized in chapters devoted mainly to one subject (soil, water, temperature, aeration, etc.), but each subject is not an independent entity. Varying one factor changes others, or changes the effects of others, in never-ending interactions. Nevertheless, it is useful to attempt to treat one subject at a time as it focuses thought upon that subject and emphasizes its importance.

Probably the most controversial chapter in this book is the last one, Evolution and Taxonomy. It may be flawed by too strong conviction (prejudice?). Sexual reproduction by microorganisms is in need of thought as to its significance and role. Most of us were taught by mycologists imbued with the beauty of the whole organism, especially of the sexual state. A plea is also made for greater respect for and emphasis upon taxonomy, with a warning of the dangers of too little precision (lumping).

Read the works of S. D. Garrett for a more philosophical treatment. His works are a pleasure to read. D. C. Norton's book on the ecology of nematodes, R. S. Russell's book on plant roots, and the "Ecology of Soil-Borne Plant Pathogens," edited by K. F. Baker and W. C. Snyder cannot be surpassed. Many good books exist, but those listed at the end of this section have special value as general references.

When looking for the references in the chapters, look at both the Key References and References lists.

Acknowledgments

I am indebted to Joyce Mikelson for typing the early drafts, and to Linda Wilson and Jane Mildren for the preparation of the final drafts.

P. B. Adams, R. Baker, C. Beckman, L. Faulkner, S. D. Garrett, R. Grogan, D. Hagedorn, B. Daniels Hetrick, M. Kainz, T. Kommedahl, R. Linderman, S. Lyda, R. Papendick, M. Stanghellini, and D. Weller were kind enough to read individual chapters, and R. Smiley read the entire manuscript. I am grateful for their helpful suggestions. The cooperation of those who supplied photographs is also appreciated.

General References

Baker, K. F. and R. J. Cook. 1974. *Biological Control of Plant Pathogens.* W H. Freeman, San Francisco, Calif.

Baker, K. F. and W. C. Snyder. 1965. *Ecology of Soil-Borne Plant Pathogens, Prelude to Biological Control.* Univ. of California Press, Berkeley.

Cook, R. J. and K. F. Baker. 1983. *The Nature and Practice of Biological Control of Plant Pathogens.* Amer. Phytopathology Soc., St. Paul, MN.

Garrett, S. D. 1944. *Root Disease Fungi.* Chronica Botanica Co., Waltham, Mass.

Garrett, S. D. 1956. *Biology of Root-Infecting Fungi.* Cambridge Univ. Press, Cambridge, England.

Garrett, S. D. 1970. *Pathogenic Root-Infecting Fungi.* Cambridge Univ. Press, Cambridge, England.

Norton, D. C. 1978. *Ecology of Plant-Parasitic Nematodes.* Wiley, New York.

Russell, R. S. 1977. *Plant Root Systems: Their Function and Interaction with the Soil.* McGraw-Hill, London, England.

Schippers, B. and W. Gams. 1979. *Soil-Borne Plant Pathogens.* Academic Press, New York.

Chapter 1

Soil

Soilborne pathogens are by definition residents in soil, for either extended or brief periods of their existence. Underground plant organs are affected directly, and above-ground plant parts indirectly, by soil conditions. Certain aspects of soil science are so important to the understanding of soilborne pathogens and the diseases they cause that brief treatment of soil will be made. Plant pathologists should view soil from several perspectives. On a grand scale is the concept of zonal soils, followed by intrazonal and azonal soils, those influenced strongly by some local condition. Soil profile, texture, and bulk density are also important. In addition, pathologists are concerned about recent treatment of the soil, such as its history of liming, drainage, cropping, fertilization, or irrigation.

The soil classification terms used in this book have been replaced by a modern system of soil nomenclature. Unfortunately, the new system of soil classification is very technical, leading Cox and Atkins (1979) to lament the adoption of the new system because it is "almost unintelligible to the nonspecialist."

ZONAL SOILS

It is apparent while driving from eastern South Dakota across the Great Plains to Glacier Park, Montana, that the color of the cultivated fields changes gradually from dark to lighter as rainfall decreases, reflecting lower amounts of total organic matter. That climate and vegetation are factors in forming soil was not easy to learn. In western Europe early scientists observed the effects of geology on soil, of glaciers, lake beds, deposition of sand and silt, of sandstone or limestone parent materials. Some countries were small, some bisected by mountain ranges, and wide travel was not common. Thus soil was studied on a limited geographic scale.

Observation of soil on a grand scale occurred in the Union of Soviet Socialist Republics (USSR). The taigas, prairies, and steppes extended over great expanses of relatively uniform topography. Soil-forming processes continued over time with minimum erosion. The temperature, in general, increased from north to south and precipitation decreased from west to east. V. V. Dokuchaev (1846–1903) was employed to classify the relative productivity of soil for the purpose of taxation (Buol et al., 1973). By good fortune he was a geologist. Dokuchaev studied soil profiles—the successive layers (horizons) of soil—from the surface downward. He observed changes in depth of top soil (A horizons), organic matter content, and characteristics of the profiles. He became aware of nonconformity between underlying rock and major characteristics of the soil. He and his students concluded that climate and vegetation were strong forces in soil formation and that, conversely, the soil is a key to the environment. The ability to see the obvious and to interpret it is rare, and soil science was greatly advanced by these concepts.

Sibirtzev, a student of Dokuchaev, expanded these observations and introduced the terms zonal and azonal into the vocabulary of soil science in 1899. *Zonal soils* are mature soils formed in place prior to tillage with little erosion. As defined in the United States Department of Agriculture Yearbook (1938), they are any one of the great groups of soils having well-developed soil characteristics that reflect the influence of the active factors of soil genesis—climate and living organisms, chiefly vegetation. Among the examples of zonal soils are Podzol, Brown Podzolic, Red and Yellow Podzolic, Laterite, Prairie, Chernozem, Chestnut, Brown, Sierozem, and Desert.

Azonal soils can vary markedly within short distances. They may reflect local drainage problems or differences due to deposition by streams or other factors of a more local nature.

Carleton (1919) was so impressed with the concept of zonal soils that, in his book on small grain cereals, he illustrated the Argentine pampas, the North American plains, and prairies and steppes in the USSR with soils reflecting climatic gradients, north to south, east to west, and west to east, respectively (moist to drier) in each case, regardless of the continent. One can be fairly confident that the soil environment of the Chernozem (chernui - black, zemla - earth) of Russia has much in common with chernozem in North America, Argentina, or anywhere else. Knowledge that the soil reflects environmental conditions is useful in surveying plant diseases.

A short description of some zonal soils will be presented to illustrate some factors of soil formation.

Grassland Soils

Prairie soils in North America, common in Iowa, south-central Minnesota, and central Illinois, formed under tall-grass vegetation where precipitation, mainly summer rainfall, just barely exceeded evapotranspiration. These soils are rich in organic matter, deep, and without an underlying deposit of calcium carbonate, Chernozems are important in eastern North Dakota, eastern South Dakota, east-central Nebraska, and central Kansas. Chernozems are darker than Prairie soils, and they are high in organic matter. The black surface layers gradually grade into lighter subsoil, under which is a layer of calcium carbonate. During Chernozem formation the vegetation was dominated by tall to mixed grasses, and the climate is cool, with summer rainfall. Potential evapotranspiration exceeded precipitation, so that salts leached from the upper horizons were deposited in the dry underlying layers at the depth of maximum common wetting, about 46–71 cm beneath the soil surface

Decomposition of organic debris in and on the soil results in the production of carbon dioxide, which forms carbonic acid with water. Calcium carbonate in the soil is made soluble as calcium bicarbonate and is leached down through the soil profile. In the Prairie, soil water is adequate to remove some of the calcium into the ground water (an open system). In the Chernozem (and similar soils in drier climates) water is inadequate to remove the calcium from the soil and it is deposited in the subsoil, a closed system (Cox and Atkins, 1979).

Westward of Chernozem soils, with increasing dryness, Chestnut (Northern Dark Brown) and then Brown soils are encountered. The total organic matter in the profile declines, and the calcium carbonate layer becomes closer to the surface with increasing dryness. The native vegetation has a corresponding change from tall grasses to mid-grasses and finally, to short or bunch grasses. In grassland soils, calcium, magnesium, and organic matter stabilize colloids, and along with the lower rainfall, minimize leaching and downward movement of colloids.

Forest Soils

Forests usually occur where moisture is abundant. Tree roots do not occur in surface horizons to the same degree that grass roots do. The grassland soils described above are slightly acid to slightly alkaline. Forest soils, in contrast, are usually slightly to strongly acid, depending upon parent material, climate, and vegetation

A Podzol is a zonal soil with an organic mat (litter) and a very thin organic-mineral layer above a gray leached layer. This gray layer has a color somewhat like ashes [in Russian "podzol" means beneath (pod) ash (zol)]. The ash-colored horizon rests on a dark brown layer enriched by humus and iron oxides and alumina from the upper horizons (U.S.D.A. Yearbook, 1938). In modern soil classification, Spodosols include most of the soils formerly called Podzols

Podzols are associated in Russia with a cool, temperate climate, especially under coniferous forests. The leaf duff of Jack pine had a pH of 4.5; of white pine, 4.8; and of maple-basswood, 6.3. The maple-basswood duff contains a ton of lime per acre, about 5 times that in the conifer duff (Weaver and Clements, 1938). In general, podzolization occurs in areas where the vegetation is low in basic elements and is acid in reaction. Rain in these areas exceeds evapotranspiration and acidified water therefore moves through the soil. Acidic podzols are unfavorable for survival of most insects, earthworms, and bacteria. Fungi assume a dominant role, particularly in the leaf litter and duff.

Disease Survey by Soil Groups

Bipolaris sorokiniana attacks the roots, crowns and sub-crown internodes of spring cereals in the Canadian Prairie Provinces so generally that Canadians refer to "common" root rot. Ledingham et al. (1973) conducted a prairie-wide survey of this disease on hard red spring wheat and utilized soil zones rather than political boundaries in interpreting results. Zone 1 is the most arid with brown soil. Zone 2 is more moist and the soil is dark brown. Zone 3 includes most of the black lands of Manitoba and parkland soils of Saskatchewan and Alberta. Zone 4 lies between parkland and boreal forest. These soil zones represent an ecological gradient from rather dry to moist, dominated essentially by common suscepts and tillage systems so that inherent soil factors as they affect common root rot should be expressed.

Disease does not differ greatly from soil zone to soil zone. The differences among zonal soils are not as important as the weather. In 1969 Zone 4 had the lowest disease loss (4.1%) and in 1970 it had the highest (9.1%). Rain was the major variable in the environment.

TOPOGRAPHY

A southern slope of 5° angle increases the amount of radiant energy absorbed by soil equivalent to that of nonsloping land a latitudinal distance 484 km away (Weaver and Clements, 1938). A northern exposure would be affected equally, but in the reverse direction. Buol et al. (1973) cite data on soil temperatures at the 2 cm depth on north and south slopes in Michigan. The south slope is 4 to 6°C warmer throughout the year, which is more than enough to influence many plant diseases.

Nyhan et al. (1972) studied the effect of a single slope (toposequence)

on the abundance of stylet-bearing nematodes in and about soybean roots in Iowa. Total nematodes per gram of root is greatest on the summit and least on the foot of the slope. Fine soil particles (silt + clay) are most abundant at the foot of the slope. The slope was 133 m long, so general climatic factors, cropping history, and the crop (soybeans) are similar. Nevertheless, the reduction of stylet-bearing nematodes from the summit down the slope to where fine soil particles accumulate is significant. Although the results of this study are not spectacular, the effect of soil toposequences on soilborne pathogens and diseases should receive more study.

SOIL PROFILE

The *soil profile* is a vertical section of the soil through all its horizons and extending to the parent material (Brady, 1974). A surface that appears excellent may lie over a hard pan or layers of accumulated minerals harmful to plant development. It is useful to think of the "surface" soil (about 15 cm depth) as about 2,240,000 kg/ha. Thus, if you added 1 part per million of something to the soil, you added about 2.2 kg/ha.

In natural, untilled situations, plant debris covers the soil surface, forming the organic (0) horizons. Although 0 horizons influence plants and pathogens, they are not part of the true soil. True soil horizons begin beneath the litter as A (surface) horizons. The bulk of plant roots, expecially of grasses, is in the 0 and A horizons, as are the highest populations of microorganisms. The horizons rich in organic matter have good structure, with the finer soil particles aggregated into granules. Roots penetrate such soil easily. The B horizons (subsoil) are lower in organic matter. Fine soil particles tend to wash from the A into the B horizons, making the horizons less porous and more difficult for roots to penetrate.

Pans (relatively impervious layers) are common beneath the A horizon. Soil particles cemented together by certain types of organic matter or silica are duripans; by sesquioxides, iron pans; or by calcium carbonates, caliche. The layers are not easy for roots to penetrate even when wet. Restrictive zones formed by accumulated clay or poor soil structure, especially structures deflocculated by high sodium content, form again if disrupted by tillage.

A pressure pan is a subsurface horizon of higher than normal bulk density that is less porous than the soils above or below. It is produced by repeated tillage with the plow sole at a certain depth or by other traffic. This type of restricted layer can be disrupted physically by chisel plows, correcting the problem.

It is important to know of the existence and characteristics of restrictive layers in the soil profile. These pans concentrate roots in the more friable upper layers in which the greatest concentration of most soilborne plant pathogens occur. Pans also concentrate water in surface soils during rainy seasons. Water accumulates above the pan, increasing root damage from heaving when soil freezes. Pans have important consequences for both microorganisms and higher plants.

Effect of an Iron Pan

Jarrah *(Eucalyptus marginata)* is an economically important tree providing 70% of the raw material of the forest industry of western Australia. In 1921 jarrah dieback was present in a few small areas in the northern part of the forest. By 1972 the disease had spread to 80,000 ha in the national forest preserve (Podger, 1972). In 1983 (Shea et al., 1983) it occurred in over 200,000 hectares. Jarrah dieback, caused by *Phytophthora cinnamomi,* is a lethal disease to several species of native trees, shrubs, and other plants in the forest, and it has changed the composition of the flora.

The commercial use of the jarrah forests began in 1850 and was intensive between 1870 and 1920. The progress of jarrah dieback provides a classic basis for questions on historical etiology. Why were its effects so long in appearing? There is no evidence that selection of resistant plants or species occurred. The disease tended to follow roads and logging operations. Podger (1972) stated that in some sites repeated disturbances have not resulted in disease. Either these sites are unfavorable to the pathogen, or the pathogen has not reached those sites. Evidence is strong that *P. cinnamomi* was introduced relatively recently.

Contributing to the extreme virulence of *P. cinnamomi* in the jarrah forests is the low level of biologic activity in the forest soils. The soils are of coarse texture, favorable for zoospore distribution and rapid wetting of the root zone. Even light rains provide much free water. Stem sections of lupine used as bait are colonized within 24 hr after wetting a previously dry soil, so prolonged wet periods are not essential, and rapid draining does not preclude infection.

The winter, when most rain occurs, is too cool for extensive zoospore production. Spread is greatest after spring, summer, and autumn showers, but once infection is established, the fungus proceeds into the roots and trunk relatively uninterrupted. Warm, wet soil is required for rapid production and release of zoospores (Zentmyer, 1980). Once infection occurs, many of the hosts, particularly *Banksia grandis*, are so susceptible that the pathogen is not confined to small feeder roots but occurs also on roots up to 10 cm in diameter and on collars.

The surface soils of jarrah forests are dry for much of the summer, and survival of *P. cinnamomi* is negligible in the surface 10 cm in stands of diseased jarrah. Rather than killing by destroying fine feeder roots in surface soils, *P. cinnamomi* attacks the roots where they pass through vertical channels in the laterite (Fig. 1.1). In all sites where the disease is severe, the compacted lateritic layer is within 1 m of the soil surface. The roots of *Banksia* and *Eucalyptus* intertwine, often occupying the same openings in the later-

SOIL

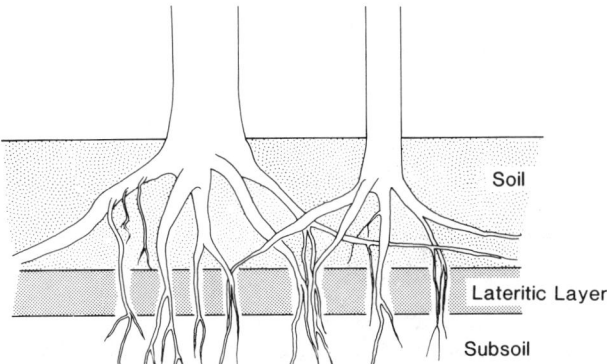

Figure 1.1. Roots of trees penetrate the rocklike lateritic layer at natural openings. Water accumulates above the laterite, facilitating concentrated lateral spread of zoospores of *Phytophthora cinnamomi*, just above the laterite. After Shea et al., 1983.

ite. Water accumulates above the laterite, favoring zoospore production and lateral movement of zoospores. The pathogen severs the roots that penetrate the restrictive lateritic layer, isolating the trees from water deep in the soil profile. Surveys show a close association between occurrences of the iron pan and severe jarrah decline.

Effect of a Pressure Pan

Fusarium solani f. sp. *phaseoli* causes a root and hypocotyl rot of common bean. It is important in irrigated sandy loams in Washington, where yields decline after several crops of beans. Part of the problem has been attributed to the formation of a tillage pan (Burke, 1968). The pathogen increases on beans; chlamydospores accumulate in the soil; the pressure pan restricts most of the roots to the surface layers above the pan, exposing roots to the maximum number of fungal propagules (Fig. 1.2). Subsoiling just prior to planting increases the yield of beans twofold. The bean roots penetrate deeply where the subsoiler fractures the tillage pan, expanding the root system, allowing deeper penetration and utilization of water deeper in the soil, reducing water stress. Subsoiling also increases yields on land not heavily infested with *F. solani*, but to a lesser extent.

Subsoiling before seedbed preparation is not beneficial, but subsoiling at planting time proves that compaction in preparation for the current season crop on some soils is sufficient to restrict root penetration beneath the tillage layer (Burke et al., 1972). Plowing barley straw into the soil without subsoiling increases the yield by 11%, subsoiling alone increases the yield by 38%. Plowing barley straw into the soil with a moldboard plow also loosens the soil and reduces root rot. Loosening the soil is beneficial in coarse-textured soils but not in a silt loam. Moisture stress develops more

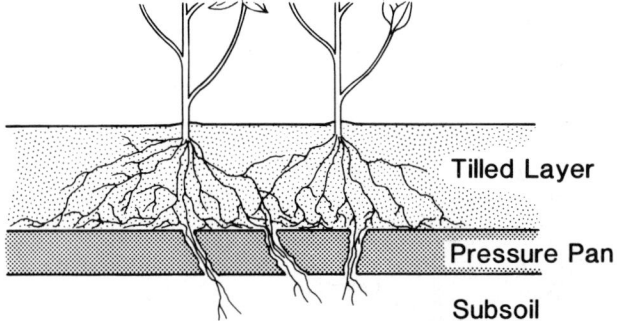

Figure 1.2. The pressure pan restricts penetration of the more dense soil by bean roots, concentrating roots in the loose tilled layer of soil. After Burke et al., 1972.

frequently in the lighter-textured soils, so roots restricted to the upper layers are more frequently stressed in soils of low moisture-holding capacity.

SOIL TEXTURE AND STRUCTURE

Texture expresses the size of the primary soil particles (sand, silt, and clay) that comprise the soil (Donohue, et al., 1971). Coarse or light refers to a sandy soil, fine or heavy refers to a clay soil. A loam is intermediate, containing a favorable mix of sand, silt, and clay particles. Sandy soils dry out and warm up the quickest and are easiest to till. Clays retain more total water, dry out, and warm up more slowly and are most difficult to till. Loams are intermediate, having good water-holding properties plus reasonably good tillage characteristics.

Structure refers to aggregations of primary particles into secondary particles. If the finer silt and clay particles are aggregated into granules (secondary structure) the soil has good structure. A crumb structure (Donahue et al., 1971) facilitates water infiltration, reduces bulk density, and facilitates penetration by roots. Both texture and structure determine the number and size of spaces (pores) between and within soil particles, primary and secondary. The pore sizes and configuration, continuous or discontinuous, affect the dispersal of motile propagules of microorganisms within the soil. Aggregation of fine particles into larger particles (flocculation) is favored by calcium, magnesium, and humus. Sodium favors deflocculation or destruction of soil structure.

Bulk Density

Bulk density refers to the grams of oven-dried soil per cubic centimeter including the pores or air space. As bulk density increases, root extension and air space are reduced. Soils of fine texture in good tilth tend to have lower bulk densities than coarse-textured soils because fine texture provides more air spaces, especially if fine soil particles are aggregated into stable granules. Organic matter reduces bulk density, not only because organic matter (humus) is lighter than mineral soil, but also because it increases aggregation of soil particles into granules. The tilled layers of soil have reduced bulk density for months after tillage, favoring extension of seedling roots. Kraft and Giles (1979) grew peas in soils infested with *Fusarium solani* f. sp. *pisi* and *Pythium ultimum* at bulk density 1.2 g/cm^3 and at 1.5 g/cm^3. Root weight was reduced and disease increased in the compact soil.

Bulk density by itself may not reflect the ability of nematodes to penetrate soil, because bulk density alone does not reveal the size of individual pores and their continuity (Jones and Thomasson, 1976). A pore of optimum size leading to a blind end is of little use for movement of a nematode (or zoospore) through soil.

Soil Texture in Relation to Nematode Movement

Root-lesion nematodes are important on tobacco, cotton, maize and peanuts in southeastern United States. Endo (1959) studied the effect of a sandy loam, loam and clay loam on the ability of *Pratylenchus zeae* to disperse. The soils were fumigated to kill all nematodes and placed in tubs with provision for drainage. A metal cylinder containing maize roots filled with *P. zeae* mixed with the same soil as in the tub was placed in the center of each tub prior to filling. After the tub was filled the cylinder was removed, leaving a core of infested debris in the center of a tub of nematode-free soil. Maize was grown in a ring about the central core. A fallow treatment (no plants) for each soil was included. No nematodes were found at 2.5 cm from the core in tubs without maize plants 4 months after the start of the experiment. When maize was present, nematode movement was greatest in the sandy loam, less in the loam, and the least in clay loam. No nematodes were found beyond 5 cm in the clay loam or beyond 10 cm in the loam, but about 50 were found at 12.5 cm in the sandy loam (Fig. 1.3). Soil texture is a major factor in nematode migration, and aeration, pore space, and particle size all influenced nematode motility.

Early observations on root-knot (*Meloidogyne* spp.) indicated that the heavier soils are unfavorable for disease development and that lighter soils are favorable. Endo's (1959) experiments with *Pratylenchus* spp. measured movement, not survival, predation, or other factors. Ability to move in soil is more critical for nematodes that lay their eggs in concentrated masses, as do root-knot nematodes, than for those that lay eggs singly.

Organic Colloids

Humus "tightens" a sandy soil, "loosens" a clay soil, and is a source of plant nutrients. An acre of fertile soil may contain 60 to 120 tons of organic

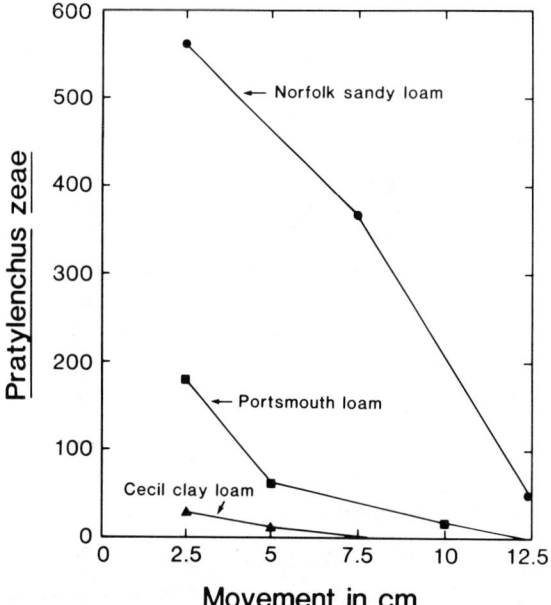

Figure 1.3. Movement of *Pratylenchus zeae* is facilitated in the sandy loam, greatly restricted in the clay loam. After Endo, 1959.

carbon, with 3 to 6 tons of nitrogen in the humus fraction, along with 36 tons of other minerals essential for plant growth (U.S.D.A. Yearbook, 1938). The resistance of humus to decomposition provides for the slow but constant release of nutrient elements. Humic acids from grassland soils may contain as much as 4 to 6% nitrogen, those from highly leached podzols may contain less than 0.3% N. Humus has a density of about 1/3 that of the mineral fraction; thus an organic content of a soil of from 1 to 15% on a dry weight basis tends to underestimate the real amount of organic matter in the soil (Weaver and Clements, 1938).

Inorganic Colloids

Clay minerals differ in their abilities to expand and contract, and this property influences plant and microbial growth. Montmorillonite clay consists of micelles (plates) bound weakly enough together to permit expansion and contraction. Kaolinite does not expand. Montmorillonite has a greater capacity than kaolinite to absorb mineral cations (K^+, Ca^{2+}, Mg^{2+}, etc.) and positively charged organic compounds. Greaves and Wilson (1973) expanded montmorillonite and found that nucleic acids could enter between the micelles. Clays with adsorbed adenine were incubated in soil. At 2 days about 20 times more bacteria were around the montmorillonite than around

SOIL   11

the kaolinite. Adenine was still present in montmorillonite after 5 weeks, but no adenine was detected on kaolinite after 1 week. The loss of adenine resulted from microbial degradation because no decline occurred when the clays were incubated in sterile soil.

Stover (in Toussoun, 1975) reviewed early efforts to determine soil factors unfavorable to Fusarium wilt of banana. Soils low in potassium favor banana wilt, but adding potassium to soils favorable to wilt did not reduce disease. Stotzky and associates, 1961–1967, studied banana soils in Central America and attributed wilt suppression to soils rich in montmorillonite clay. Montmorillonite clay favors growth of bacteria which may have suppressed growth of the pathogen.

Vertisols Vertisol, a new soil order, is of particular importance to plant pathologists. A vertisol is rich in expanding clays. It swells when wet and shrinks when dry, forming cracks that extend deeply into the soil. Bits of surface soil fall and wash into the cracks. Through the years this process results in a churning of the soil. Vertisol gets its name from the Latin, *verto,* to turn, and *sol,* from *solum,* soil. Normal surface horizons did not develop because of the churning. The Houston black clay, famous in relation to Phymatotrichum root rot, is a vertisol.

SOIL pH

Soil acidity is common in all regions of high precipitation (Brady, 1974) where appreciable amounts of exchangeable bases have been leached from the surface horizons of soil. Calcium, magnesium, potassium, and sodium ions are called *exchangeable bases,* and these four cations are measured to determine the percent base saturation (Cox and Atkins, 1979). The soil pH estimates the abundance of these cations in soil. Over 99% of the available cations are held by adsorption on the surface of soil colloids. The capacity of the soil to hold cations in a form available to higher plants is termed the *cation exchange capacity,* and it is expressed as milligram equivalent (meq) per 100g of dry soil; meq states the number of milligrams of hydrogen ions that would be present if all the adsorption sites were occupied by this cation. The cation valence and atomic weight determine the actual amount of the mineral capacity of the system. One meq of hydrogen (valence 1, atomic weight 1) would hold 20 mg of calcium (valence 2, atomic weight 40) per 100 grams of soil.

Organic colloids (humus) have the highest exchange capacity. The cation exchange capacity of humus is strongly influenced by soil pH (Cox and Atkins, 1979), and is about 90 at pH 4 and over 200 at pH 8. The cation exchange capacity of inorganic colloids (clay minerals) is less affected by pH.

The pH of a soil can vary considerably, sometimes being changed by 1 pH unit with rainfall, season, and vegetation (Weaver and Clements, 1938).

Drying soil at high temperature can increase its acidity (Brady, 1974). The pH tends to decrease during summer and increase during winter and spring. Soil pH can also differ within very small distances (microsites) within soil.

When pH is determined electrometrically, the active acidity (or the concentration of hydrogen ions in the soil solution) is determined. Hydrogen ions held by soil colloids constitute the reserve or exchange acidity (Brady, 1974). The active acidity is probably only 1/1000 of the reserve acidity of a sandy soil, and about 1/50,000, or less of that of a clay soil. When lime is added to soil to elevate the pH, most of the lime is used to reduce the reserve acidity. Because the exchange capacity of a sandy soil is less than that of a clay soil, more lime must be added to the clay soil to attain a given correction in pH than is needed for the sandy soil.

Soil reaction influences the stability and activity of enzymes and antibiotics, the adsorption of substances and bacteria to soil colloids, the balance between total bacteria and fungi as well as qualitative differences in bacterial and fungal populations, and the availability of mineral nutrients to higher plants and microorganisms. Low pH makes iron, manganese, zinc, copper and cobalt more available (Brady, 1974). Below pH 5.5 nitrogen, calcium, magnesium, phosphorus, potassium, sulfur, and molybdenum may be less available. Above pH 7 copper and zinc become less available, and high pH's often lead to shortages of iron, manganese, and cobalt, in addition to copper and zinc. Soils with pH values between 5.6 and 7 are best for most crop plants. Take-all *(Gaeumannomyces graminis)* of winter wheat is favored by alkalinity for several reasons, and Reis et al. (1982) obtained evidence that reduction of available minor elements at higher pH's may predispose wheat to greater losses.

Measurements

Measurements of soil pH in water are influenced by salt content, but measurements in 0.01 M $CaCl_2$ solution are little affected by salts (Smiley and Cook, 1972); differences in soil pH in water and in calcium chloride solution vary from 0 to 0.9 units. Dobson et al. (1983) believe taking pH by different methods may contribute to differing results in experiments with *Plasmodiophora brassicae* and other pathogens sensitive to soil pH.

Influence of pH on Diseases

Walker (1950) listed Fusarium wilt of cotton (race 1) and club root of cabbage as favored by acidity and Phymatotrichum root rot and potato scab as favored by alkalinity. It is interesting that Walker noted (p. 554) that club root does not occur in heavily limed soil kept continuously moist, but that if the soil water content is allowed to fluctuate, or if aerated, infection occurs. He suggested that active roots may reduce the pH of soil in their immediate proximity, favoring infection. He speculated that the reduced pH in the rhizosphere permits spores to germinate and infection to occur and that after infection had occurred, soil conditions probably have no effect. There is no doubt of the significance of pH, yet we do not understand how soil pH affects many pathogen-host interactions.

Dobson et al. (1983) reviewed the literature on the success or failure of raising the soil pH to control club root of cabbage (*Plasmodiophora brassicae*). They attributed the failures to acidic microsites within the bulk soil resulting from the use of coarse lime and lack of thorough mixing with the soil. Calcium carbonate changes the soil pH more rapidly than magnesium carbonate, and the size of particle is critical (Brady, 1974). After 3 months calcite particles in the 20-mesh screen size had about a 16% reaction, the 30-mesh size had about 30% reaction, and the 100-mesh size had about 80% of the calcite reacted with the soil. Equivalent amounts of $Ca(OH)_2$ and laboratory $CaCO_3$, both fine powders, raised the soil pH to 7.4, and no cabbage plants contracted clubroot (Dobson et al., 1983). Agricultural flour limestone gave a bulk soil pH of 7.3 and 8% infected plants. Fine limestone (<0.5 mm) gave a pH of 7.3 and 27% infection. Medium limestone (0.5 to 1.0 mm) gave a bulk soil pH of 7.2 and 48% infection, coarse limestone a pH of 7.1 and 80% infection, and no lime, pH 5.8 and 100% infection. All pH's were of bulk soil, and the soil was heavily infested by the pathogen. In soil with coarse limestone, acidic microsites within the soil were abundant, and poor control resulted, even after thorough mixing. All liming trials involved planting within 7 days of liming, so the effect of particle size was strongly expressed. Bremer (in Macfarlane, 1952) observed that up to 37% of resting spores germinate in wet, acid soil, and only 1 to 2% germinate in alkaline soil. Macfarlane obtained information, supporting Bremer, that few resting spores of *P. brassicae* germinated spontaneously in alkaline soil. Control by liming may result from inhibiting spore germination.

In contrast to the club root example, it is not always practical to control diseases by modifying soil pH, even though this practice is effective for disease control per se. Black shank of tobacco occurs on a wide variety of soils ranging in pH from 5 to 6 in North Carolina and in soils derived from limestone with pH up to 7.4 in Kentucky (Lucas, 1975). Kincaid and Gammon (1954) adjusted a fine sandy loam to pH's of 4.5 to 6.2 and recorded plants killed by *Phytophthora parasitica* var. *nicotianae* on three dates. The time sequence provided graphic illustration of the effect of soil pH (Table 1.1). One month after transplanting the effect of pH was pronounced. Three months after transplanting there was no difference in mortality of the susceptible cultivar at soil pH values between 5.1 and 6.2. The resistant cultivar did well over the entire pH range. Acidity retards disease, but from a practical standpoint it makes little difference. Death at 3 months is a greater economic loss than death at 1 month, because more expense has gone into the crop and the chances of planting a different type of plant to salvage the rest of the growing season are reduced. In addition, this is an impractical

Table 1.1. Death of tobacco plants killed by *P. parasitica* var. *nicotianae* as influenced by pH of Marlboro fine sandy loam (Kincaid and Gammon, 1954) (transplanting was done on 26 March)

Soil pH	Mortality, susceptible cultivar			Resistant cultivar
	26 April	25 May	26 June	
4.5	24	38	58	1
5.1	67	87	100	1
5.5	74	92	95	2
5.8	83	97	100	2
6.2	88	99	100	3

control strategy, because lime is applied to acid soils to approach a pH of 6.0 to improve the leaf quality. Disease control procedures must be in harmony with agronomic principles. Liming to pH 6 may have brought the soil to the pH most favorable for black shank, but resistance and crop rotation have solved the problem (Kincaid, et al., 1970).

Isolates of *P. parasitica* var *nicotianae* from the burley-producing areas of Kentucky, Tennessee, and western North Carolina differ from those from flue-cured areas of North Carolina and adjacent states (Apple, 1957), not only culturally but in response to pH. Burley isolates grow fastest in culture near pH 5.5; flue-cured isolates grow most rapidly at about pH 6-6.5. Differences in response to pH among isolates of a species, especially of this magnitude, are not common.

COMMENTS

Soil properties give clues as to the nature of the environment. The failure of the great soil groups (zonal soils) in Canada to be highly correlated with the prevalence of *Bipolaris sorokiniana* on wheat is evidence that the gradual changes reflected in soil formation over many years is not as important to the pathogen-host interaction as seasonal variations, in this case, of rainfall. The most dramatic effects of soil on disease development result from tillage practices, such as development of a pressure pan, and responses to azonal factors (wet spots, changes in soil texture, alluvial deposits by streams, or other factors). The iron pan and pressure pan are both presented, because one results from geologic forces, the other from man.

How soil pH affects so many diseases in such diverse ways requires cooperative experimentation by soil chemists and plant pathologists. The lowering of soil pH through use of some fertilizers, especially of anhydrous ammonia, has already had important effects. Baker and Cook (1974, p. 287) present the effects of several nitrogenous fertilizers on bulk soil pH.

Diseases caused by soilborne plant pathogens often vary in incidence

SOIL

and severity within a field, even within a field that has had the same cropping history and management for many years. These local differences in all probability reflect differences in the soil itself. Study of soils in the areas with severe and slight disease within a field can contribute to understanding the soil factors involved.

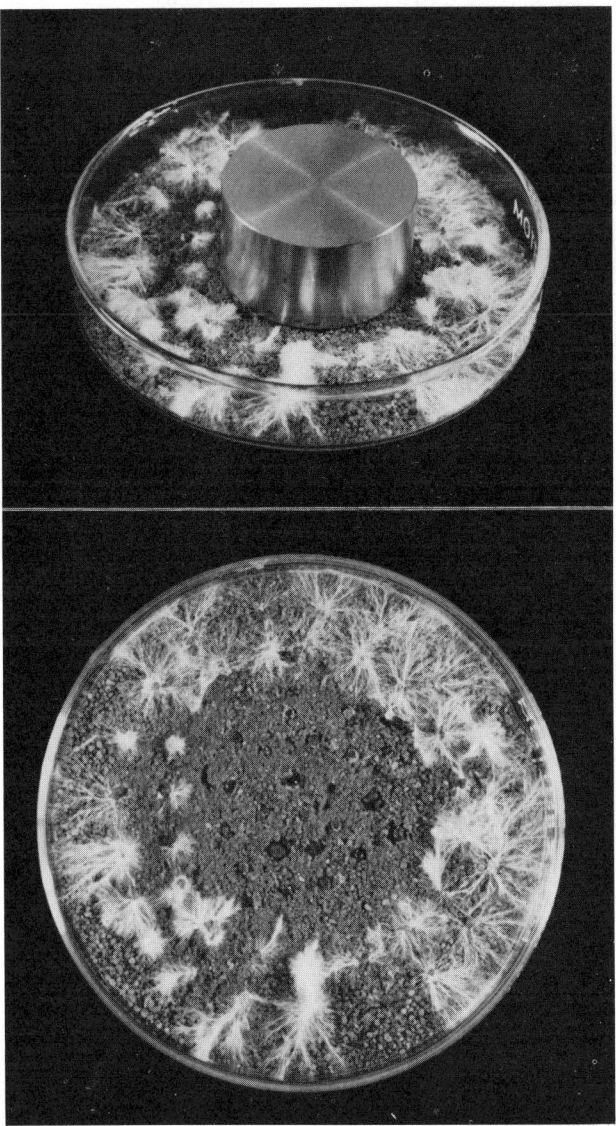

Figure 1.4. Sclerotia of *Sclerotium rolfsii* failed to germinate when subjected to pressure. After Punja and Jenkins, 1984.

ADDENDUM

The effect of pressure on propagules within soil has been little studied. Punja and Jenkins (1984) placed a weight on sclerotia of *Sclerotium rolfsii* on the surface of moist soil and found that leakage of solutes from the sclerotia increased and that germination of the sclerotia was inhibited (Fig. 1.4). This aspect of life within soil should be investigated further. Much life occurs within the pore spaces within soils, but what happens where the weight of the overburden is sustained?

KEY REFERENCES

Burke, D. W., D. E. Miller, L.D. Holmes, and A. W. Barker. 1972. Counteracting bean root rot by loosening the soil. *Phytopathology* **62**:306-309.

Dobson, R. L., R. L. Gabrielson, A. S. Baker, and L. Bennett. 1983. Effects of lime particle size and distribution and fertilizer formulation on club root disease caused by *Plasmodiophora brassicae*. *Plant Dis.* **67**:50-52.

Endo, B. Y. 1959. Responses of root-lesion nematodes, *Pratylenchus brachyurus* and *P. zeae*, to various plants and soil types. *Phytopathology* **49**:417-421.

Shea, S. R., B. L. Shearer, J. T. Tippett, and P. M. Deegan. 1983. Distribution, reproduction and movement of *Phytophthora cinnamomi* on sites highly conducive to jarrah dieback in south western Australia. *Plant Dis.* **67**:970-973.

REFERENCES

Apple, J. L. 1957. Pathogenic, cultural and physiological variation within *Phytophthora parasitica* var. *nicotianae*. *Phytopathology* **47**:733-740.

Baker, K. F. and R. J. Cook. 1974. *Biological Control of Plant Pathogens*. Freeman, San Francisco.

Brady, N. C. 1974. *The Nature and Properties of Soils*. Eighth ed. Macmillan, New York.

Buol, S. W., F. D. Hole and R. J. McCracken. 1973. *Soil Genesis and Classification*.Iowa State Univ. Press, Ames.

Burke, D. W. 1968. Root growth obstruction and Fusarium root rot of beans. *Phytopathology* **58**:1575-1576.

Carleton, M. A. 1919. *The Small Grains*. Macmillan, New York.

Cox, G. W. and M. D. Atkins. 1979. *Agricultural Ecology*.Freeman, San Francisco.

Donahue, R. L., J. C. Shickluma, and L. S. Robertson. 1971. *Soils. An Introduction to Soils and Plant Growth*. Prentice-Hall, Englewood Cliffs, N. J.

Fitzpatrick, E. A. 1971. *Pedology. A Systematic Approach to Soil Science*. Constable, Edinburgh, Scotland.

Greaves, M. P. and M. J. Wilson. 1973. Effects of soil micro-organisms on montmorillinite-adenine complexes. *Soil Biol. Biochem.* **5**:275-276.

Jones, F. G. W. and A. J. Thomasson. 1976. Bulk density as an indicator of pore space in soils usable by nematodes. *Nematologica* **22**:133-137.

Kincaid, R. R. and N. Gammon, Jr. 1954. Incidence of tobacco black shank directly related to soil pH. *Plant Dis. Reptr.* **38**:852-853.

Kincaid, R. R., F. G. Martin, N. Gammon, Jr., H. L. Breland, and W. L. Pritchett. 1970. Multiple regression of tobacco black shank, root knot and coarse root indexes on soil pH, potassium, calcium, and magnesium. *Phytopathology* **60**: 1513–1516.

Kraft, J. M. and R. A. Giles. 1979. Increasing green pea yields with root-rot resistance and subsoiling. Pages 407–413 in *Soilborne Plant Pathogens*. B. Schippers and W. Gams, eds. Academic Press, New York.

Ledingham, R. J., T. G. Atkinson, J. S. Horricks, J. T. Mills, L. J. Piening, and R. D. Tinline. 1973. Wheat losses due to common root rot in the prairie provinces of Canada, 1969-1971. *Can. Plant Dis. Surv.* **53**:113–122.

Lucas, G. B. 1975. *Diseases of Tobacco*. Third ed. Biological Consulting Associates, Raleigh, N. C.

Macfarlane, I. 1952. Factors affecting the survival of *Plasmodiophora brassicae* Wor. in the soil and its assessment by a host test. *Ann. Appl. Biol.* **39** :239–256.

Nyhan, J. W., L. R. Frederick, and D. C. Norton. 1972. Ecology of nematodes in Clarion-Webster toposequences associated with *Glycine max* (L.) Merrill. *Soil Sci. Soc. Am. Proc.* **36**:74.

Podger, F. D. 1972. *Phytophthora cinnamomi,* a cause of lethal disease in indigenous plant communities in Western Australia. *Phytopathology* **62**:972–981.

Punja, Z. K. and S. F. Jenkins. 1984. Influence of temperature, moisture, modified gaseous atmosphere, and depth in soil on eruptive sclerotial germination of *Sclerotium rolfsii. Phytopathology* **74**:749–754.

Reis, E. M., R. J. Cook, and B. L. McNeal. 1982. Effect of mineral nutrition on take-all of wheat. *Phytopathology* **72**:224–229.

Smiley, R. W. and R. J. Cook. 1972. Use and abuse of the soil pH measurement. *Phytopathology* **62**:193–194.

Toussoun, T. A. 1975. Fusarium-suppressive soils. Pages 145–151 in *Biology and Control of Soilborne Plant Pathogens*. G. W. Bruehl, ed. American Phytopathology Society, St. Paul, Minn.

U.S.D.A. Yearbook. 1938. *Soils and Men*. U.S. GPO, Washington, D.C.

Walker, J. C. 1950. *Plant Pathology*. McGraw-Hill, New York.

Weaver, J. E. and F. E. Clements. 1938. *Plant Ecology* Second ed. McGraw-Hill, New York.

Zentmyer, G. A. 1980. *Phytophthora cinnamomi* and the diseases it causes. American Phytopathology Society, St. Paul, Minn.

Chapter 2

Water

Higher plants use enormous quantities of water, and water stress is one of the greatest constraints on yield. In deciduous forests over half the energy of the plant is expended below ground to obtain water and form new roots (Lange et al., 1976). Comparative estimates for mesic meadows are about 50%, for short grass prairies about 75%. Water stress increases the proportion of roots to shoot. It also increases permeability of membranes and leakage of solutes from roots. Water stress sufficient for pathologic effects disrupts intracellular compartmentation, releasing latent acid hydrolases to cause direct injury. The abundance of water in soil is important to host plants, to microorganisms, and to host-pathogen interactions.

WETNESS

In most of the older literature water content of soil was expressed in volumetric terms, as percentage of total water-holding capacity of the soil or in terms of percentage of field capacity. Most modern workers express water in relation to its energy status, reflecting the energy required to remove water from the soil. Free, pure water has 0 energy potential; water at moisture-holding capacity of many soils is at -0.3 bars, and at the permanent wilting point, near -15 bars water potential. Other expressions of energy relationships include pF values ($pF_1 = -0.01$ bar, $pF_3 = -1$ bar, $pF_5 = -100$ bars, etc.—a logarithmic relationship between bars and pF values), or the equivalent suction exerted by columns of water or mercury of various heights. Soil physicists express water in soil in relation to its energy status (thus the term *water potential*). The lower the water potential the less available water becomes.

Solution chemists used the term *water activity* to express the status of water. They work with osmotic factors, not the matric forces that dominate in soil. Scott (1956) studied the preservation of foods by dehydration (removal of water) and by addition of salts (reducing the water activity a_w). An a_w of 0.96 is equivalent to a relative humidity of 96%. *Staphylococcus aureus* grew in synthetic media, nutrient broth, milk, meat, and soup at a_w 0.86 to 0.88, regardless of the actual water content of the substrate (it varied from 16 to 375% of the dry weight of the substrates he studied). In another experiment, Christian and Scott (in Scott, 1962) grew a bacterium in brain-heart infusion either dehydrated to varying degrees or to which electrolytes were added to obtain different water activities. Whether water was removed by dehydration or the water content was held constant and its activity reduced by added electrolytes, the water activities obtained by the two methods had similar (but not identical) effects on the test organism. At a water activity of 0.96 both methods gave bacterial reproduction of 0.5 divisions per hour, in spite of removal of half the actual water by dehydration to attain the 0.96 activity of water.

Most workers test several solutes, because some organisms may not tolerate certain salts. Molal (not molar) solutions of potassium chloride are most commonly used because this salt is easy to use and most fungi tolerate it in high concentrations. When sucrose is used as the osmoticum, the final volume of the solution increases greatly if low water potentials are desired. The medium also becomes viscous and is difficult to dispense.

Papendick and Campbell (1981) give factors for converting water potential in bars to other units. The water potential at a given relative humidity within the soil can be determined by subtracting 1.00 from the humidity and then multiplying by 1350. A relative humidity of 96 equals $.96 - 1.00 = -0.04 \times 1350 = -54$ bars.

Components of Water Potential

Total soil water potential includes gravitational, matric, osmotic, and other factors, but for our purposes, only the matric and osmotic forces are important. The *matric* potential represents the adsorptive force by which soil particles attract water plus capillary forces, including surface tension of water. The *osmotic* potential results from ions and molecules dissolved in the soil solution. Matric potential limits many biological processes in soil out of proportion to the volume of water actually present. The spermosphere and rhizosphere shrink as water films become thinner, due in part to physical restriction of diffusion of solutes within thin water films.

Loams are generally the most productive agricultural soils because they contain a favorable mix of soil particles of differing sizes, giving a favorable balance between water retention and aeration. Water in a fine sand at field capacity is at about -0.1 bar water potential, and in loams and silt loams, at about -0.3 bar water potential. Except in saline soils or in unusual situations, water potentials below -0.6 bar are due mainly to matric forces.

Within the soil, matric forces usually predominate; within the plant, osmotic forces.

Matric versus Osmotic Potential

Zentmyer (1980), in his monograph on *Phytophthora cinnamomi*, stated that soil moisture is the "primary environmental factor in development of Phytophthora root rot," that zoospores are responsible for most of the severe disease outbreaks, and that infection from direct germination is important only in sustaining populations under less favorable conditions. *Direct germination* is germination by means of a germ tube rather than by production of zoospores (indirect germination). Sterne et al. (1977) studied the effects of matric and solute water potential on root rot of avocado incited by *Phytophthora cinnamomi*. The osmotic levels of this study approximated the range of those present in some avocado groves in southern California. The total water potential in the soil was the sum of the saturation extract of that soil (-0.37 bars), added salts, and the matric potential. Total water potentials ranged from -0.37 to -2.62 bars. The treatments consisted of combination of matric potentials from 0 to -0.25 bars plus osmotic potentials of from -0.37 to -2.00 bars, including the constant -0.37 potential of the saturation soil water extract. Disease severity was controlled almost completely by matric water potential. Highest disease levels occurred at matric potentials of 0 to -0.05 bars. Disease severity on inoculated plants was greatest at -0.37 bars (0 matric plus -0.37 saturation extract) and least at -2.62. Roots of noninoculated plants were healthy over this range of water potentials after 12 days. The authors concluded that zoospores were the major infective propagule and that matric water potentials above -0.10 were required to keep enough water in the larger soil pores to facilitate zoospore movement.

Salinity

In normal agricultural soils matric forces predominate and osmotic forces are of minor consequence. Salinity is important in local situations, however. When plants are grown on ridges and irrigated from ditches, water moves up the ridge by capillarity and is lost by evaporation and transpiration. Salts accumulate in the zones of greatest water evaporation or removal by the plant. Cook and Papendick (1972) calculated that, in a situation described for a cotton field in the Southwest, when the average matric potential was -0.3

bar, the total water potential in the soil varied from about −3 to −45 bars. In this example osmotic (solute) forces were dominant. These concentrations of salts greatly altered water relations of roots or microorganisms exposed to these varied conditions under a single row of cotton.

Temperature

Not only solutes within water affect its availability, but also its temperature. The viscosity of liquid water decreases about 50% between 25 and 0°C (Kramer, 1949). In ice, some liquid water remains, so at temperatures below freezing, humidity is influenced by the vapor pressure of both ice and water. Ice at 0°C has a relative humidity of 100%; at −5°C, 95.3%; at −10°C, 90.7% and at −15°C, 86.4% (Scott, 1962). Several snow mold fungi grow at temperatures below freezing, at least to −5°C, so these relationships have significance in plant pathology, in addition to food preservation. D. A. Gaudet (personal communication) found that *Coprinus psychromor bidus* attacked wheat leaves between +2 and −10°C.

Strong interactions between temperature and water potential often occur. Manandhar and Bruehl (1973) found that *Verticillium dahliae* did not grow on cornmeal dextrose agar (about −2 bars) at 35°C, but with KCl added to achieve −30 to −50 bars, it made appreciable growth (Fig. 2.1). It is common for reduced water potential to change the optimum temperature.

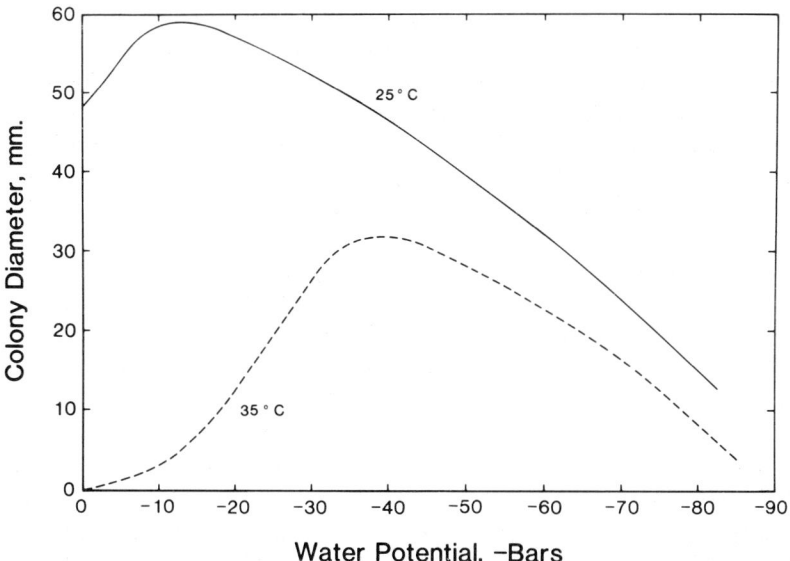

Figure 2.1. *Verticillium dahliae* does not grow at high water potential at 35°C, but grows well at that temperature near -40 bars osmotic water potential. At 25°C its optimum osmotic water potential is near -15 bars, illustrating interaction between temperature and osmotic water potential. After Manandhar and Bruehl, 1973.

Sclerotinia borealis, a psycrophile, grows very slowly on laboratory media at 5°C and is considered difficult to isolate. If sucrose is added to the medium to produce −20 to −30 bars water potential, it grows as fast in culture as *Fusarium nivale*, a fungus easily isolated on standard media. *Sclerotinia borealis* is most pathogenic at about −3°C, and it grows most rapidly on culture media at close to −30 bars. Most laboratory media, unamended by osmotica, have water potentials of about −1 to −3 bars. Knowledge of water relations can be of value in isolating pathogens.

CLASSES OF WATER IN SOIL

Gravitational Water

Gravitational water is water in excess of that which is held by soil particles. It passes through the soil to deep within the profile, and it may reach the water table. Gravitational water displaces the soil atmosphere. If it is retained or percolates too slowly through the soil, many organisms may be harmed by lack of oxygen. Just after the gravitational water has drained away, the soil is at field capacity, at moisture-holding capacity for many soils, or at a matric water potential of about −0.3 bar compared with pure free water at 0 bar water potential.

Movement of Propagules by Gravitational Water Most microorganisms are present in the surface horizons of soil, yet small propagules are moved downward by water in soils with large pores. Some fungal propagules (hydrophobic) are repelled by water, and others, particularly those borne in slimy masses that wet easily (hydrophilic), are adapted to downward movement by water. In experiments in which nematodes, fungal spores, or bacteria are added to the soil surface in water, the soil matrix often restricts their movement into the soil. This may greatly influence the interpretation of results, if the inoculum was assumed to be uniformly distributed throughout the soil matrix.

During the Irish potato famine, tubers were infected by *Phytophthora infestans*. DeBary in 1863 (in Large, 1962) proved that spores produced on foliage could be washed into the soil. Great losses have sometimes followed harvest when the tops were diseased but still alive. Potato vines killed by desiccants several days before harvest reduces this danger. Lacey (1965) used potato tuber slices to assay the infectivity of sporangia in soil and estimated the number of sporangia in soil under diseased plants in England. He found 402 sporangia/ml of soil at 0 depth, 69 sporangia/ml at 5 cm, 48 sporangia/ml at 10 cm and 35 sporangia/ml at 15 cm. Soil infestation and actual tuber infection in the field are highly erratic, but Lacey's experiments proved that sporangia produced aerially are carried to shallow depths. Most of the sporangia were retained within the top 5 cm, but zoospores penetrated more deeply. Lacey stressed that rain runs down stems of erect potato plants, leading to high concentrations of sporangia and zoospores in local areas. The force of water moving through soil is greater than that of motile zoospores (Wilkinson et al., 1981).

Entrapment within the soil reduced the numbers of bacteria carried through the soil to a greater extent than their adsorption to the soil particles (Wilkinson et al., 1981). In a discussion of the possible role of cations in fungistasis, Ashworth et al. (1976) reviewed literature on sorption of bacteria and macromolecules to soil colloids. T. C. Peele in 1936 tried to wash all bacteria from soil to get accurate plate counts. Al^{3+} and Fe^{3+} favor bacterial adsorption. Soils saturated with these cations retain bacteria, and respiration is greatly reduced. In one soil *Bacillus cereus* var. *mycoides* was 96% adsorbed, *B. mesentericus* 41%, and *Esherichia coli*, 12%. Sorption of substances to soil colloids was reviewed by Marshall (1971).

Wakefield and Bisby (1941) classified *Gliomastix* in the Gloiosporeae (spores borne in slime) and *Penicillium* in the Xerosporae (spores borne dry). Spores borne in slime are normally water or insect-dispersed; those borne dry are usually airborne. Burges (1950) made a suspension of spores of *Penicillium cyclopium* and of *Gliomastix convoluta*, dispensed it on the surface of a wet sand column, and then poured excess water onto the surface of the column. *Pencillium* spores were mostly trapped within the first 5 cm of sand. Most of the spores of *Gliomastix* were in the 15 to 40 cm depths. Hydrophobic spores (Xerosporae) were not moved far; easily wetted spores (Gloiosporeae) were readily moved. Burges commented that most fungi isolated from deep in soil have wettable spores.

If, as suggested by Burges (1950), wettable spores were readily carried downward by water, they should, at least in some soils, accumulate at a restricted soil horizon beneath the soil surface. Hepple (1960) found a concentration of *Mucor ramannius* spores at 20 to 30 cm in the B horizon, or at the humus pan of a podsol in England. This fungus usually grows in forest litter on the soil surface. The spores were washed through the sandy A_1 and A_2 horizons to accumulate on the more compact B_1 horizon. Hepple believed that the mucilaginous coating of wettable spores, except under rapid movement of water, permits adhesion to the sand grains, that adhesion is favorable to survival because most substrates are present in surface soil layers, and that adhesion to soil particles (or roots) would maintain higher propagule numbers in the surface 5 to 10 cm of soil. Deep washing is unnecessary, because roots penetrating through surface layers generally carry some of their surface flora with them.

Fungal propagules that lack motility, such as those of *Fusarium*, *Verticillium*, and *Cephalosporium*, depend on gravitational water for penetration into and within soil.

Capillary Water

Capillary water is water held by soil particles as films coating the soil particles and filling some angles between them. It is the only class of soil water important in sustaining mesophytic plants. It flows from higher to lower water potential. Capillary water moves from bulk soil toward the root surface when water is transpired by the plant. As the water films about soil particles become thinner, water is held by increased force. When the water film becomes extremely thin, capillary movement ceases. Capillary water is generally that between -0.3 and -31 bars. Around -15 bars, depending on root system morphology, soil structure and texture, and atmospheric demand, plants cannot obtain enough water to sustain life, and the permanent wilting point has been reached. At -15 bars the soil atmosphere is near 98.9% relative humidity, favorable for many activities by some soil microorganisms.

During the day, when water use is high, matric potential is lower at the root surface than a few millimeters away, reflecting water removal and creating a gradient favoring water movement toward the root. When the soil mass is at near -5 bars matric potential, it can be about -8 bars 1 mm from the root surface and about -14 bars at the root surface. At night, water in the deficit zone is replenished by capillary movement. During the day, water potential in the rhizosphere often differs considerably from that of bulk soil.

The thickness of water films about soil particles and aggregates also affects soil aeration within microsites. Oxygen diffuses 10,000 times more slowly in water than in air (Baver et al., 1972, in Papendick and Campbell, 1981). Anaerobic microsites develop within aggregates surrounded by thick water films.

Soil Pores The size, continuity, and distribution of pores within soil, governed by soil texture and structure, affect the infiltration of water and gases and the movement of zoospores and nematodes within soil. Many motile propagules exhaust themselves and die within soil pores that do not lead to a substrate, which is why pore continuity is important. Estimates of the size of pores required for movement of propagules vary. Water-filled pores 40 to 60 μm in diameter (Stolzy et al., 1965), or 24 μm in diameter (Young et al., 1979) are required by zoospores of *Phytophthora* spp. Small nematodes maneuver well within 30 to 60 μm pores; bacteria, within pores 1 to 1.5 μm in diameter. Carlile (1983) and Duniway (1983) reviewed motility of zoospores of *Phytophthora* spp. in soil. Data of several workers with various organisms are presented by Papendick and Campbell (1981).

As saturated soil dries, the largest pores drain first, and increasingly smaller pores drain as drying proceeds.

Size of Swimming Propagules Primary zoospores of *Plasmodiophora brassicae* produced from resting spores infect cabbage root hairs. Sporangia are then produced in the root hairs. Zoospores from these sporangia are liberated into the soil solution, from which they infect other root hairs or act as gametes and fuse. The larger fusion "zoospores" are capable of cortical and stellar invasions which cause clubbed roots. Root-hair infections by primary zoospores do not result in clubs and are essentially symptomless. This life cycle, proposed by Ingram and Tommerup (1972), presented so many technical difficulties it could not be proved.

Dobson et al. (1982) obtained cortical infections down to −0.1 bar, and root-hair invasion down to as low as −0.8 bar. After determining that root-hair invasion could occur in soil drier than that at which cortical infections cease, Dobson and Gabrielson (1983) analyzed the proposed roles of primary and fused zoospores: In very wet soil (−0.02 bar), favorable for both root-hair and cortical infections, root-hair infections occurred within 1 day, mature sporangia were present within 2 days, and cortical infections occurred within 3 days. If the water tension was increased to −0.16 bar or higher after 1 or 2 days, no clubs developed on the cabbage plants, even though infection of root hairs by primary zoospores had occurred. Dobson and Gabrielson concluded that the primary zoospores were smaller, capable of traversing smaller soil pores, and thus able to infect in drier (down to −0.8 bar) mineral soil. Further, they concluded that fused secondary zoospores were larger, unable to infect below −0.15 bar, and required larger water-filled soil pores, evidence that fused secondary zoospores were required for cortical infections. The precision of this study showed that swimming spores of the same species differed in relation to motility in soil.

Hygroscopic Water

Hygroscopic water is water at below −31 bars. It is so tightly held by soil colloids that it does not move by capillarity. At −31 bars the relative humidity within soil is about 97.8, so many microbial activities occur in the presence of hygroscopic water.

Air-dry soil (loam and sandy loams) usually contains 2 to 4% water by weight. Air drying kills many organisms, and on remoistening, a flush of microbial activity occurs in response to nutrients released by dead microbes within the soil. Soil is dried at 100 to 110°C for 24 hr to determine the amount of water in it.

CLASSES OF ORGANISMS

In wet soil bacteria dominate; in drier soil actinomycetes and fungi dominate. Kouyeas (1964) studied microorganisms on buried glass slides. He observed a sharp decrease in bacteria when moisture fell below about −1 bar. Actinomycetes increased as the moisture level declined to the lowest levels he used (−3.3 bars).

Kouyeas placed culture disks of various fungi on glass slides and incubated them over water and salt solutions to study the range of relative humidities (RH) at which hyphal growth from the inoculum disks occurred. *Saprolegnia ferax* grew at 100% RH but not at 99.5%. *Mortierella alpina* grew best at 100% RH and only slightly at 98%. *Pythium ultimum* grew fastest at 100% RH and had slight growth at 95%. *Fusarium culmorum* grew at

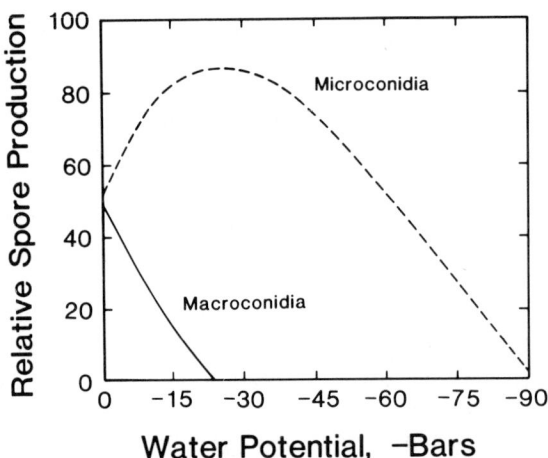

Figure 2.2. Macroconidia of *Fusarium moniliforme* are produced in greatest quantity at high water potential, microconidia at lower osmotic water potentials. After Sung, 1984.

100% RH, better at 98%, and had only a trace of growth at 92%. *Aspergillus nigricans* grew only slightly at 100% RH, well at 94%, and extremely slowly at 86%. These experiments illustrate differences in the ability of hyphae to grow through moist air from a food base.

Walter (1955) divided fungi into three groups: hygrophilic species grew to 95% RH, mesophilic species to 90%, and xerophilic species to 85%. Kouyeas agreed with Walter that important ecological groups in relation to water exist among fungi. The xerophytes could not thrive under wet conditions, and the hygrophilic fungi could not thrive under xerophytic conditions. *Aspergillus repens*, a xerophyte, grew best at RH 94% (−80.6 bars).

Fusarium moniliforme is atypical in epidemiology among the fusaria pathogenic on cereals. It is seedborne in maize to an unusual degree under low rainfall, suggesting that microconidia produced in chains are airborne, and are therefore not dependent on water-splash dispersal. Sung (1984) reported the optimum water potential for macroconidium production as −1.4 bars and for production of microconidia as −30 bars (Fig. 2.2). Macroconidia were not produced on media drier than −30 bars. Micoconidia were produced in small amounts to −90 bars. The relatively xerophytic nature of microconidium production lends support to the role of microconidia as airborne propagules. Sung commented that microconidia are produced most abundantly on dead tissue, which is usually drier than living-host tissues.

Dry Situations

Seed in Dry Soil Farmers in semi-arid regions sometimes seed into soils too dry to support good germination. Wallace (1960) attempted to determine the amount of water required to sustain germination of small grain seeds. Soil moisture, on an air-dry basis, was 3, 6, 9, 12, 15, and 20%. No seedlings emerged in 18 days at water content of 12% or less. Emergence at 15% water was 70% and at 20% water was 100%. Ungerminated seeds were recovered from the drier soils and weighed. At subemergence water levels the seeds imbibed water in direct proportion to the water content of the soil. Ungerminated seeds recovered from the dry soils were then incubated on moist filter paper. Lowest germination (36%) occurred in seeds from the 9% moisture treatment. Highest germination (90%) occurred in seeds from the driest (3% water) soil. In a second experiment Wallace incubated seeds in dry soil (8% water) for up to 17 days. The soil was then moistened to adequate levels. Germinability decreased with length of exposure in dry soil. Seeds incubated in sterilized or unsterilized soils responded equally. Reduced germination following incubation in below adequate moisture correlated with injuries to the seed coat, especially near the embryo. Response to seed treatment was proportional to the extent of seed-coat injuries.

In moist soil the dominate fungi on seeds included *Helminthosporium* and *Alternaria* species present in the soil. In dry soil, seedborne *Penicillium* spp. dominated. Treating seed with formalin prior to exposure to dry soil increased germination. Formalin leaves no residual protectant, so destruction of seedborne *Penicillium, Mucor, Aspergillus*, and *Rhizopus* spores was beneficial in dry soil. "Field" fungi were favored in moist soil, "Storage" fungi in dry soil.

Lindstrom et al. (1976) studied the effect of soil water content on germination and emergence of wheat seed in soils of water contents varying from -0.4 to -20 bars. The seed eventually germinated at -20 bars, but the rate of emergence was more sensitive to low moisture than was the germination. For all practical purposes they concluded that the minimum soil water content to sustain economically acceptable germination in the field was about -6 to -7 bars.

Water Stress within the Plant Winter wheat seeded early in the dryland-summer fallow region of central Washington tillers profusely in late summer and early autumn and develops a vigorous root system. Moisture is optimum in early spring, but soon becomes inadequate (the plant has tillered for 67 quintals/ha of wheat, but there is water for only 27 to 40 quintals/ha), and water stress develops within the plant. The plant draws moisture from deep in the soil so that a gradual adjustment of internal water relations is accomplished with a minimum of damage to the plant. Increased growth resulting from the application of nitrogen increases the stress. Cook and Papendick (1970) reported leaf water potentials of about -23 bars with no added N, -31 bars with 60 lb of N, and -37 bars with 120 lb of N per acre. Heavy tillering and fertilizer combined to expose the plant to severe water stress predisposing winter wheat to disease caused by *Fusarium culmorum* during the summer.

Wet Situations

Seeds in Wet Soil *Pythium ultimum* attacks peas, and Kerr (1964) studied the effect of soil moisture on damping-off and root rot. Kerr soaked pea seeds in water, then measured the loss in weight and the amount of sugar in the soak water. The amount of sugar lost by the seeds was proportional to the duration of soaking, and it accounted for nearly half the weight lost. Weight loss was greatest in the sand and less in light sandy loam and loam. Kerr than tested pea seed in soils differing in water content and found a direct relationship between disease and weight loss by the seed. The loss in seed weight increased with increasing bulk density of the soil. Diffusion of solutes in soil was proportional to moisture per unit volume of soil, and disease incidence increased with increased bulk density (see Fig 8.1). Disease was correlated with the amount of sugar lost per seed in all trials. Papendick and Campbell (1981) state that diffusion of solutes in water films is directly related to the cross-sectional area for flow.

Effects on the Host Gingrich and Russell (1956) studied the relationship between oxygen and water on growth of maize seedlings. Maize seedlings with radicles 11 to 14 mm in length were transplanted to silt loam with moisture tensions of about -0.3, -0.5, -1, -3, -6, -9, and -12 bars, each in atmospheres of 0.26, 2.10, 5.25, 10.5, and 21.0% oxygen by volume at 25°C. After 24 hr the seedlings were removed, the endosperms excised from the seedlings, and root growth and dry weight of the seedlings determined.

Low oxygen produced watery seedlings (99% water at 0.26% oxygen, 96% water at 2.10% oxygen) and oxygen levels of 5.25 to 21.00 produced seedlings of equal water content (94% water). Gingrich and Russell noted that enzyme activity, translocation, and deposition of hydrolysates from the endosperm were favored by adequate oxygen. With a high oxygen atmosphere seedlings made the greatest growth in the wettest soils, indicating that high water content per se did not reduce growth. Increase in dry weight was more sensitive to low oxygen than water uptake (Table 2.1), but both depend on oxygen.

Sporangium Formation and Zoospore Release The precise relationships of the responses of soilborne *Phytophthora* spp. to water have only recently begun to be known. Remember that saturated soil has a water potential of 0, ignoring the solutes in the soil solution. Water free to move downward by gravity extends from 0 to about -0.2 bar, and field capacity is usually near -0.3 bar. Thus any figure between 0 and -0.3 bar represents very wet soil.

Table 2.1. Relative water-uptake, gain in dry weight, and dry matter content of maize seedlings with differing amounts of oxygen[a]

Oxygen, %	Relative water uptake, %	Relative dry matter, %	% dry matter
0.26	18	2	1
2.10	43	29	4
5.25	67	64	6
10.50	83	74	6
21.00	100	100	6

After Gingrich and Russell, 1956.
[a]Average of all soil moisture tensions from 0.3 to 12 bars.

P. megasperma established in alfalfa seedling radicles produced the greatest number of sporangia in saturated soil (0 matric potential) (Pfender et al., 1977), very few at -0.6, and none at -2.8 bars. *P. cinnamomi* produced sporangia between -0.01 and -2.5 bars (Gisi et al., 1980), and unlike *P. megasperma*, not in saturated soil. Like *P. cinnamomi*, *P. dreschleri* did not produce sporangia in saturated soil, but they were abundant at -0.3 bar, and decreased to very few at -4 bars and none in soils drier than that (Duniway, 1975). Sporangia of *P. cactorum* were abundant at -0.1 and -0.3 bar and few at -3 bars (Sneh and McIntosh, 1974). Some species do not form sporangia in saturated soil, but all produced them abundantly from near-saturation to about field capacity.

Zoospores are released in an even narrower soil water matric potential range. Zoospores are released most abundantly at 0 matric potential, declining to near zero release at -0.02 bar in both *P. cryptogea* and *P. megasperma* (MacDonald and Duniway, 1978). Zoospores of *P. megasperma* are released in greatest abundance at 0 bar, and some sporangia germinate directly by germ tube at -0.05 bar and by germ tube only at water potentials below -0.1 bar (Pfender et al., 1977). Zoospores are released only when soil pores are essentially filled with water.

The above data are all based on matric forces only. Water availability governed by osmotic forces (ψs) provides different results. Zoospore release by *Aphanomyces euteiches* was more inhibited by -0.01 bar matric potential than by -2.0 bars solute potential (Hoch and Mitchell, 1973). In sucrose solutions *P. capsici* released zoospores to a solute potential of -8 bars, below which the sporangia germinated directly (Katsura, 1971). Matric forces of -0.02 bar stopped zoospore released by *P. cryptogea* and *P. megasperma*, but release continued in solutions of various osmotica to -4.5 to -9 bars ψs (MacDonald and Duniway, 1978). Matric and osmotic forces are not comparable in their effect on zoospore release. Zoospore formation and release mechanisms are much more sensitive to liquid volume than to water activity.

INTERACTIONS

Germling Survival

Kerr (1964) speculated that fungi favored by dry soil might be hindered by antagonism from bacteria when exudates were abundant, as around pea seeds in wet soil. Direct support of this hypothesis was obtained by Cook and Papendick (1970) in their study of *Fusarium culmorum*. Chlamydospores of *F. culmorum* germinated well in two silt loams amended with glucose and ammonium sulfate down to -50 to -60 bars, with some germination at -80 to -85 bars in 72 hr. At water potentials of -10 bars or higher, however, germ tubes either lysed or converted into replacement chlamydospores within 48 to 72 hr. In the drier soil the germlings survived and grew as hyphae for at least 6 days. Bacteria increased 200 to 300 times as much in soils -5 bars or wetter than in soil -14 to -17 bars or less. When antibacterial antibiotics were added with the nutrients, survival of *F. culmorum* germlings was high, even at water potentials greater than -1 bar. Cook and Papendick concluded that soils wetter than -10 bars were unfavorable to infection because of the susceptibility of *F. culmorum* germlings to bacterial antagonism.

Kerr (1964) worked with *Pythium ultimum*, which germinates and grows rapidly in wet soil in the presence of host exudates. It penetrates host tissues within a few hours. It escapes serious antagonistic effects from soil bacteria because infection occurs before bacterial reproduction becomes well established. *P. ultimum* invades seeds in the food-rich environment of seeds which exude nutrients into wet soil. *F. culmorum* can not compete in this situation.

Competition among Bacteria

Streptomyces scabies was controlled on potato tubers in England by irrigation which maintained soil water above -0.4 bar from tuber initiation to about 4 to 6 weeks later (Lapwood and Adams, 1975). Lenticels are susceptible for only a short period, but, to prevent scab, each successive period of lenticel development must be kept moist (Fig. 2.3). If the tuber was scabbed at the attachment end, irrigation had not been started soon enough. Scab is favored in soils drier than -0.4 bar.

B. G. Lewis believed that in wet soil true bacteria colonized the lenticel area more quickly than *S. scabies* (Lapwood and Adams, 1975). Labruyere (1971) speculated that if bacteria colonized the area first, they would utilize the leaking nutrients, protecting the lenticel area from other organisms. In drier soil, at -1 bar, soil pore necks down to less than -1.5 μm in diameter would be drained. True bacteria could not move rapidly to the developing lenticels, and *S. scabies*, which has filamentous characteristics, would have the advantage. Lapwood and Adams (1975) concluded that occupancy of

Figure 2.3. Common potato scab is favored by dry soil. The bands of scab in the top row of tubers correspond to periods of dry weather during tuber formation; the scab-free bands in the lower row of tubers correspond to periods of wet soil during tuber development. From Lapwood and Adams, 1975.

the lenticel area by nonpathogenic bacteria in wet soil protected tubers from attack by *S. scabies*.

Effect of Oxygen and Water on Hatching of Eggs of *Meloidogyne* spp.

Eggs of *M. javanica* hatched most rapidly in air or in 20% oxygen plus 80% nitrogen and less rapidly as oxygen decreased and carbon dioxide increased (Wallace, 1968a). Only the rate of hatch decreased and not total hatch when oxygen was reduced from 20% to 5%. Free eggs and eggs in egg sacs responded similarly to the various gasses; thus the egg sac matrix did not impede oxygen diffusion, nor did respiration of eggs within the egg sac have any appreciable effect (Wallace, 1968b).

Wallace exposed eggs to anaerobic conditions (100% N_2) for 2 to 10 days and then attempted to hatch them in aerated water. Two days without oxygen did not reduce hatch, but longer exposure did. After 4 days in nitrogen, about 50% hatched; after 6 or 8 days about 39% hatched.

Oxygen enters soil very rapidly as the pores drain. Penman in 1940 (in Wallace 1968a), stated that when air porosity in soil exceeds 10% by volume, the atmosphere within the soil comes to equilibrium with the surface atmosphere to depths of 39 cm within 1 hr. As soil drains and oxygen increases, hatch is stimulated. When it drains beyond a certain point, the egg sac shrinks, and hatch is restricted. If water is sufficient to support hatching, it is sufficient for larval motility. Hatching and emergence of larvae from the egg sac have similar requirements.

Wallace (1968b) determined the evaporative loss of water by egg sacs and by drops of water of the same size. Water was lost by egg sacs and free water at the same rate, and thus the egg sac has no special ability to reduce water loss. Egg sacs were incubated at 98% RH for varying numbers of days, and the number of second-stage larvae within the eggs increased for about 5 days, but no increase was noted after that. Therefore, larval development is possible in soil near the permanent wilting point, at least for a few days.

Eggs in the egg sac were protected from rapid desiccation even though Wallace's previous experiments showed that egg sacs had no special ability to reduce water loss. When egg sacs and free eggs were exposed to 98% RH and then placed in water, hatch from egg sacs was good, even after 30 days at 98% RH. In contrast, hatch from free eggs declined after only 0.5 hr at 98% RH, and none hatched after 10 days at 98% RH. Individual eggs have a larger surface-to-volume ratio than egg sacs. Water droplets the size of eggs evaporate quickly. Eggs on the inside of the egg sac were protected from the rapid desiccation that occurred in free eggs. Free larvae were as susceptible to desiccation as free eggs.

The work with oxygen and with water enabled Wallace to devise hypothetical hatching curves for a soil with a coarse and a fine texture. The pores in a coarse soil permitted oxygen to enter the soil while tension on the water was still low. Oxygen stimulated hatching, and ample water supported it. In the fine soil, oxygen did not enter the pore space until more water had been lost, and by that time the egg sac was subjeced to shrinkage, reducing hatch (Fig. 2.4). The well-aerated, coarse soil favored hatch, the finer-textured soil did not. This is probably an important factor affecting the success of root-knot nematodes in sandy soils. Wallace cautioned that eggs of nematodes that are singly distributed in soil, not in egg masses, could hatch in soil at much higher suctions because they would not be subject to the compression that occurs when egg masses are subjected to slight desiccation.

COMMENTS

This chapter is labeled water. It is impossible to treat water, or any other environmental factor, by itself, because varying one factor affects others. High soil water leads to low soil oxygen. Water interacts with temperature relations, as illustrated by the growth of *Verticillium dahliae* in culture. Interactions are seen throughout the book, but treating water as a separate subject emphasizes water and concentrates thinking about it, so the effort is justified.

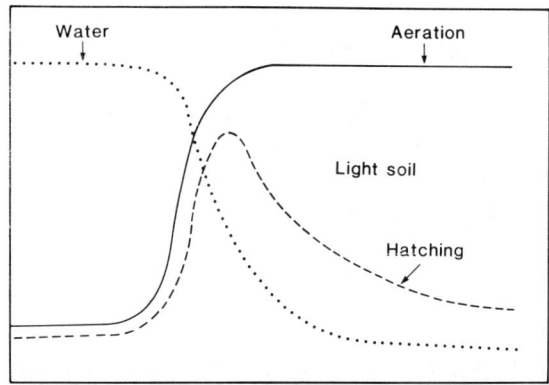

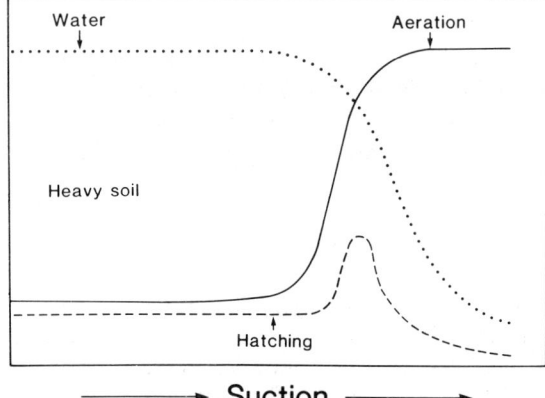

Figure 2.4. Hatching of eggs of *Meloidogyne javanica* eggs within the egg sac is very sensitive to oxygen and shrinkage of the egg sac. Hatching is greatest in a light-textured soil which is penetrated rapidly by oxygen while water is still adequate (upper) and least in a heavy soil in which oxygen diffusion is restricted while water is still adequate to prevent shrinkage of the egg sac (lower). After Wallace, 1968b. 14:223–230.

How plants respond to water shortage or root damage varies greatly with transpirational demand. Burstrom (1965) cites root mutilation experiments in which any reduction in root surface resulted in reduced top growth. In contrast, F. J. Newhook (personal communication) observed trees exposed to constant high humidities that had roots devastated by *Phytophthora cinnamomi*, yet the trees did not wilt. The relationship of root disease to actual loss is not necessarily in direct proportion. A wheat stem under little water stress may survive a lesion of given size at its base, but under severe stress that same lesion may cause death.

Another aspect of organism-enviroment interactions is the need to study different stages of a life cycle independently. This is illustrated by the effect of water on formation of macroconidia and microconidia of *Fusarium moniliforme* and by the differing responses of sporangium formation and zoospore discharge by *Phytophthora* spp. Each stage in the life cycle of an organism probably responds differently to many factors, adding to adaptability within a species.

The velocity and volume of water moving within the plant is critical to vascular parasites that spread by propagules carried upward in the plant by water in the xylem vessels. Bacteria and small fungal spores are sufficiently buoyant to be carried. Burstrom (1965) diagrammed the speed of water movement in relation to the cross-sectional areas involved. He calculated the surface areas of root hairs and roots of a 5-day wheat plant to be 690 mm^2, and the rate of water movement at 2.3 μm/min as it moved across the epidermis. Water movement through the cortex, endodermis, and within the xylem vessels was 7, 16, and 34,000 μm/min respectively. As a given volume of water, collected by a large surface, moved through increasingly smaller channels, its velocity increased.

Water quality can also be a factor. Lipps (1980), working with *Pythium iwayamai*, a snow-rot pathogen of wheat, found that snow melt water stimulated the production of more sporangia on wheat leaf disks in the water than distilled water or any salt mix he tried. The snow melt water had a water potential of about −0.035 bar.

KEY REFERENCES

Cook, R. J. and R. I. Papendick. 1970. Effect of soil water on microbial growth, antagonism, and nutrient availability in relation to soilborne fungal diseases of plants. Pages 81–88 in *Root Diseases and Soilborne Pathogens*. T. A. Toussoun, R. V. Bega and P. E. Nelson, eds. Univ. of Calif. Press, Berkeley.

Dobson, R. L. and R. L. Gabrielson. 1983. Evidence that clubroot development requires infection by both primary and secondary zoospores of *Plasmodiophora brassicae*. *Phytopathology* **73**:559–561.

Duniway, J. M. 1975. Limiting influence of low water potential on the formation of sporangia by *Phytophthora drechsleri* in soil. *Phytopathology* **65**:1089–1093.

Lapwood, D. H. and M. J. Adams. 1975. Mechanisms of control of common scab by irrigation. Pages 123–129 in *Biology and Control of Soilborne Plant Pathogens*. G. W. Bruehl, ed. Am Phytopathol. Soc., St. Paul, MN.

Sterne, R. E., G. A. Zentmyer, and M. R. Kaufmann. 1977. The effect of matric and osmotic potential of soil on Phytophthora root disease of *Persea indica*. *Phytopathology* **67**:1491–1494.

Wallace, H. R. 1968a. The influence of aeration on survival and hatch of Meloidogyne javanica. *Nematologica* **14**:223–230.

Wallace, H. R. 1968b. The influence of soil moisture on survival and hatch of *Meloidogyne javanica*. *Nematologica* **14**:231–242.

REFERENCES

Ashworth, L. J., Jr., O. C. Huisman, R. G. Grogan, and D. M. Harper. 1976. Copper-induced fungistasis of microsclerotia of *Verticillium albo-atrum* and its influence on infection of cotton in the field. *Phytopathology* **66**:970–977.

Burges, A. 1950. The downward movement of fungal spores in sandy soil. *Trans. Brit. Mycol. Soc.* **33**:142–147.

Burstrom, H. G. 1965. The physiology of roots. Pages 154–169 in *Ecology of Soil-borne Plant Pathogens*. K. F. Baker and W. C. Snyder, eds. Univ. of Calif. Press, Berkeley.

Carlile, M. J. 1983. Motility, taxis, and tropism in *Phytophthora*. Pages 95–107 in *Phytophthora, Its Biology, Taxonomy, Ecology and Pathology*. D. C. Erwin, S. Bartnicki-Garcia, and P. H. Tsao, eds. American Phytopathology Society, St. Paul, MN.

Cook, R. J. and R. I. Papendick. 1970. Soil water potential as a factor in the ecology of *Fusarium roseum* f. sp. *cerealis* 'Culmorum'. *Plant Soil* **32**:131–145.

Cook, R. J. and R. I. Papendick. 1972. Influence of water potential of soils and plants on root disease. *Ann. Rev. Phytopathol.* **10**:349–374.

Dobson, R. L., R. L. Gabrielson, and A. S. Baker. 1982. Soil water matric potential requirements for root-hair and cortical infection of Chinese cabbage by *Plasmodiophora brassicae*. *Phytopathology* **72**:1598–1600.

Duniway, J. M. 1983. Role of physical factors in the development of *Phytophthora* diseases. Pages 175–187 in *Phytophthora, Its Biology, Taxonomy, Ecology, and Pathology*. D. C. Erwin, S. Bartnicki-Garcia and P. H. Tsao, eds. American Phytopathology Society., St. Paul, MN.

Gingrich, J. R. and M. B. Russell. 1956. Effect of soil moisture tension and oxygen concentration on the growth of corn roots. *Agron. J.* **48**:517–520.

Gisa, U., G. A. Zentmyer, and L. J. Klure. 1980. Production of sporangia by *Phytophthora cinnamomi* and *P. palmivora* in soils at different matric potentials. *Phytopathology* **70**:301–306.

Hepple, S. 1960. The movement of fungal spores in soil. *Trans. Brit. Mycol. Soc.* **43**:73–79.

Hoch, H. C. and J. E. Mitchell. 1973. The effects of osmotic water potentials on *Aphanomyces euteiches* during zoosporogenesis. *Can. J. Bot.* **51**:413–420.

Ingram, D. S. and I. C. Tommerup. 1972. The life history of *Plasmodiophora brassicae*. *Proc. Roy. Soc. Lond., Ser B*.**180**:103–112.

Katsura, K. 1971. Some ecological studies on zoospore of *Phytophthora capsici* Leonian. *Rev. Plant Prot. Res.* **4**:58–70.

Kerr, A. 1964. The influence of soil moisture on infection of peas by *Pythium ultimum*. *Aust. J. Biol. Sci.* **17**:676–685.

Kouyeas, V. 1964. An approach to the study of moisture relations of soil fungi. *Plant Soil* **20**:351–363.

Kramer, P. J. 1949. *Plant and Soil Water Relationships*. McGraw-Hill, New York.

Labruyere, R. E. 1971. Common scab and its control in seed-potato crops. *Versl. Landouwk. Onderz. Ned.* No. 767.

Lacy, J. 1965. The infectivity of soils containing *Phytophthora infestans*. *Ann. Appl. Biol.* **56**:363–380.

Lange, O. L., L. Kappen, and E. D. Schulze. 1976. *Water and Plant Life: Problems and Modern Approaches*. Springer-Verlag, Berlin.

Large, E. C. 1962. *The Advance of the Fungi*. Dover, New York.

Lindstrom, M. J., R. I. Papendick, and F. E. Koehler. 1976. A model to predict winter wheat emergence as affected by soil temperature, water potential and depth of planting. *Agron J.* **68**:137–141.

Lipps, P. E. 1980. The influence of temperature and water potential on asexual reproduction by *Pythium* spp. associated with snow rot of wheat. *Phytopathology* **70**:794–797.

MacDonald, J. D. and J. M. Duniway. 1978. Temperature and water stress effects on sporangium viability and zoospore discharge in *Phytophthora cryptogea* and *P. megasperma*. *Phytopathology* **68**:1449–1455.

Manandhar, J. B. and G. W. Bruehl. 1973. *In vitro* interactions of Fusarium and Verticillium wilt fungi with water, pH, and temperature. *Phytopathology* **63**:413–419.

Marshall, K. C. 1971. Sorption interactions between soil particles and microorganisms. Pages 405–445 in *Soil Biochemistry*, Vol. 2. A. D. McLaren and J. J. Skujins, eds. Marcel Dekkker, New York.

Papendick, R. I. and G. S. Campbell. 1981. Theory and measurement of water potential. Pages 1–22 in *Water Potential Relations in Soil Microbiology*. J. F. Parr, W. R. Gardner, and L. F. Elliott, eds. Soil Science Society of America.

Pfender, W. F., R. B. Hine, and M. E. Stanghellini. 1977. Production of sporangia and release of zoospores by *Phytophthora megasperma* in soil. *Phytopathology* **67**:657–663.

Scott, W. J. 1956. Water relations of food spoilage microorganisms. *Advan. Food Res.* **7**:83–127.

Scott, W. J. 1962. Available water and microbial growth. Pages 89–105 in *Proceedings, Low Temperature Microbiology Symposium, Camden, NJ, 1961*. Campbell Soup Co.

Sneh, B. and D. L. McIntosh. 1974. Studies on the behavior and survival of *Phytophthora cactorum* in soil. *Can. J. Bot.* **52**:795–802.

Stolzy, L. H., J. Letey, L. J. Klotz, and C. K. Labanauskas. 1965. Water and aeration as factors in root decay of *Citrus sinesis*. *Phytopathology* **55**:270–275.

Sung, J. M. 1984. The soilborne diseases of the major crops in Korea. Pages 1–16 in *Soilborne Crop Diseases in Asia*. Food and Fertilizer Technology Center for the Asian and Pacific Region, Taipei, Taiwan.

Wakefield, E. and G. R. Bisby. 1941. List of Hyphomycetes recorded for Britain. *Trans. Brit. Mycol. Soc.* **25**:49–126.

Wallace, H. A. H. 1960. Factors affecting subsequent germination of cereal seeds in soils of subgermination moisture content. *Can. J. Bot.* **38**:287–306.

Walter, H. 1955. The water economy and the hydrature of plants. *Ann. Rev. Plant Physiol.* **6**:239–252.

Wilkinson, H. T., R. D. Miller, and R. L. Millar. 1981. Infiltration of fungal and bacterial propagules into soil. *Soil Sci. Soc. Am. Proc.* **45**:1034–1039.

Young, B. R., F. J. Newhook, and R. N. Allen. 1979. Motility and chemotactic response of *Phytophthora cinnamoni* zoospores in 'ideal soils'. *Trans. Br. Mycol. Soc.* **72**:395–401.

Zentmyer, G. A. 1980. *Phytophthora cinnamomi and the Diseases It Causes*. American Phytopathology Society, St. Paul, MN.

Chapter 3

Temperature

Temperature directly affects metabolism and growth of higher plants, plant pathogens and other microorganisms in the soil, and host plant pathogen-microbiota interactions. DeCandolle in 1855 observed that higher plants tended to be distributed around the world in east-to-west zones reflecting temperature. Kelman (1953) reported that *Pseudomonas solanacearum* occurred only within the warm, moist climatic zones of the earth. Taylor and Sasser (1978) described the general distribution of the major root-knot nematodes. *Meloidogyne javanica, M. incognita, M. arenaria* and *M. hapla*, with respect to temperature zones.

The golden period of temperature studies in relation to soilborne diseases began at the University of Wisconsin under the leadership of L. R. Jones (see Jones et al., 1926 and Walker, 1950 for details). The impetus originated with the discovery of the significance of temperature in the development of Fusarium wilts, black root of tobacco, and Fusarium seedling blights of wheat and maize.

A classic approach to the study of temperature on plant disease was used by Ball in Egypt between 1905 and 1908, in his study of sore shin of cotton caused by *Rhizoctonia solani* (in Young, 1928). Ball found that *R. solani* grew fastest in culture near 23°C; that cotton seedlings were most vigorous between 30 and 35°C, with the optimum near 32°C; and that the disease was severe in soil at 23°C or cooler and slight at 30 to 35°C. The pathogen was most vigorous in relatively cool conditions, the host in warm conditions, and the disease in cool conditions. Ball conceived the fundamental principles of temperature relations necessary for disease (pathogen, host, pathogen × host interaction).

When the disease curve follows the curve of the growth of the pathogen in culture, the natural conclusion is that the pathogen dominates the interaction. The susceptible host in these cases has little real resistance, since the pathogen progresses in the host proportional to its development on agar media. Fusarium wilt in susceptible cabbage, flax, and tomato roughly follows the growth rates of the pathogens in culture (Jones et al., 1926).

When the disease curve departs markedly from the growth curve of the pathogen in culture, the host plays a greater role in the interaction. The host possesses significant resistance, but adverse temperature weakens it and acts as a predisposing factor. The study by Dickson (1923) of seedling blight of wheat and corn caused by *Fusarium graminearum* exemplifies predisposition. The same isolates of the pathogen are more destructive on maize in cool soil and on wheat in warm soil. Maize, a warm-weather plant, is more resistant in warm soil. Wheat, a cool-weather plant, is more resistant in cool soil. These plants possess significant natural resistance overcome by conditions adverse to the host (predisposition).

When the disease-temperature curve does not follow the pathogen-temperature curve, interaction with the soil microflora may be involved. Take-all, caused by *Gaeumannomyces graminis*, was most severe on wheat in natural soil at 12 to 16°C, but in sterilized soil the disease was more severe at 18°C. Henry (1932) concluded that in natural soil the pathogen was inhibited more by the soil microflora at higher soil temperatures, inleuencing the host × pathogen interaction indirectly, and establishing a temperature × pathogen (which grows best in culture near 25°C) × host × soil microflora × disease interaction.

True vascular wilt pathogens are usually confined to xylem vessels during systemic invasion, with invasion of surrounding parenchyma occurring during and after host decline. They are not classed as cortical rotters, yet if environmental conditions cause plant stress, some become typical cortical rotters.

Environmental conditions unfavorable to the host are particularly important in cortical rots. Stress of the host, whether due to water, temperature, oxygen deficiency, malnutrition, or host fruiting, reduces the level of resistance. Stress is not as simple a concept as may appear, however. Wounds should be distinguished from stress. All nematodes that enter roots make wounds, but unless the numbers are high, little host stress may result. Root-knot nematodes, in contrast, may make few wounds but cause changes in host physiology that destroy or greatly reduce resistance to several pathogens. A wounded seed imbibes water too rapidly, damaging membranes within the seed. If the intercellular spaces within the seed fill with water, the wound leads to oxygen deficiency within the embryo.

Teachers, of necessity, make many generalities, but generalities make it possible to impart more understanding in limited time than would otherwise be possible. That the severity of Fusarium wilts follows the growth curve of the pathogens on culture media is more true than false but some host-pathogen interactions depart from this generality.

FUSARIUM WILTS

All formae speciales of *F. oxysporum* have in vitro temperature optima near 28°C, evidence of a basic physiologic similarity that persists after pathogenic specialization within different suscepts develops. The Fusarium wilt fungi almost surely descended from a common progenitor. Association with cool-weather crops (flax, peas, cabbage) has not altered the basic physiology of the fungus.

The Fusarium wilts, in general, are warm soil diseases, being most devastating near 28°C or near the optimum temperature for growth of the pathogens in culture. Wilts of muskmelon and watermelon differ, in that at low temperatures serious preemergence and postemergence seedling blights can occur, aggravated by water stress. On these hosts serious losses occur over a range of temperatures, from as low as the host can grow to that where the soil is so warm it is past the optimum temperature for growth of the fungus. Another exception is pea wilt, which is as serious as a true wilt at 21 to 22°C, well below the pathogens' optimum for growth in culture, 28°C. With low-temperature crops (cabbage, flax, peas) early seeding is an advantage, in that plants make considerable growth before temperatures favoring maximum virulence are reached.

Bell and Mace (1981) generalized that temperature has a greater effect on Fusarium wilt severity than host genotype; that resistance to cabbage yellows, whether mono or polygenic, breaks down at temperatures much above 24°C; and that resistance to pea wilt likewise breaks down above 28°C, at which temperature severe cortical rot destroys all genotypes. This is not a serious problem in peas and cabbage, however, because these plants are adapted to cool conditions and they are produced commercially in climates in which the resistance is effective.

Disease Proportional to In Vitro Growth of the Pathogen

Flax, like cabbage, is a cool-weather crop. *F. oxysporum* f. sp. *lini* grows in culture with an optimum between 25 and 28°C. Tisdale (1917) reported that wilt developed most aggressively at 24 to 28°C, and that susceptible plants wilted at 20°C. As in cabbage yellows, flax wilt severity roughly followed the rate of growth of the pathogen in culture. Grossman (in Kommedahl et al., 1970) reported that at 27°C both resistant and susceptible flaxes wilted.

Resistance in flax and cabbage both fail in warm soil. In the field, flax wilt severity was proportional to the number of hours at which the soil temperature exceeded 21°C (Kommedahl et al., 1970). Water stress increased wilt in plants with moderate resistance.

Tomato and cotton are warm-weather plants, but apparently this makes no fundamental difference in the host-pathogen interaction when compared to that of cabbage and flax. The optimum for *F. oxysporum* f. sp. *lycopersici* in culture is 28°C, wilt of tomatoes is most severe at about 28°C, and disease is inhibited below 21°C (Clayton, 1923). Cotton wilt (*F. oxysporum* f. sp. *vasifectum*) severity in an inoculated loamy fine sand corresponds closely with the growth curve of the fungus in culture (Young, 1928).

Temperature × Type of Resistance Interactions

Cabbage yellows, caused by *Fusarium oxysporum* f. sp. *conglutinans*, became devastating near Racine, Wisconsin. Much of the best cabbage soil, after having been repeatedly sown to cabbage, became "cabbage sick" and unproductive (Jones and Gilman, 1915). Cabbage yellows is severe in warm, dry seasons and mild in cool, wet seasons. Gilman grew cabbage in infested soil in a cool greenhouse (12 to 16°C) and in a warm house (23 to 26°C). The cabbage remained healthy in the cool house and was severely diseased in the warm house. Plants moved from the cool house to the warm house became diseased.

F. oxysporum f. sp. *conglutinans* grows best in culture near 28°C. Maximum wilt develops at about 26°C. Wisconsin Hollander is wilt resistant, but in warm, dry seasons it suffers from wilt. Wisconsin All Seasons is resistant even in warm seasons. Walker (1950) and his students found that at 22°C all cabbages lacking resistance died, and those with appreciable resistance lived. At 24°C, Wisconsin All Seasons lived, Wisconsin Hollander died. If the temperature was maintained much above 24°C, Wisconsin All Seasons also failed. Wisconsin Hollander has polygenic resistance (type B) and Wisconsin All Seasons has an additional single dominant gene that confers greater resistance (type A). Polygenic resistance is valuable, because cabbage is typically well developed by the time soil temperatures rise sufficiently to render this resistance ineffective.

Temperature × Race of the Pathogen Interactions

Fusarium oxysporum f. sp. *pisi* grows fastest on solid media at about 28°C. Race 1 produces typical wilt most rapidly at about 21 to 23°C. Race 2 wilts peas most rapidly at about 24 to 28°C. Race 2 was not discovered until peas resistant to race 1 were widely grown. Disease produced by race 1 is called *pea wilt*; disease produced by race 2 is called *near-wilt*. Near-wilt is important only to peas resistant to the other races and during warm seasons. The severity of pea wilt departs significantly from the growth

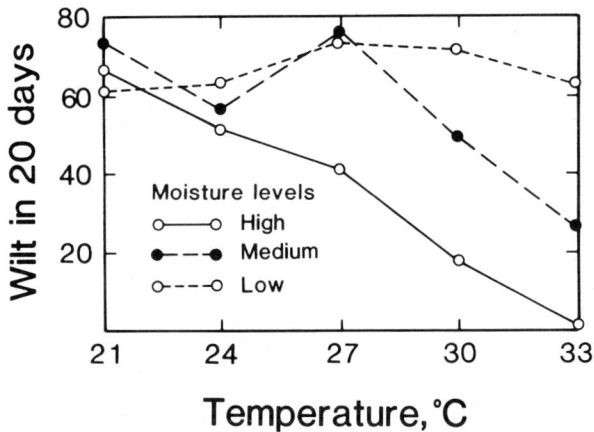

Figure 3.1. The response of muskmelon to *Fusarium oxysporum* f. sp. *melonis* in temperature × water interaction studies is typical of a cortical rot rather than of a vascular wilt. Muskmelon is susceptible over a wide temperature range when subjected to moisture stress. After Miller, 1945.

curve of the fungus in culture, but near-wilt follows the in-vitro growth curve closely (Walker, 1950).

Temperature × Moisture Interactions

Muskmelon, a warm-weather plant, has been grown repeatedly in Minnesota and southern Ontario on sandy soils that warmed early in the spring. Such soils quickly have become heavily infested with *F. oxysporum* f. sp. *melonis*, and muskmelons have been severely attacked from seeding time onward (Leach and Currence, 1938). The fungus grows in culture between 6 and 37°C, with the optimum near 27°C. Wilt is worst from 27°C downward and even susceptible cultivars are resistant from 30°C upward. When Miller (1945) grew muskmelons in infested soil, maximum damage occurred at 18°C. In moist sandy loam at 45% moisture-holding capacity (MHC), disease declines from a maximum at 18 to 21°C down to no disease at 33°C. In dry soil (27% MHC), muskmelons are affected by disease at all temperatures tested (18 to 33°C). Water stress negated the temperature effect over the entire temperature range (Fig. 3.1).

VERTICILLIUM WILTS

Fusarium wilts were so destructive where they were adapted that they forced each host to economic ruin. All cultural methods of control, though helpful, were inadequate. These wilts were controlled economically only by development of resistant cultivars. Breeding for disease resistance received much impetus from successes with the Fusarium wilts. In every case in which a concerted effort was made, resistance was found. Verticillium wilts, in contrast, are usually not as devastating, resistance is generally not as clearly defined, and crop management is more important in reducing losses in annual crops.

The two major Verticillium wilt pathogens are *V. albo-atrum* and *V. dahliae*. The species epithet *albo-atrum* describes cultures that are light-colored (albo) at first, becoming dark (atrum) with age in culture. *V. albo-atrum*, with its dark, thickened mycelium, became known as DM (dark, mycelial type). *V. dahliae* produces abundant microsclerotia, or tiny pseudosclerotia, on most media, and it became known as the PS (pseudosclerotial) or MS (microsclerotial) form of *V. albo-atrum* by those who did not recognize *V. dahliae* as a valid species. Most modern workers refer to *V. albo-atrum* as those isolates that do not form microsclerotia and to *V. dahliae* as those isolates that form microsclerotia in abundance. When the description by the author is adequate, dark mycelial forms are called *V. albo-atrum* and microsclerotial forms are called *V. dahliae*, regardless of the terminology of the cited author.

Verticillium wilts are prevalent in temperate climates, with *V. albo-atrum* important in the cooler, humid northern areas and *V. dahliae* extending further south, particularly on irrigated crops (Snyder and Smith, 1981). Both fungi are unimportant or absent in the true, humid tropics and very humid semitropics. The wider distribution of *V. dahliae* is due to its adaptation to slightly warmer conditions and to its production of microsclerotia, making it more difficult to control by rotation.

Effect of Temperature on Geographic Distribution

V. albo-atrum and *V. dahliae* differ in relation to temperature (Edson and Shapavalov, 1920). Potato isolates from West Virginia produce microsclerotia abundantly in culture and grow at 30°C (*V. dahliae*). Potato isolates from Maine produce few or no sclerotia and little or no growth at 30°C (*V. albo-atrum*). Ludbrook (1933) studied cultures from parts of the United States, Europe, and Australia. *V. albo-atrum* comes only from cool areas; *V. dahliae* from cool and warm areas. *V. albo-atrum* from all sources fails to grow at 30°C. Isaac (1949) and Robinson et al. (1957) found the same temperature relationships. To date, *V. albo-atrum*, though virulent on cotton, has not been isolated from cotton in the major cotton-producing areas of the United States (Schnathorst, 1973). *V. dahliae* is an important pathogen of cotton in Mississippi, Arkansas, the high plains of Texas, and across the arid Southwest to the San Joaquin Valley in California. *V. albo-atrum* is more virulent than *V. dahliae* on cotton, so relative virulence is not responsible for their difference in geographic range. The small difference in response to temperature is important.

Both species are pathogenic on potatoes from 12 to 28°C, but *V.*

dahliae alone is still pathogenic at 30° (Ludbrook, 1933). Brinkerhoff (1973) reviewed temperature studies of *V. dahliae* in 19 host species. In all cases the disease curves closely follows the growth curve of the fungus in culture, indicating that the pathogen dominates the host-pathogen interaction.

Caution is needed in reading the literature (Brinkerhoff, 1973). Verticillium wilt (*V. dahliae*) of mint in Michigan is reported to be most severe in hot dry seasons (Nelson, in Brinkerhoff); whereas Verticillium wilt of cotton is most severe in cool, wet summers in Arkansas (Young et al., 1959). June-August air temperatures in the "hot" season in Michigan averaged 21 to 24°C; temperatures during the same period in the "cool," wet season in Arkansas averaged 24 to 25°C. The "hot" season in Michigan was cooler than the "cool" season in Arkansas.

Soil and Air Temperature

When soil and air temperatures differ, they interact, and the effects could affect pathogens in the vascular system disproportionately. Edgington and Walker (1957) found that the effect of soil temperature was more influential than air temperature, but both were important (Table 3.1). Water from cool soil (20°C) reduces the temperature of a tomato stem in warm air (28°C) during periods of active transpiration. Cool water reduces the stem temperature significantly 4 cm above the soil line and has some effect even at 30 cm (Fig. 3.2). The temperature of the lower part of the stem is significant in

Table 3.1. Effect of soil and air temperature upon severity of verticillium wilt of tomato

Air temperature	Soil temperature		
	20	24	28
V. albo-atrum			
16	40	33	7
20	39	40	7
24	36	35	6
28	22	20	6
V. dahiae			
16	37	38	39
20	37	38	39
24	40	41	46
28	41	44	44

Source: Edgington and Walker, 1957.

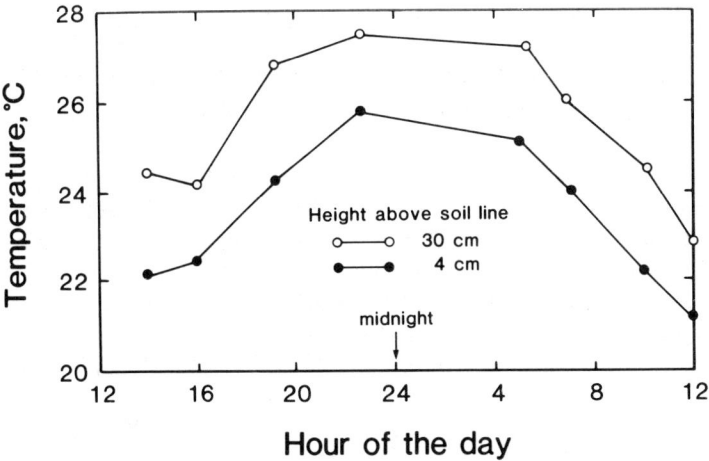

Figure 3.2. Soil colder (20°C) than the air (28°C) causes discrepancies between stem tissue and air temperature during periods of intense sunlight because of the cooling effect of water on internal stem tissue, particularly near the soil line. This effect could be important in vascular wilts. After Edgington and Walker, 1957.

wilts, and this experiment illustrates that soil temperature (which equals water temperature) affects aerial portions of the plant to some degree.

Irrigation Methods Guayule plants in a semi-desert suffered the greatest disease (77%) from *V. dahliae* with weekly irrigations, and less with less frequent waterings (62% with irrigation every 2 weeks, 42% every 4 weeks, and 8% in unirrigated plots). Schneider (1948) attributed disease increase with many irrigations to reduced soil temperature. In New Mexico Leyendecker (1950) found that four irrigations on cotton caused an average of 32% wilt and seven irrigations caused 48% wilt. Water can not be withheld as a means of control, however, because water deficit affects yield more than disease. Overirrigation wastes water and increases Verticillium wilt by cooling the soil. Raney (1973) recommended irrigating before planting to fill the soil profile and the use of as few heavy irrigations as practicable for maximum yields. Since each irrigation reduces soil temperature, the use of frequent, light irrigations leads to the highest disease incidence.

Planting on ridges can elevate the soil temperature. Leyendecker (1950) tested three methods of planting and irrigating cotton. Conventional rows planted on flat land and flood irrigated result in the most disease and lowest yield. Cotton seeded on beds raised 20 cm with furrows for water between pairs of two rows on the beds leads to intermediate levels of disease and yield. Cotton seeded in paired rows on beds raised 37 cm with 130 cm between the furrows results in the least disease and the highest yield. Reduced disease in the high-ridge system is attributed to elevated soil temperature at the 15-cm soil depth during the growing season. The soil averaged 27°C under the cotton on ridges, 24°C in the low ridges, and 21 to 23°C under cotton under flat tillage. Young et al. (1959) grew cotton in Arkansas on soil with and without raised ridges. In Arkansas the soil temperature is essentially the same in both management systems, and no wilt reduction is achieved by ridging.

The cotton in Arkansas was rain fed, not irrigated. Note that cultural methods play little role in Fusarium wilts but that in Verticillium wilts, workers have found significant responses to management practices that influenced soil temperature.

Recovery In regions of high summer temperatures plants can recover, at least temporarily, from Verticillium wilt. At Shafter, California, Garber and Presley (1971) isolated *V. dahliae* from all leaves of an infected cotton plant in June 1967. In July and August leaf isolations dropped to 30 to 40%. It is presumed that these leaves had been infected but that high air temperatures led to the death of the fungus in many leaves. Air temperatures reached daily maxima above 35°C on 49 days during July and August. Leaf infection returned to 100% with cooler weather in September and October following reinvasion from below. Wilhelm and Taylor (1965) found the same phenomenon in olive trees in California. With sustained cool weather there is no temperature-mediated recovery.

Tests for Resistance

All cottons tested by Bell and Presley (1969) are susceptible at 22°C and all are resistant at 32°C (*V. dahliae*). Differentiation between resistant and susceptible cottons occurs between 25 and 29°C. Varieties intermediate in resistance are susceptible at 25°C, tolerant at 27° and resistant at 29°. In Texas, where Verticillium and Fusarium wilts are both important, cottons in the breeding program are tested for resistance to both diseases at 28°C (Abdel-Raheem and Bird, 1968). Fusarium wilt is severe between 28 and 32°C, and Verticillium wilt is severe from 20 to 28°C, so their ranges overlap.

Bell (1973) stated that "dropping the temperature from 30°C to 22°C has more influence on the resistance of cotton to *Verticillium dahliae* than the total genetic variability known to occur in the genus *Gossypium*." The fungus is more virulent at lower temperatures. Bell recommends testing cottons at 26 to 28°C when working with the virulent, defoliating strains and at 23 to 25°C when working with the mild strains.

CORTICAL ROTS

Fusarium graminearum, the anamorph of *Gibberella zeae*, causes a severe seedling blight of maize at temperatures below 20°C. Above 24°C maize is resistant. Wheat seedlings resist *F. graminearum* at 8°C and become susceptible at 12°C and above. These reactions occur reciprocally with the same isolate; the difference is due to host reactions. Dickson (1923) credited B. Koehler with unpublished evidence that at soil moisture levels below 45 to 50% MHC wheat seedlings are predisposed to *F. graminearum*, even at low temperatures. Dickson confirmed this relationship. At 60% MHC, no wheat seedlings become blighted at 8°C, but 44% are blighted at 45% MHC and 72% blighted at 30% MHC. Moisture stress negate the effects of a favorable temperature in this combination, as it did in Miller's 1945 study of Fusarium wilt of watermelon.

Thielavopsis basicola causes a serious root rot of tobacco in cold soil. The dry weight of White Burley tobacco grown in naturally infested soil at 17 to 18°C averages 10% of the growth in clean soil, 21% at 20 to 21°C, 23% at 23 to 24°C, 44% at 25 to 26°C, 48% at 28 to 29°C, and 97% at 31 to 32°C (Johnson and Hartman, 1919). Tobacco resists the fungus increasingly as the temperature rises, even though *T. basicola* grows fastest in culture at 28 to 30°C. Both the fungus and the host are warm-weather organisms. A short, cool period after transplanting results in stunting but if warm weather comes quickly, recovery can occur. The main effect of cool soil is in predisposing tobacco, a warm-climate plant.

TEMPERATURE AFFECTS GEOGRAPHIC DISTRIBUTION

Temperature plays a major role in plant geography, and it also delimits the geographic range of many important plant pathogens. Among the widely distributed root-knot nematodes, *Meloidogyne javanica* is truly tropical. The southern root-knot nematode *M. incognita* is more subtropical; *M. arenaria* is of temperate adaptation; and *M. hapla*, the northern root-knot nematode, occurs as far north as Washington State and parts of southern Canada (Taylor and Sasser, 1978). The range of *Pseudomonas solanacearum* distribution is primarily between 45°N and 45°S latitude in moist areas (Kelman, 1953). *P. solanacearum* has been introduced into northern areas repeatedly, but under cooler conditions it does not persist. *Phymatotrichum omnivorum* is limited on the north by temperature (Ezekiel, 1945 and Fig. 3.3). *Verticillium albo-atrum* is more common in cooler regions that *V. dahliae* (Snyder and Smith, 1981). The above pathogens, all with wide host ranges, illustrate the effect of temperature on geographic distribution because lack of a suitable host does not affect their geographic distribution.

In the northern United States onions are planted in cool soil in the spring, in the southern United States they are planted in summer or early autumn in very warm soil. Teliospores of *Urocystis cepulae* germinate from about 6°C to nearly 30°C, but at high temperatures fewer spores germinate and the promycelia are weak (Walker and Wellman, 1926). Infection is high from 10 to 12°C, about the lower limit for onion seedling development. Onion smut has not been reported south of Kentucky (Walker, 1950), and temperature is the controlling factor.

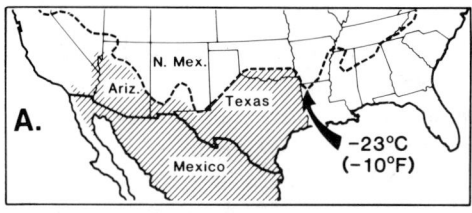

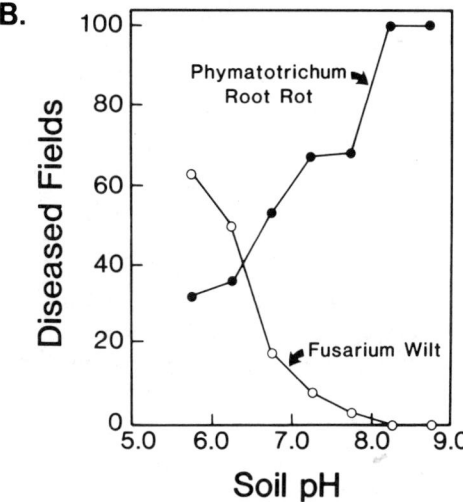

Figure 3.3. Two classics of ecology, the limitation of the survival of *Phymatotrichum omnivorum* by low minimum winter temperatures (A, after Ezekiel, 1945) and by soil pH upon its distribution (B, after Taubenhaus et al., 1928). This pathogen, in spite of its wide host range, has not spread to regions of cold winters (North) and it has not spread into warm regions with acid soils (East).

Transition Zones

Observations in transitional zones are of great value in determining critical factors in the distribution of an organism. Black shank of tobacco (*Phytophthora parasitica* var. *nicotianae*) is serious every year in North Carolina when susceptible cultivars are grown without rotation (Apple, 1963). In contrast, Wills (1964) reported considerable year-to-year variation in black shank in Virginia, the state immediately to the north of North Carolina. Using uniform soil, cultivar, and standard cultural practices, he compared disease severity with weather records. In Virginia disease severity increases following a wet autumn with a late killing frost and decreases following a dry autumn with an early killing frost. The date of the first autumn freeze is the single most important weather factor. Wills concluded that frost kills the vegetative hyphae but not chlamydospores, that a prolonged autumn enables the pathogen to form many chlamydospores before the tobacco is killed, that an early frost curtails production of chlamydospores, thus reducing inoculum for the next crop.

The hypothesis of Wills explains "why (1) soil infestation in black shank nurseries is more uniform from year to year south of Virginia than in Virginia, (2) why black shank has never been and probably never will be serious north of Virginia, and (3) why in Virginia it has not been possible to depend upon field tests for indexing tobacco for black shank resistance."

In contrast, Kenerley and Bruck (1983) observed in a forest seedling nursery that propagules of *P. cinnamomi* in nonrhizosphere soil are abundant before onset of cold, but that most are killed by the cold. Viable chlamydospores of *P. cinnamomi* decrease rapidly when soil temperatures at the 7.5 cm depth fall to freezing or below. The literature review in Kuske and Benson (1983) indicates that *P. parasitica* and *P. cinnamomi* are limited in their northward adaptation by susceptibility of the resting structures to cold.

Escape from Climatic Rigors

Temperature seemingly plays a minimal role in the geographic distribution of some pathogens. *Fusarium solani* f. sp. *phaseoli* follows beans relentlessly into widely different areas, and *F. oxysporum* f. sp. *lini* follows flax. These pathogens are well adapted to hosts, developing when the host develops. After maturation or death of the host, they survive by means of dormant chlamydospores in the soil until favorable conditions return. Distribution maps of these pathogens are really distribution maps of the hosts.

Latitude and Altitude

Distribution maps, superficially examined, can be misleading. The cyst-nematodes of potato (*Solanum tuberosum*) are cool-temperature pathogens, yet distribution maps show them to be present in "warm" countries. Potatoes are grown at high elevations in northern South America and in Central America. Potato is a cool-weather plant, and *Globodera rostochiensis* and *G. pallida* are cool-weather nematodes. Latitude and altitude both change climate.

The zonal soil concept (Chap. 1) has been of greatest value to me in regions of rapidly changing elevation rather than in the gradually changing Great Plains. In a small county (Asotin) of Washington, in a distance of about 35 km, the soil changes from a light brown to a deep black prairie soil on a sloping plateau from the Snake River to the foot of the Blue Mountains. As the elevation increases, moisture increases and the temperature decreases. Within this short distance the soil change is equal to that from the western edge of Minnesota to the Rocky Mountains, or equal to the changes in the entire Canadian prairie. Elevation, rapidly changing, requires alertness among pathologists dealing with soilborne plant pathogens.

PLANTING DATE

In climates with wide annual fluctuations between winter and summer temperatures, soil temperature undergoes an annual cycle, cold to cool in early spring, warm to hot in summer, and warm to cool in autumn. These annual cycles offer advantages and disadvantages for disease management. Warm-weather plants are stressed when seeded in cool soil early in the spring, as in Pythium seed rot and seedling blight of maize or in seedling disorders of cotton. Potatoes in part of California can be grown during the winter on land infested with root-knot nematodes, harvested, and sold on the early market without blemish, because they are dug before temperatures are warm enough for nematode activity. Vegetables are grown in southern Texas, Arizona, New Mexico, and northern Mexico during winter months with no danger of attack by *Phymatotrichum omnivorum*. Common bunt (*Tilletia caries* and *T. foetida*) can be minimized by seeding winter wheat early when the soil is warm. Flag smut of wheat (*Urocystis agropyri*) in contrast, is favored by seeding early in warm, relatively dry soil. To the extent that seeding date can be economically altered, a farmer has some control over soil temperature, at least during certain critical periods of host or pathogen development.

Epps and Chambers (1963) studied the effect of seeding date (soil temperature, day length) on the production of soybeans on land infested with the soybean cyst-nematode, *Heterodera glycines*. Soybeans yielded best when seeded May 1–15. The authors found no way to prevent losses by manipulating the seeding date of soybeans. The host is a relatively warm-season crop, and the nematode is most active in warm soil.

SEQUENCE OF PATHOGENS

An excellent example of the effect of temperature on sequence of pathogens is provided by the wheat rusts. *Puccinia striiformis* is adapted to cool temperatures, and it starts to increase early in the spring. *P. recondita* is intermediate as to temperature requirement, developing later in the spring, and *P. graminis* is restricted to warm weather. *P. striiformis* (stripe rust) develops slowly over a long period and it can occupy the leaves to the exclusion of *P. recondita* (leaf rust). If the wheat is resistant to stripe rust but susceptible to leaf rust, *P. recondita* can occupy the leaves and exclude P. graminis from the leaves. *P. graminis*, which develops last, can still attack the stem proper, but it will be restricted to tissues not already occupied by the other rusts. Thus the differing temperature relations of these three rusts affect competitive relationships.

Similar but less easily seen sequences exist among soilborne pathogens. *Pythium ultimum* is serious on maize seeded in cold wet soil; *P. arrhenomanes*, though capable of rotting corn seed, is significant later in the season as a root rotter when soil temperatures are higher.

DIURNAL FLUCTUATIONS

In the early studies by University of Wisconsin scientists and by others using Wisconsin soil temperature tanks, soil temperatures fluctuating about a mean were compared with constant temperatures. In general, fluctuations about a mean give the same result as a constant temperature at that mean, so studies at constant temperatures have been accepted as the standard practice. Smith (1966), however, found that fluctuating temperature increased the severity of charcoal rot (*Macrophomina phaseolina*) on sugar pine in forest nurseries in California. Charcoal rot on most hosts is dependent on host stress, probably accounting for the strong effect of diurnal temperature fluctuations. Soils in the forest nurseries have a daily fluctuation in temperature of about 10°C at the 2.5 cm depth and 3.3°C at the 15 cm depth. Smith grew sugar pine seedlings in a sterilized sand-vermiculite mixture inoculated with sclerotia of *M. phaseolina*. He chose 25 and 30°C as main temperatures and varied the daily temperature fluctuations about these means 0, ±4.5, and ±8.9°C. Mortality of seedlings was lowest at 25°C, but it increased as the amplitude of the fluctuations about this mean increased. Mortality was high at a mean of 30°C, but the speed of death increased greatly as the daily temperature fluctuations increased (Fig. 3.4).

M. phaseolina is a high-temperature fungus, and Smith (1964) studied the effect of daily temperature fluctuations on its growth in culture. Most rapid hyphal growth occurred at 30°C ±2.2°C. Growth at 30°C ±8.9°C was slower. The increased mortality (Smith, 1966) of pine seedlings at 30°C

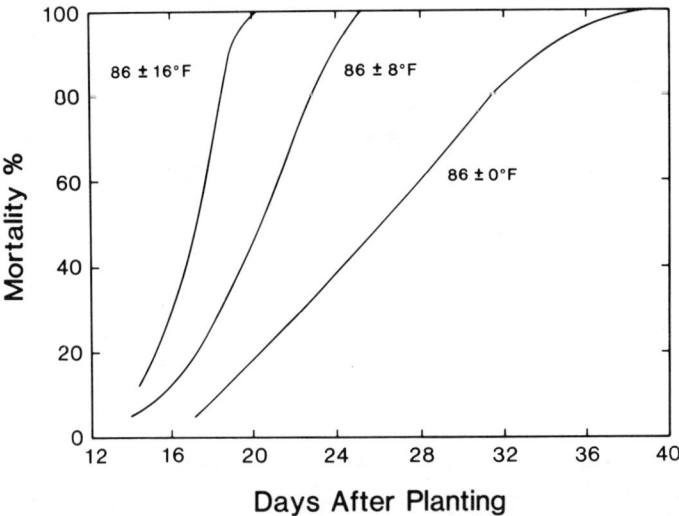

Figure 3.4. Strong fluctuations in diurnal temperature in surface soils increase the severity of *Macrophomina phaseoli* charcoal rot of sugar pine. After Smith, 1966.

±8.9°C was not because the fungus grew faster in that regime; rather, host stress must have been important.

SOME EFFECTS OF TEMPERATURE ON NEMATODES

Larval Survival

Eggs of the cereal cyst nematode, *Heterodera avenae*, in most of its geographic range, hatch in a seasonal cycle which includes an inactive period in winter (Kerry and Jenkinson, 1976). In Europe and North America rapid escape of larvae from cysts has been associated with previous exposure to low temperatures. Cotten (1962) and Fushtey and Johnson (1966) consider 8 weeks at 7°C necessary to stimulate hatching. In Britain larvae begin to emerge in autumn, and larvae survive all winter in the absence of a host when soil temperatures are low (but not severe), and die within a few weeks when soil temperatures rise (Kerry and Jenkins, 1976). Second-stage larvae become more active as the soil warms in the spring, resulting in exhaustion and death when no host is available.

Kerry and Jenkins (1976) placed newly formed cysts in small plastic cylinders with the ends covered with silk through which larvae could easily emerge. These containers (hatching tubes) were placed in moist sand and incubated outdoors. All larvae in the sand came from the cysts. The larval concentration increased througout the autumn and into winter in the absence of host exudates at Rothamsted, England. The hatching tubes were removed in late January. The number of active larvae remained high until the end of March, after which a sharp decline occurred. Removal of the cysts in the hatching tubes ensured that all larvae recovered after that were "overwintering" larvae. Larvae from an autumn or winter hatch could infect a winter host or accumulate and remain alive long enough to infect an early-seeded spring host. They could not survive the summer free in soil.

Vertical Migration

Florida citrus are grown on deep, sandy soils, and *Radopholus similis* are frequently abundant deep in the soil. Forida nematologists believe the main factor governing the depth of nematode activity is temperature, not moisture (Norton, 1978). Fibrous roots between 25 and 76 cm depth sustain about 25 to 30% damage, and 90% of those below 76 cm are killed within a few weeks. The surface soils are apparently so hot that the nematode migrates to cooler, deeper soil.

Sex Ratios

Laughlin et al. (1969) inoculated bermuda grass roots with larvae of *Meloidogyne graminis*, incubated the plants at 27°C for 2 days to facilitate infection, and then placed the plants at 16, 21, 27, and 32°C constant temperatures. About 1% of the larvae became males at 16°C, 2% at 21°C, 4% at 27°C, and 80% at 32°C. If plants were inoculated with larvae at 32°C, held at this temperature for 2 days, and then transferred to 27°C, essentially all became females. If held 3 or more days at 32°C before transferring to 27°C, practically all became males. Laughlin et al. believe that temperature itself rather than indirect effects through the host, are responsible for these dramatic changes in sex ratio. Incubation for 3 days or more at 32°C initiates an irreversible differentiation toward the male sex.

Number of Generations

Ichinohe (1959) studied the effect of temperature upon development of the soybean cyst-nematode, *Heterodera glycines*, in Hokkaido, the northernmost island of Japan. The soybean cyst-nematode causes severe damage in soil that is frozen to depths up to 27 cm from November to April. Eggs of this species in cysts are cold-resistant. The threshold temperature for nematode development is 10°C, and Ichinohe calculated the accumulated number of degree-days required for a life cycle in Hokkaido at between 304 and 320, with an average of 313. Development occurs between 10 and 34°C. In the Sapporo District the nematode develops from June 1 to October 10, coinciding with the vegetative period of the host. At 5 and 30 cm deep, the average season contains 1190 and 1069 degree-days, respectively. In general, the soybean cyst nematode requires 24 or 40 days at 23 and 18°C, respectively, to complete a generation. At 5 and 30 cm the average season would, therefore, sustain 3.8 and 3.4 generations (23°C average daily temperatures $-10 = 13 \times 24$ days $= 312$ total degree-days). Three to six generations are possible in the United States depending upon the location.

SOIL SOLARIZATION

Katan (1980) described *soil solarization* as the process of increasing the heat in soil by covering it with clear polyethylene plastic, 25 to 30 μm thick, to reduce pathogen populations. Solarization is more effective if the soil is moist. Moist soil is covered during the hottest part of the year. Soil temperatures reach 40 to 44°C at 20 cm depth in Israel. The method is more effective than the amount of heat alone would be, and it is probable that adverse factors other than heat, such as an altered soil atmosphere, are involved.

As many as 85% of pistachio nut trees have had to be replaced within 6 years when trees have been planted in a grove infested with *Verticillium dahliae* in California. Ashworth and Gaona (1972) attempted to arrest the advance of the disease to healthy trees and possibly to assist infected trees by spreading clear plastic mulch over the soil surface in July, with and without preirrigation. They were concerned that the trees would be damaged by mulching or even be made more susceptible to the pathogen. Microsclerotia occur 120 cm deep in this soil, probably beyond significant soil heating. Also, soil temperatures might not rise very high in shade under the tree. The soil was irrigated to the depth of the soil profile in one treatment prior to mulching, and in another treatment the soil had only the residual moisture from the winter. The trees were irrigated by drip, so outside the reach of the drip irrigation the soil was below the permanent wilting percentage to a depth of 1.2 m before the experiment began. The soil surface was completely covered by plastic sheets for 2 months during summer. Microsclerotia per gram of soil were reduced to nondetectable levels under the tarp in direct sunlight and reduced greatly in the shaded areas.

The following year wilt developed in 6.3% additional trees in the control, 1.7% with "dry" mulch, and in 1.5% with moist mulch. Ashworth and Gaona did not believe that increased soil temperature below the 30 cm depth accounted for sclerotial death. They hypothesized that chemicals in the soil environment released the microsclerotia from fungistasis and that suicidal germination may have occurrd. The maximum soil temperature reached with mulch at 30 cm depth was 42°C, and that was for a short period of time. The maximum temperature at 120 cm was about 33°C, and that was for only 1 week. Under laboratory conditions all microsclerotia survived in moist soil at 35°C for 8 weeks. They expected a "solarization", direct kill by heat but encountered a more complex, unknown soil condition from restricted aeration.

Moorman (1982) grew eggplant on soil infested or noninfested with *Verticillium dahliae,* a cool soil pathogen, with and without black plastic spread as a mulch over the soil surface. The plastic raised the soil temperature, but unlike solarization treatments, the plastic was left in place the entire growing season. On noninfested land, the plastic mulch increased yields 13%. On infested land, yields were increased 40%. Raising the soil temperature in Massachusetts favored eggplant, but it reduced the severity of Verticillium wilt even more.

FIRE

Hardison (1976 and 1980) has discussed the use of fire in a general way, with emphasis on its use in grass seed production. Fire is particularly beneficial in production of grass seed. It reduces ergot (*Claviceps purpurea*), blind seed (*Gloeotinia temulenta*), silver top (sterile panicles—etiology somewhat obscure—*Fusarium poae,* thrips. mites), seed gall nematodes (*Anguina* spp.), and miscellaneous leaf spots. It removes residues that accumulates after threshing. Residue control is important, because large accumulations lead to growth of rhizomes and roots within the residues. Shallow-rooted plants become very susceptible to changes in environmental conditions and to residue-inhabiting pathogens and insects.

Ascospores of *Gaeumannomyces graminis* are airborne but do not infect plant foliage. If they reach exposed roots and rhizomes formed above ground, they could infect these organs with little microbial competition or antagonism. The take-all fungus is favored in wet places, and residue accumulation could result in root and rhizome development, above the soil, as described above. Before tillage, ascospores could have played an important role in dissemination of this fungus, because accumulated residues must have been important, resulting in roots and rhizomes above the soil surface. In cultivated crops, however, burning has little effect on severity of take-all.

Burning wheat straw reduces Cephalosporium stripe. Because the systemic pathogen, *Cephalosporium gramineum*, invades the entire stem and awns, much potential inoculum is destroyed by burning. In contrast, *Pseudocercosporella herpotrichoides*, which is limited to local lesions near the soil surface, is not materially reduced by burning (see Fig. 11.1). Much of this fungus, at or just below the soil line, escapes. *Fusarium culmorum* has been only slightly reduced in infested stubble by burning in Washington. Wheat of the 1963–1964 season was burned after harvest, and stubble was collected the following September and October, 1965. *F. culmorum* was isolated from 10% of the charred segments, 32% of stem segments just below the charred segments, and 45% of the segments further from the charred ends. Whether the heat was insufficient to eradicate the fungus or whether the fungus grew from below into the heated ends is not known.

Natural burns are sufficient to reduce straw to manageable amounts in rice fields, and burning aids in controlling rice-stem pathogens. Reducing the amount of straw also aids in limiting problems resulting from anaerobic fermentation in some rice soils. Smoke gases and condensates are fungistatic to several fungi (Zagory and Parmeter, 1984), and smoke may augment the sanitizing effects from burning. However, it seems doubtful that smoke from a natural burn would penetrate soil.

COMMENTS

Temperature is a major factor determining the severity of some vascular wilts, mainly by affecting the growth rate of the pathogens. Fusarium and Verticillium wilts tend to follow this generality, as does southern bacterial wilt. *Pseudomonas solanacearum* grows fastest in culture between 30 and 35°C, and disease develops most severely in tobacco at 30 to 35°C (Lucas, 1975). The wheat rusts (*Puccinia graminis*, *P. recondita*, and *P. striiformis*) differ in relation to temperature, not because of predisposition of the host but because the causal fungi differ in relation to temperature. Less fastidious pathogens, such as cortical rotters, tend to be most destructive when the host is under stress (temperature, water, nutrition).

Although the number of generations per season of many pathogens is determined by temperature, this is not always so. *Typhula idahoensis* produces one crop of sclerotia per year under snow, and if conditions are favorable, only one crop of spores before snowfall. Its life cycle is relatively fixed. Some organisms, like *Tilletia caries* that causes stinking smut of wheat, produce only one crop of spores per growing season, because in such an organism its life cycle depends on the host's completing its life cycle. It sporulates in the inflorescence.

It is fortunate that most host-pathogen interactions are strongly affected by water and temperature. Usually, weather favors one or only a few pathogens at a time. Warm, dry weather favors fusarium foot rot of wheat, and cool, moist weather favors strawbreaker foot rot of wheat. All diseases are not severe at the same time on a genetically susceptible plant because of these and other environmental influences.

KEY REFERENCES

Fushtey, S. G and P. W. Johnson. 1966. The biology of the oval cyst nematode, *Heterodera avenae* in Canada. I. The effect of temperature on the hatchability of cysts and emergence of larvae. *Nematologica* **12**:313–320.

Katan, J. 1980. Soil solarization of soils for disease control: Status and prospects. *Plant Dis.* **64**:450–454.

Miller, J. J. 1945. Studies on the Fusarium of muskmelon wilt. II. Infection studies concerning the host of the organism and the effect of environment on disease incidence. *Can. J. Res. C.***23**:166–187.

Smith, R. S., Jr. 1966. Effect of diurnal temperature fluctuations on the charcoal root disease of *Pinus lambertiana*. *Phytopathology* **56**:61–64.

Walker, J. C. 1950. *Plant Pathology*. McGraw-Hill, New York.

Wills, W. H. 1964. Autumn weather in relation to subsequent occurrence of tobacco black shank in Virginia. *Plant Dis. Reptr.* **48**:32–36.

REFERENCES

Abdel-Raheem, A. and L. S. Bird. 1968. The interrelationship of resistance and susceptibility of cotton to *Verticillium albo-atrum* and *Fusarium oxysporum* f. sp. *vasinfectum* as influenced by soil temperature. *Phytopathology* **58**:725 (Abstract).

Apple, J. L. 1963. Persistence of *Phytophthora parasitica* var. *nicotianae* in soil. *Plant Dis. Reptr.* **47**:632–634.

Ashworth, L. J. Jr. and S. A. Gaona. 1982. Evaluation of clear polyethylene mulch for controlling Verticillium wilt in established pistachio nut groves. *Phytopathology* **72**:243–246.

Bell, A. A. 1973. Nature of resistance. Pages 47–62 in *Verticillium Wilt of Cotton*. C. D. Raney, ed. U.S. Dept. Agric. Publ. ARS-S-19.

Bell, A. A. and M. E. Mace. 1981. Biochemistry and physiology of resistance. Pages 431–486 in *Fungal Wilt Diseases of Plants*. M E. Mace, A. A. Bell, and C. H. Beckman, eds. Academic Press, New York.

Bell, A. A. and J. T. Presley. 1969. Temperature effects upon resistance and Phytoalexin synthesis in cotton inoculated with *Verticillium albo-atrum*. *Phytopathology* **59**:1141–1146.

Brinkerhoff, L. A. 1973. Effects of environment on the pathogen and the disease. Pages 78–88 in *Verticillium Wilt of Cotton*. C. D. Raney, ed. U.S. Dept. Agric. Publ. ARS-S-19.

Clayton, E. E. 1923. The relation of temperature to the Fusarium wilt of the tomato. *Am. J. Bot.* **10**:71–88.

Cotten, J. 1962. The effect of temperature on hatching in the cereal root eelworm. *Nature* **195**:308.

Dickson, J. G. 1923. Influence of soil temperature and moisture on the development of the seedling-blight of wheat and corn caused by *Gibberella saubinetii*. *J. Agric. Res.* **23**:837–870.

Edgington, L. V. and J. C. Walker. 1959. Influence of soil and air temperature on Verticillium wilt of tomato. *Phytopathology* **47**:594–598.

Edson, H. A. and M. Shapalov. 1920. Temperature relations of certain potato-rot and wilt-producing fungi. *J. Agric. Res.* **18**:511–524.

Epps, J. M. and A. Y. Chambers. 1963. Influence of planting date on yield of soybeans in fumigated and untreated soil infested with *Heterodera glycines*. *Plant Dis. Reptr.* **47**:589–593.

Ezekiel, W. N. 1945. Effect of low temperatures on survival of *Phymatotrichum omnivorum*. *Phytopathology* **35**:296–301.

Garber, R. H. and J. T. Presley. 1971. Relation of air temperature to development of Verticillium wilt on cotton in the field. *Phytopathology* **61**:204–207.

Hardison, J. R. 1976. Fire and flame for plant disease control. *Ann. Rev. Phytopathol.* **14**:355–379.

Hardison, J. R. 1980. Role of fire for disease control in grass seed production. *Plant Dis.* **64**:641–645.

Henry, A. W. 1932. The influence of soil temperature and soil sterilization on the reaction of wheat seedlings to *Ophiobolus graminis*. *Can. J. Res.* **7**:198–203.

Ichinohe, M. 1959. Studies on the soybean cyst nematode *Heterodera glycines* and its injury to soybean plants in Japan. *Plant Dis. Reptr. Suppl.* **260**:239–248.

Isaac, I 1949. A comparative study of pathogenic isolates of Verticilliium. *Brit. Mycol. Soc. Trans.* **32**:137–157.

Johnson, J. and R. E. Hartman. 1919. Influence of soil environment on the root-rot of tobacco. *J. Agric. Res.* **17**:41–86.

Jones, L. R. and J. C. Gilman. 1915. The control of cabbage yellows through disease resistance. *Wis. Agr. Expt. Sta. Res. Bull.* 38.

Jones, L. R., J. Johnson, and J. G. Dickson. 1926. Wisconsin studies upon the relation of soil temperature to plant disease. *Wis. Agr. Expt. Sta. Res. Bull.* 71.

Kelman, A. 1953. The bacterial wilt caused by *Pseudomonas solanacearum*: A literature review and bibliography. *N. C. Agric. Expt. Sta. Tech. Bull.* 99.

Kenerley, C. M. and R. I. Bruck. 1983. Overwintering and survival of *Phytophthora cinnamomi* in Fraser fir and cover cropped nursery beds in North Carolina. *Phytopathology* **73**:1643–1647.

Kerry, B. R. and S. C. Jenkinson. 1976. Observation on emergence, survial and root invasion of second-stage larvae of the cereal cyst-nematode. *Nematologica* **22**:467–474.

Kommedahl, T., J. J. Christensen, and R. A. Frederiksen. 1970. A half century of research in Minnesota on flax wilt caused by *Fusarium oxysporum*. *Minn. Agric. Expt. Sta. Tech. Bull.* 237.

Kuske, C. R. and D. M. Benson. 1983. Overwintering and survival of *Phytophthora parasitica* causing dieback of rhododendrum. *Phytopathology* **73**:1192–1196.

Laughlin, C. W., A. S. Williams, and J. A. Fox. 1969. The influence of temperature on development and sex differentiation of *Meloidogyne graminis*. *J. Nematol.* **1**:212–215.

Leach, J. G. and T. M. Currence. 1938. Fusarium wilt of muskmelons in Minnesota. *Minn. Agr. Expt. Sta. Tech Bull.* 129.

Leyendecker, P. J., Jr. 1950. Effects of certain cultural practices on Verticillium wilt of cotton. *New Mex. Expt. Sta. Bull.* 356.

Lucas, G. B. 1975. *Diseases of Tobacco.* Biological Consulting Associates, Raleigh, NC.

Ludbrook, W. V. 1933. Pathogenicity and environal studies on Verticillium hadromycosis. *Phytopathology* 23:117-154.

Moorman, G. W. 1982. The influence of black plastic mulching on infection rates of Verticillium wilt and yield of eggplant. *Phytopathology* 72:1412-1414.

Norton, D. C. 1978. *Ecology of Plant Parasitic Nematodes.* Wiley, New York.

Raney, C. D. 1973. Cultural control. Pages 98-104 in *Verticillium Wilt of Cotton.* C. D. Raney, ed. U.S. Dept. Agric. Publ., ARS-S-14.

Robinson, D. B., R. H. Larson and J. C. Walker. 1957. Verticillium wilt of potato in relation to symptoms, epidemiology and variability of the pathogen. *Univ. Wisc. Res. Bull.*

Schnathorst, W. C. 1973. Nomenclature and physiology on *Verticillium* species, with emphasis on the *V. albo-atrum* versus *V. dahliae* controversy. Pages 1-19 *Verticillium Wilt of Cotton.* C. D. Raney, ed. Agricultural Research Service, S-19.

Schneider, H. 1948. Susceptibility of guayle to Verticillium wilt and influence of soil temperature and moisture on development of infection. *J. Agric. Res.* 76:129-143.

Smith, R. S., Jr. 1964. Effect of diurnal temperature fluctuations on linear growth rate of *Macrophomina phaseoli* in culture. *Phytopathology* 54:849-852.

Snyder, W. C. and S. N. Smith. 1981. Current status. Pages 25-50 in *Fungal Wilt Diseases of Plants.* M. E. Mace, A. A. Bell, and C. H. Beckman, eds. Academic Press, New York.

Taubenhaus, J. J., W. N. Ezekiel and D. T. Killough. 1928. Relation of cotton root-rot and Fusarium wilt to the acidity and alkalinity of the soil. *Tex. Agric. Exp. Sta. Bull.* 389.

Taylor A. L. and J. N. Sasser. 1978. Biology, identification and control of nematodes (*Meloidogyne* species). Coop. Publ., Dept. Plant Pathol., N. C. State Univ. and U.S. Agency Int. Dev., Raleigh.

Tisdale, W. H. 1917. Relation of temperature to the growth and infecting power of *Fusarium lini. Phytopathology* 7:356-360.

Walker, J. C. and F. L. Wellman. 1926. Relation of temperature to spore germination and growth of *Urocystis cepulae. J. Agric. Res.* 32:133-146.

Wilhelm, S. and J. B. Taylor. 1965. Control of Vericillium wilt of olive through natural recovery and resistance. *Phytopathology* 55:310-316.

Young, V. H. 1928. Cotton wilt studies. I. Relation of soil temperature to the development of cotton wilt. *Ark Agr. Expt. Sta. Bull.* 226.

Young, V. H., N. D. Fulton, and B. A. Waddle. 1959. Factors affecting the incidence and severity of Verticillium wilt disease of cotton. *Ark. Agr. Ext. Sta. Bull.* 612.

Zagory, D. and J. . Parmeter, Jr. 1984. Fungitoxicity of smoke. *Phytopathology* 74:1027-1031.

Chapter 4

Soil Aeration

Because most plant roots are in the upper layers of soil, most of the organic matter and microorganisms are also in surface layers. These general relationships hold in a wide range of soils occurring in humid to arid and tropical to arctic regions. Root systems are responsive to soil conditions; they are not fixed in an immutable architecture. They predominate in surface layers because these layers are most favorable, and the main factor responsible appears to be the relative abundance of oxygen. The roots of higher plants are aerobic (Burstrom, 1965). The great advantage in energy conversion conferred by aerobic over anaerobic respiration (about 19 times) is a factor in determining dominant life forms.

QUANTITY OF OXYGEN REQUIRED

The majority of feeder roots in normal soils are within the top 20 to 40 cm. Brady (1974) quoted the work of Boynton and his colleagues on the influence of oxygen on apple tree roots. At least 3% of the soil atmosphere must be oxygen to keep an apple root alive. From 5 to 10% oxygen permits the growth of existing roots, but new roots are not initiated unless at least 12% oxygen is present. This experiment, if applicable to many mesophytic plants, would explain why most of the feeder roots are in aerated layers of soil. Winter wheat roots attain depths of 180 to 240 cm in deep soils of good texture, but at lower depths individual roots have few branches.

The volume of oxygen consumed by a densely rooted species such as cereals at their height of development is surprising. Ten liters of oxygen per square meter of surface per day are consumed (Drew and Lynch, 1980). Russell (1977) cites the work of Currie, which reported that kale in July used 12.4 g of oxygen per m^2 per day. The oxygen content of soil air seldom falls below 10 to 12% in surface layers, but oxygen consumption is great. Clark and Kemper (1967) calculated microbial consumption of oxygen for a hectare of fallow soil at 27 to 54 kg per day, the hectare with crop residue at 54 to 107 kg/dy, and for the hectare with some residue and a growing crop at 81 to 161 kg of oxygen per day. The latter figures equal the amount of nitrogen added in fertilizer per hectare per year for some crops. Oxygen is not a minor element!

Soil ranges from 40 to 60% pore space, depending on soil texture and structure, and its air content varies inversely with the water content. If all the pores are filled with water, air is excluded. Oxygen deficiencies usually develop when porosity is reduced to near 10% or less for some time, particularly in warm weather and if decomposable organic matter is present in the soil. Many papers deal with CO_2 toxicity when CO_2 is present in high quantities, such as above 10%, and many deal with O_2 deficiencies. Russell (1977) and Smith and Griffin (1971) question the importance of CO_2 toxicity in most soils. They conclude that O_2 deficiency is in most cases far more significant than CO_2 toxicity.

Oxygen Diffusion

Most oxygen enters the soil by diffusion in the gaseous state. Soil scientists measure the oxygen diffusion rate within soil. A rate of $20g \times 10^{-8}/(cm^2)(min)$ is minimal for root development and about $30-40g \times 10^{-8}/cm^2/(min)$ for normal top growth (Brady, 1974); oxygen deficiency in soil is expressed in the foliage before it is detected below ground. The rate at which oxygen diffuses within soil can be estimated by the electric current that flows between two electrodes placed in the soil. Oxygen is reduced at one of the electrodes, and the current reflects the rate of oxygen replenishment at that electrode by diffusion. The oxygen diffusion rate decreases nearly linearly in a deep soil from 10 cm below the surface to 95 cm, being reduced slightly more than half at 95 cm. Subsoils become deficient in oxygen because they are further from the source, because microorganisms and roots higher in the profile consume oxygen as the air passes by, because in most soils fine particles have filtered downward, and because the structure of subsoil is usually poorer.

Russell (1977) stressed the existence of anaerobic conditions inside soil crumbs within an otherwise aerobic soil. These crumbs are said to contain anaerobic microsites. Since oxygen moves 10,000 times faster as a gas than in water, a relatively small water film around a soil aggregate profoundly reduces aeration within it. Russell emphasized that toxic products of anaerobic decomposition can occur in an otherwise aerated soil. Roots of mesophytes receive small amounts of oxygen from the shoots, and roots in an oxygen-rich microsite can supply some oxygen to the same root in a nearby oxygen-poor microsite. These limited exchanges could be significant in meeting the oxygen requirements of a sedentary, endoparasitic nematode incapable of changing its feeding position. Russell also illustrated asymmetrical root systems resulting from asymmetric distribution of fertilizers. These experiments demonstrate the pliable nature of root systems. They depart from symmetry in response to soil conditions. In wet seasons the proportion of total roots in surface layers increases. In dry seasons the proportion of roots deeper in the soil increases, but drying is accompanied by increased aeration at greater depths.

Oxygen Gradients Relative to Saprophytic Survival

The speed with which organic matter decomposes is strongly influenced by aeration and temperature, as well as by pH and nitrogen. In warm, well-aerated, moist soils, organic matter decomposes rapidly. Cultivation hastens organic decomposition, largely through increased aeration. These factors are important to microorganisms that exist in an active state in debris or in residues of diseased plants. Rapid decomposition of substrate reduces their longevity.

Macer (1961) placed wheat straws containing *Pseudocercosporella herpotrichoides* on the soil surface and buried some at depths of 2.5, 15, and 45 cm. Sporulation by this fungus is high on straw exposed to light under cool, moist conditions. Macer began the experiment in September in England, when natural sporulation would begin. Sporulation was profuse on surface straws for about 34 weeks; it declined abruptly by 56 weeks and ceased within 120 weeks. Sporulation exhausted the energy reserves available to the fungus in straw on the soil surface. When the buried straws were removed from the soil and exposed to conditions favoring sporulation, survival, as judged by ability to sporulate, was stronger on straws from 15 than from 2.5 cm deep in the soil. Longer life at 15 cm can be attributed to reduced substrate utilization due to reduced aeration at 15 cm as compared to 2.5 cm deep in the soil (Table 4.1).

Oxygen Effects on Rhizomorph Development

Rhizomorphs develop thin, wefty areas, "breathing pores," when they encounter oxygen-rich microsites within an otherwise oxygen-poor region of the soil. Oxygen diffuses through the hollow medulla of the rhizomorph to the growing tip (Smith and Griffin, 1971). The outer layer of the mature

SOIL AERATION

Table 4.1. Survival of *Pseudocerocosporella herpotrichoides* in wheat straw on the soil surface and buried

Time in weeks	Month	Survival, %			
		Depth of burial, cm			
		0	2.5	15.0	45.0
0	Sept., 1953	100	100	100	100
28	March, 1954	94	96	100	100
34	May	84	95	98	97
56	October	10	85	100	98
79	March, 1955	14	76	100	—
120	December	0	33	88	72
156	Sept., 1956	0	20	76	70

Source: Macer, 1961.

rhizomorph is dark, and its hyphae are closely packed, almost impervious to O_2. They established *Armillaria elegans* in wood blocks and incubated them in such a way that only air from the wood block (the source) was available. Maximum growth of a rhizomorph near the surface of the wood block was possible with 6.4% oxygen at the source. When the rhizomorph was 2 cm long, 6.9% oxygen at the source was required to maintain maximum growth. At 6 and 10 cm, 8.3% and 9.3% of oxygen were required at the source. Thus, the greater the diffusion distance, the higher the concentration of oxygen required at the source to sustain rapid growth.

Melanin develops only when the growing rhizomorph is exposed to 4% or more oxygen. The browning results from deposition of melanins in the intercellular spaces of the rhizomorph rind, restricting both ingress and egress of materials. p-Diphenol oxidase activity in the presence of 4% or more of oxygen results in the browning of the rhizomorphs. If melanin forms too rapidly at the tip, growth is retarded. When the rhizomorph enters a high oxygen atmosphere, its growth rate is restricted by changes in the exterior layers. Smith and Griffin believe that, in aerated soil, water films about the growing tip restrict O_2 diffusion, permitting maintenance of a significant rate of growth. Motta (1969) found a mucilaginous sheath around the apices of rhizomorphs, and Smith and Griffin (1977) suggest this may suffice to protect the tip from excess oxygen in short air pockets in soil through which the rhizomorph must pass.

Oxygen Requirements of Nematodes

Sweet orange trees and the citrus nematode, *Tylenchulus semipenetrans*, have been exposed to a gradient of oxygen (Stolzy et al., 1963). The nematodes are more sensitive to low oxygen than the host. If oxygen enters the

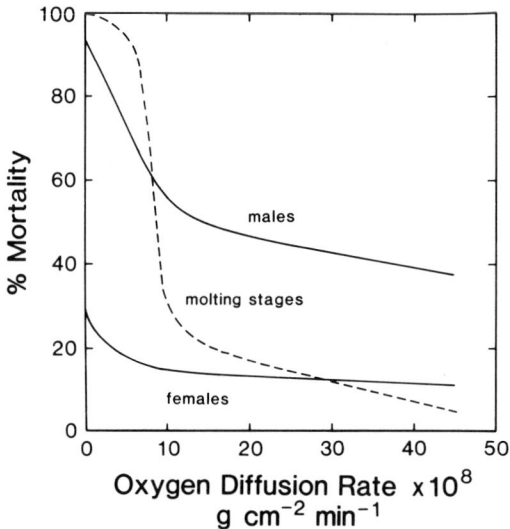

Figure 4.1. Females of *Hemicycliophora arenaria* tolerate low oxygen better than males, and molting stages are highly susceptible to reduced oxygen. Van Gundy and Stolzy, 1963.

roots from the shoots, the host root tissue has first access to this oxygen. Between diffusion rates of 10 and 100 $\mu g/(cm^2)(min)$, a straight-line relationship between oxygen and reproduction by *T. semipenetrans* results. This nematode is highly aerobic.

A sheath nematode, *Hemicycliophora arenaria*, forms galls on roots of rough lemon nursery stock in the Coachella Valley of southern California in a soil consisting of 75% sand, 9% silt, and 16% clay (Van Gundy and Stolzy, 1963). Exposure of soil to oxygen diffusion rates below 20 $\mu g/(cm^2)/(min)$ for 1 month lead to a decline in nematode numbers. Events (molting and egg production) requiring the greatest energy are most oxygen-dependent (Fig. 4.1). Males die before females, and females and fourth-stage juveniles are most tolerant to oxygen deficiency. Reproduction rates increase linearly with increasing oxygen at oxygen diffusion rates between 10 and 40 $\mu g/(cm^2)/(min)$.

OXYGEN IN RELATION TO BEAN ROOT ROT

Dry beans are grown with furrow irrigation in central Washington. Even though the fields are "leveled" to attain a constant slope for a steady flow of runoff, water collects temporarily in low areas. Miller and Burke (1975) observed that root rot caused by *Fusarium solani* f. sp. *phaseoli* is often severe in the low areas. Beans on land not infested by the pathogen withstand periods of temporary saturation well. They confirmed, in greenhouse studies, that periods of anoxia of 1 to 3 days duration predispose beans to root rot, increasing in severity with length of the period without oxygen. Beans are uninjured by the short exposures to anoxia when the soil is free of pathogens. When the gas above the soil consists entirely of nitrogen, oxygen is essentially exhausted within the soil within 24 hr. At 8 cm, the soil atmosphere was 0.9% oxygen; at 24 cm it was 2.7% within 24 hr. The loam was held at -0.15 bar, so even when quite wet but not saturated, gas diffusion was rapid. They also found that anoxia greatly reduces the ability of roots exposed to the pathogen to penetrate compact soil.

INFLUENCE OF OXYGEN ON SOFT ROT OF POTATO TUBERS

Erwinia carotovora ssp. *carotovora* and ssp. *atroseptica* and *Clostridium* spp. rot potato tubers under conditions of reduced aeration. Reduced aeration and insects, particularly the larval stage of the seed-corn maggot (*Hylemyia cilicrura* = *Delia platura*), are important in black leg and tuber rot of potatoes in heavy, wet soil (Leach, 1940). Under favorable conditions the wounded tuber (as when cut into pieces preparatory to planting, being alive and rich in energy) has great ability to heal wounds. It forms a corky layer several cells thick behind the cut surface, an effective barrier against soft-rot bacteria. When oxygen is deficient, the potato cannot heal wounds. *Erwinia* spp. are common on potato tubers, so lack of inoculum is unlikely. In addition, maggots are often born contaminated with *Erwinia carotovora*. The adults lay eggs near or on the cut tuber or young potato shoots, and the maggots feed on the tuber. They make wounds with their larval hooks faster than the host can respond, so healing cannot occur. Before adequate insectides were available, losses were sometimes heavy, especially when heavy rains occurred after planting.

Burton and Wiggington (in Lund and Kelman, 1977) reported that a thin film of water about the tuber quickly rendered the tuber anaerobic. They calculated that the tuber is anaerobic after 2.5 hr at 21°C and in 6.5 hr at 10°C, when surrounded by a water film 0.03 mm in thickness. Lund and Kelman took advantage of this phenomenon and studied the influence of harvest and storage procedures on soft rot of potato by incubating potatoes in a mist chamber. The water film is about 0.09 mm thick immediately after washing, about 0.088 mm after 4 hr in the mist chamber, and 0.10 mm after 5 days in the mist chamber. The potatoes absorb water during misting, gaining from 0.6 to 1.0% in weight in 5 days. The water film reduces oxygen diffusion into the tuber and hydrates the tissues, both of which favor bacterial soft rot.

The significance of anoxia is emphasized by the presence of pectolytic anaerobic clostridia in rotting tubers. Campos et al. (1982) reported that anaerobic bacteria were associated with rotting potatoes by van Tieghem as early as 1884, that Rudd-Jones and Dowson in 1950 found that *Clostridium* sp. caused decay when *Erwinia carotovora* was present, and that Lund and Nicholls in 1970 found that clostridia alone could rot tubers. Campos et al.

obtained rotting tubers from commercial storage and found that pectolytic clostridia were present in 22%, *Erwinia carotovora* ssp. *carotovora* in 13% and ssp. *atroseptica* in 45% of the samples. Clostridia without association with *Erwinia* were obtained from 9% of the tubers. In inoculation experiments using the mist technique, clostridia were virulent. Failure to employ anaerobic techniques led early scientists to overlook the importance of closteridia in the United States. *E. carotovora* is usually considered a facultative anaerobe, whereas *Clostridium* spp. are anaerobes.

Macerating enzymes produced by *E. carotovora* injected cell-free into tubers produce soft rot in anaerobic potato tubers but not in tubers containing oxygen (Maher and Kelman, 1983). This resistance to enzyme action in aerobic tubers differs from the cork-formation mechanism. The manner in which these enzymes are rendered harmless by aerated potato tissues is not known. Activity is still present when the enzyme is recovered from injected, aerobically incubated tissue, so lack of rotting is not due to destruction of the enzyme. Tuber slices are widely used in experiments, and Maher and Kelman caution that, being well-aerated, results of some types of trials could differ from results with intact tubers.

OXYGEN IN RICE FIELDS

Most rice is produced on flooded soil (paddy rice). This is possible because rice tolerates low levels of oxygen. Cortical lacunae connect shoot and large roots, permitting oxygen diffusion from the shoot to the roots. Iron becomes soluble when reduced, and Turner and Chen (1981) found a positive correlation between the yields of rice and the amount of iron precipitated on the roots. This serves as an indirect measure of the amount of oxygen present at the root surface of paddy rice. The reduced iron in solution is precipitated on the root surface by oxygen escaping from the roots. Turner and Chen calculated that 31,500 l of air per ha were required to remove this iron from solution, so at least 6615 l oxygen must have diffused from roots into the rhizosphere.

The highest oxidizing power of rice roots occurs in the presence of balanced mineral nutrition. The least occurs when potassium is deficient (Trolldenier, 1977). Plants low in oxygen suffer from iron toxicity, because reduced iron is soluble and is absorbed by roots (Table 4.2). Once anaerobic conditions are established in the rhizosphere, root exudation accelerates, promoting further microbial growth and oxygen consumption (Trolldenier, 1979).

Hydrogen Sulfide in Rice Fields

Hirschmaniella oryzae, the rice root nematode, is common in rice fields of Louisiana, but the standard flooding of fields reduces its numbers. Hollis

Table 4.2. Effect of potassium on growth and iron content of rice plants grown in potassium-deficient soil

Exchangeable K, meq/100 g of soil	Shoot dry weight, g	K, % of dry matter	Fe ppm, dry matter
0.08	16.9	0.45	520
0.15	20.7	0.93	400
0.25	25.3	1.41	380
0.40	27.7	2.00	330

After Trolldenier, 1977

(1967) presented nematode numbers per pint of top soil in a Louisiana field seeded April 30 and flooded May 14. On April 2, before seeding, there were 3639 nematodes/pint of soil. On May 20, 6 days after flooding, there were 5326. The numbers declined regularly with each successive sampling and reached 149 by August 7. Flooding reduces the nematode population during the period of greatest host development. Within about 6 weeks after flooding, the top soil is considered to be in a reduced state of oxidation. Nitrates and manganic manganese are reduced first, iron next, and sulfates last. Hydrogen sulfide is toxic to rice roots and nematodes. It is produced by reduction of sulfate. Nematode populations in the reduced soil zone decline rapidly about 6 weeks after flooding, and this is attributed to H_2S toxicity more than to oxygen deficiency.

Fortuner and Jacq (1976) exposed three rice-infecting nematodes to varying concentrations of H_2S. All three were equally susceptible. They believe that aerobic microsites along roots favor the nematodes. Increasing anaerobic conditions and sulfide production during rice production to control nematodes is impractical, because H_2S is toxic to rice. The authors suggest that creating anaerobic conditions when the field is not producing rice could achieve biological control of the soilborne nematodes by means of H_2S.

TOLERANCE OF FUNGI TO CARBON DIOXIDE

Bisby et al. (1935) sampled fungi in five virgin Canadian soils at different depths in the soil profile for their ability to grow in the presence of elevated CO_2 and reduced O_2. They incubated dilution plates in normal atmosphere and in a chamber flushed with CO_2 and expressed the results in terms of the percentage of fungal colonies developing in the altered atmosphere compared with that in a normal atmosphere. In the A_0 horizon, 0.8% grew in reduced O_2 versus air; in the A_1 horizon, 4%; in the A_2 horizon, 18%; in the B horizon, 35%; and in the C horizon, 39%. Relatively more fungal propagules from deep in the soil grew in the atmosphere with less oxygen and enhanced carbon dioxide.

Marked differences in tolerance to CO_2 exist among fungi (Burges and Fenton, 1953). The growth of several is reduced in 3% CO_2, while some, including *Trichoderma viride* grow well in 10% CO_2. *Penicillium nigricans* is sensitive to 3% CO_2, and it is essentially restricted to the surface 5 cm of an English soil. *Zygorhynchus vuillemini* tolerates 10% CO_2, and it is rare in soil above 10 cm deep. Carbon dioxide varies between 3.5 and 9.2% at the 40 cm depth in the soil with which Burges and Fenton worked.

Durbin (1959) described variable CO_2 sensitivity within a single fungus species, *Rhizoctonia solani*. This species includes many variants, some virulent pathogens of foliage above the soil, some attacking primarily at the soil surface, and some primarily within the soil. Durbin assembled 11 aerial isolates, 11 soil-surface isolates, and 11 isolates from subterranean habitats. On potato dextrose agar in air the aerial isolates grew fastest and those from the soil slowest. Durbin then grew them in an atmosphere containing 20% CO_2. The growth rate of aerial isolates was reduced 80%, the soil surface isolates 52%, and those from within the soil, 31%. Thus tolerance to CO_2 is reflected in the natural habitat of these fungi.

CO_2 Effects on pH

The quantity of CO_2 required to raise the CO_2 partial pressure changes with soil pH (Griffin, 1972). Assuming quantity X is required to increase partial pressure 0.1 at pH 4, then at pH5, 6, 7, 8, and 9, CO_2 must be increased approximately 1×, 1.5×, 5.3×, 45×, and 460×, respectively, to achieve the 0.1 increase in CO_2 partial pressure. If the respiration rate of soils is relatively independent of pH, then partial pressures of carbon dioxide will decline with increasing pH.

Garrett (1937) grew wheat seedlings inoculated with *Gaeumannomyces graminis* var. *tritici* in sterilized soil adjusted to pH 5.9 or to pH 8.0. He measured the distance the ectotrophic runner hyphae traversed the seminal root surface with and without forced aeration. Aeration increased fungal growth at both pH levels. In loosely plugged tubes, *G. graminis* grew 16 mm at pH 5.9 and 33 mm at pH 8.0. In aerated tubes growth was 32 mm at both pHs. Improved aeration negated the adverse effect of low pH on growth of *G. graminis*. Garrett concluded that CO_2 accumulates in acid soils at the root surface but to a much less extent in alkaline soils and that *G. graminis* is sensitive to CO_2. Garrett did not know at that time that CO_2 diffuses through water 23 times as fast as O_2 (Greenwood, 1970), so that the water film surrounding the root surface and the surface of the ectotrophic hyphae would be more subject to O_2 starvation than to CO_2 toxicity. Ferraz (1973) confirmed that either a shortage of O_2 or volatile substances other than CO_2 were responsible for restricting ectotrophic growth by *Gaeumannomyces graminis* var. *tritici*.

Morphogenetic Effect of CO_2

Phymatotrichum omnivorum, like *Gaeumannomyces graminis*, grows ectotrophically along the root surface of its hosts, and its strands penetrate soil beyond the root system, so part of it is exposed directly to the soil atmosphere. *P. omnivorum* is well adapted to Houston black clay, a vertisol with swelling montmorillonite clay that restricts aeration when wet. This soil is rich in calcium carbonate and is alkaline in reaction. *P. omnivorum* survives by means of sclerotia that are most abundant 30 to 75 cm deep within the soil or in regions of reduced aeration.

Lyda and Burnett (1971) grew *P. omnivorum* on sterilized Houston black clay with sorghum seeds on the surface. The fungus was grown in sealed jars, vented jars, and jars containing KOH to remove CO_2. In the sealed jar the CO_2 concentration rose rapidly (within 1 week), peaked at 2 weeks and then dropped. Sclerotium production was initiated at the time of highest CO_2 concentration, and more sclerotia were produced in sealed than in vented jars containing KOH. They believe that high CO_2 concentrations favor sclerotium formation, contributing to their formation deep in the soil.

Lyda and Burnett (1975) monitored CO_2 levels in the field to a depth of 180 cm for 3 years at Temple, Texas. CO_2 reached its greatest concentration between 30 and 60 cm in June and then declined as the soil dried out. Maximum CO_2 at 90 cm was reached in July and deeper than 90 cm in August and September. If sufficient rain fell to fill the shrinkage cracks, the CO_2 concentration increased. The maximum CO_2 concentrations encountered in the field ranged from 3 to 5%.

Phymatotrichum omnivorum occurs primarily in soils in which CO_2 in the soil solution is primarily in the HCO_3^- form. At low pH, CO_2 is in the H_2CO_3 form which does not prevent hyphal growth but which prevents sclerotium formation. Too much HCO_3^-, as at pH 9, restricts sclerotium formation. Lyda and Burnett (1975) concluded that the $HCO_3^-:H_2CO_3$ or $HCO_3^-:CO^2$ ratios are critical, having a morphogenetic, regulatory action on sclerotium formation.

Effects on Saprophytic Activity

In most arable soils the CO_2 concentrations, though higher than in air, are not great; CO_2 concentrations greater than 10% generally occur only under unusual conditions (Papavizas and Davey, 1962). For instance, when green organic matter is plowed under, CO_2 may reach 18% for a short time. All the papers reviewed by Papavizas and Davey state that high CO_2 concentrations inhibit growth but do not kill and that the inhibition is due primarily to CO_2 rather than to shortage of O_2.

Papavizas and Davey incubated mature soybean stem segments in soil inoculated with *Rhizoctonia solani*. The soils were infiltrated with air and air

+ CO_2 mixtures at 10, 20 and 30% CO_2 content for 4 days as follows:

	Air, %	Added CO_2 to air, v/v		
CO_2	0.03	10°	20°	30°
O_2	21	19	17	15
N_2	79	71	63	55
Colonization	84	61	19	2

Colonization of the soybean stem baits in soil infiltrated with air was high (84%) and declined rapidly as the amount of CO_2 was increased. After the permeation period, fresh baits were added to each treatment, and all soils were incubated under air. Colonization was about the same (32 to 41%) in all four treatments, indicating that exposure of soil to very high CO_2 levels for 4 days did not permanently reduce the vigor of the fungus.

When seed was planted 7 days after soil permeation with 10, 20, and 30% CO_2 had ceased, fewest seedlings emerged in the 30% CO_2 treatment. Papavizas and Davey suspected that 30% CO_2 altered microbial activity to such an extent that *R. solani*, a rapid-growing fungus, recolonized the altered soil advantageously. Early workers had shown that high CO_2 suppresses vegetative development of fungi (fungistatic), but that when the excess CO_2 is removed, fungal activity returns to normal (not fungicidal). The work of Papavizas and Davey in natural soil corroborated these findings. High concentrations of CO_2 that may exist for short periods in agricultural soils would be expected to have little lasting effect on the microbial balance in soil.

COMMENTS

In recent years a fascination with interorganism interactions has led to a relative neglect of physical factors in the ecology of soilborne plant pathogens. Water, temperature, and the soil atmosphere still require study. Griffin (1972) discussed the response of various stages of fungi to CO_2 and to oxygen and stated that the optimum conditions for mycelial growth and spore germination differ. He concluded from his review of the literature that oxygen at 3 to 6% is adequate for most fungal activities, but below that, oxygen deficiencies are commonly expressed. The dangers of generalizations about oxygen-CO_2 relationships are demonstrated by the differing responses of *Verticillium dahliae*, as reported by Ioannou et al. (1977), Fig. 4.2, and by the tolerance of different stages of the same nematode to oxygen deficiency (Fig. 4.1). Nematodes are more sensitive to oxygen deficiency than most plants.

High CO_2 concentrations for short duration are fungistatic, not fungicidal, and the soil microflora returns to normal soon after aerobic condi-

SOIL AERATION

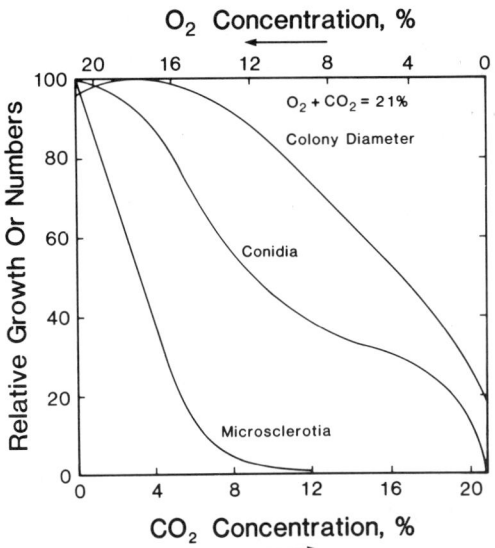

Figure 4.2. Hyphal growth, conidium production, and formation of sclerotia by *Verticillium dahliae* respond differently to varying concentrations of oxygen and carbon dioxide. After Ioannou et al., 1977.

tions return. I was impressed by the speed and depth at which oxygen diffuses into soils with reasonable texture and structure. The reduction of pathogen numbers that often follows incorporation of organic matter is apparently not due to induced oxygen deficit or excess CO_2.

My account of the oxygen-CO_2 relations of *Gaeumannomyces graminis* is incomplete. Smith and Noble (1972) found this species highly sensitive to CO_2, as reported by Garrett (1937). In contrast, Garrett (1981) accepted the work of Ferraz (1973), that *G. graminis* is highly sensitive to low oxygen. I cannot resolve this argument, but there is no doubt that good aeration favors the activity of this fungus.

Many pathogens are controlled in paddy rice by anaerobic or near-anaerobic conditions in soil resulting from flooding during warm weather. Flooding alone as a control practice has not been widely used. In rice culture a valuable plant is grown during flooding, so any resulting control of pathogens is free.

Covering the soil with clear plastic during warm weather (solarization) pasteurizes soil partially through heat, partially because of changes in the soil atmosphere. Both effects are involved in death of fungal propagules (see Chap. 3). The soil atmosphere is much more complex than indicated in this chapter. Many substances result from life and death within the soil, some appearing in the soil solution, some in the soil atmosphere (Chap. 9), some in both.

KEY REFERENCES

Lyda, S. and E. Burnett. 1971. Changes in carbon dioxide levels during sclerotial formation by *Phymatotrichum omnivorum*. *Phytopathology* **61**:858–861.
Maher, E. A. and A. Kelman. 1983. Oxygen status of potato tuber tissue in relation to maceration by pectic enzymes of *Erwinia carotovora*. *Phytopathology* **73**:536–539.
Miller, D. E. and D. W. Burke. 1975. Effect of soil aeration on Fusarium root rot of beans. *Phytopathology* **65**:519–523.
Papavizas, G. C. and C. B. Davey. 1962. Activity of *Rhizoctonia* in soil as affected by carbon dioxide. *Phytopathology* **52**:759–766.
Van Gundy, S. D. and L. H. Stolzy. 1963. The relationship of oxygen diffusion rates to the survival, movement and reproduction of *Hemicycliophora arenaria*. *Nematologica* **9**:605–612.

REFERENCES

Bisby, G. R., M. I. Timonin, and N. James. 1935. Fungi isolated from soil profiles in Manitoba. *Can. J. Res.* **13**:47–65.
Brady, N. C. 1974. The Nature and Properties of Soils. Eighth ed. Macmillan, New York.
Burges, A. and E. Fenton. 1953. The effect of carbon dioxide on the growth of certain soil fungi. *Trans. Brit. Mycol. Soc.* **36**:104–108.
Burstrom, H. G. 1965. The physiology of plant roots. Pages 154–166 in *Ecology of Soil-Borne Plant Pathogens*. K. F. Baker and W. C. Snyder, eds. Univ. of Calif. Press, Berkeley.
Campos, E., E. A. Maher, and A. Kelman. 1982. Relationship to pectolytic closteridia and *Erwinia carotovora* strains to decay of potato tubers in storage. *Plant Dis.* **66**:543–546.
Clark, F. E. and W. D. Kemper. 1967. Microbial activity in relation to soil water and soil aeration. Pages 472–480 in *Irrigation of Agricultural Lands*. R. M. Hagan, H. R. Haise, and T. W. Edminster, eds. Am. Soc. Agron, Madison, Wisc.
Drew, M. C. and J. M. Lynch. 1980. Soil anaerobiosis, microorganisms, and root function. *Ann. Rev. Phytopathol.* **18**:37–66.
Durbin, R. D. 1959. Factors affecting the vertical distribution of *Rhizoctonia solani* with special reference to CO_2 concentration. *Am. J. Bot.* **46**:22–25.
Ferraz, J. F. P. 1973. Influence of soil atmosphere on spread of *Ophiobolus graminis* along wheat roots. *Trans. Brit. Mycol. Soc.* **61**:237–249.
Fortuner, R. and V. A. Jacq. 1976. *In vitro* study of toxicity of soluble sulphides to three nematodes parasitic on rice in Senegal. *Nematologica* **22**:343–351.
Garrett, S. D. 1937. Soil conditions and the take-all disease of wheat. II. The relation between soil reaction and soil aeration. *Ann. Appl. Biol.* **24**:747–751.
Garret S. D. 1981. Introduction. Pages 1–11 in *Biology and Control of Take-All*. M. J. C. Asher and P. J. Shipton, eds. Academic Press, New York.
Greenwood, D. J. 1970. Distribution of carbon dioxide in the aqueous phase of aerobic soils. *J. Soil Sci.* **21**:314–329.
Griffin, D. M. 1972. Ecology of Soil Fungi. Syracuse Univ. Press, Syracuse, New York.

Hollis, J. P. 1967. Nature of the nematode problem in Louisiana rice fields. *Plant Dis. Reptr.* **51**:167–169.

Ioannou, N., R. W. Schneider, and R. G. Grogan. 1977. Effect of oxygen, carbon dioxide, and ethylene on growth, sporulation, and production of microsclerotia by *Verticillium dahliae*. *Phytopathology* **67**:645–650.

Leach, J. G. 1940. Insect Transmission of Plant Diseases. McGraw-Hill, New York.

Lund, B. M. and A. Kelman. 1977. Determination of the potential for development of bacterial soft rot of potatoes. *Am. Potato J.* **54**:211–225.

Lyda, S. and E. Burnett. 1975. The role of carbon dioxide in growth and survival of *Phymatotrichum omnivorum*. Pages 63–68 in *Biology and Control of Soilborne Plant Pathogens*. G. W. Bruehl, ed. Am. Phytopathol. Soc., St. Paul, Minn.

Macer, R. C. F. 1961. The survival of *Cercosporella herpotrichoides* Fron in wheat straw. *Ann. Appl. Biol.* **49**:165–172.

Motta, J. J. 1969. Cytology and morphogenesis in the rhizomorph of *Armillaria mellea*. *Am. J. Bot.* **56**:610–619.

Russell, R. S. 1977. *Plant Root Systems: Their Function and Interaction with the Soil*. McGraw-Hill, London.

Smith, A. M. and D. M. Griffin. 1971. Oxygen and the ecology of *Armillariella elegans* Heim. *Aust. J. Biol. Sci.* **24**:231–262.

Smith, A. M. and D. Noble. 1972. Effects of oxygen and carbon dioxide on the growth of two varieties of *Gaeumannomyces graminis*. *Trans. Brit. Mycol. Soc.* **58**:499–503.

Stolzy, L. H., S. D. van Gundy, C. K. Labanauska, and T. E. Szieskiewicz. 1963. Response of *Tylenchulus semipenetrans* infected citrus seedlings to soil aeration and temperature. *Soil Sci.* **96**:292–298.

Trolldenier, G. 1977. Mineral nutrition and reduction processes in the rhizosphere of rice. *Plant Soil* **47**:193–202.

Trolldenier, G. 1979. Effects of mineral nutrition of plants and soil oxygen on rhizosphere organisms. Pages 235–240 in *Soilborne Plant Pathogens*. B. Schippers and W. Gams, eds. Academic Press, New York.

Turner, F. T. and C. C. Chen. 1981. Iron coating on rice roots. *IRRN* **6**:18–19.

Chapter 5

Host Development and Stress

Host physiology in relation to disease development is such a comprehensive subject that only certain aspects of special interest to soilborne plant pathogens are discussed. Hypersensitivity and phytoalexins, important in specific host-pathogen reactions, have been dealt with extensively by others. Resistance derived from these reactions are little influenced by the general health of the plant. In nonspecific host-pathogen interactions, factors that affect general plant vigor such as host age, senescence, structure, nutrition, food reserves, and water, and temperature stresses are important. So, too, are interorgan relationships responsive to hormonal control such as macrophysiology or physiology of the roots versus the tops, roots versus fruiting, and stalk rot versus fruiting. Many fungal pathogens profit from stressed hosts, but nematodes, in contrast, profit from good plant health.

Plant physiology can be studied on an interorgan, organ, tissue, cell, organelle, or molecular level. All levels are important, but at present more emphasis at the interorgan level is needed in plant pathology.

EFFECT OF HOST DEVELOPMENT ON SUSCEPTIBILITY

Juvenile Susceptibility

Seedling age and cell wall development are particularly important to seed rot and seedling blight. Halpin and Hanson (1958) inoculated small-seeded legumes with mycelium of five species of *Pythium* from 0 to 56 days after seeding. The exact time of host-pathogen contact is not known, but the fungi contacted the seeds and seedlings rapidly after inoculation. Zoospores of *P. debaryanum, P. irregulare, P. splendens,* and *P. paroecandrum* were recovered from the drainage water within 24 hr of inoculation. *P. ultimum* also reached all parts of the sand within 24 hr, apparently by hyphal growth.

When introduced at the time of seeding, *P. irregulare, P. ultimum,* and *P. debaryanum* rotted some seeds and killed almost all seedlings, either before or after emergence. When the pathogens were introduced 1 day after planting, seed-rot was absent but some preemergence damping-off occurred. When introduced 2 days after seeding, no seed-rot and no pre-emergence blight occurred. When introduced 3 days after seeding, all host species were unaffected by the pathogens. Red clover and Ladino white clover developed complete resistance in 2 days; alfalfa and sweet clover developed most of their resistance in 2 days. By 3 days all were resistant, and no disease developed. The age of the seedling at the time of exposure to the pathogen is critical. Halpin and Hanson cautioned that these results would not hold over a wider range of environmental conditions. They pointed out that excised corn embryos at 24°C were resistant after 3 days but that embryos of the same strains of corn required 15 days to become resistant at 12°C (Hooker and Dickson, 1952).

Tissue Maturation

By tissue maturation I mean development of secondary thickening of cell walls. Cell walls of warm-weather plants in cold, wet soils are often watery (see Table 2.1) and rich in wall substances hydrolysable by *Pythium* spp. Ample potassium and phosphorus hasten tissue maturation; excess nitrogen retards it. The importance of tissue maturation is illustrated by the short period of susceptibility of legume seeds and seedlings to attack by *Pythium* spp. It is less dramatically seen in attack of wheat seedlings by *Gaeumannomyces graminis* var. *tritici*.

Robertson (1932) studied Marquis spring wheat and noted an increase in lignification during the first 40 days. Lignification is most pronounced in the xylem, pericycle, fibers of the vascular bundle sheaths and in the subepidermal layers of the cortex of the crown roots. Robertson inoculated wheat with *G. graminis* at planting time and at 10-day intervals until maturity. Disease development declined dramatically at 40 days and later, the time when lignification is essentially complete. In 10-day-old seminal roots the endodermis was well developed, but lignification had not occurred, except for possibly a trace in the spiral thickenings of the protoxylem vessels. The endodermis of the subcrown internode is well developed in plants 20 days old and sclerenchyma is present, so resistant mechanical tissues develop quickly.

Fellows and Ficke (1934) grew wheat in pots with inoculum of *G.*

graminis at different depths in the soil. Severe damage developed only when the inoculum was 7.5 cm or less from the seed, but seedling blight is atypical in the field. When inoculum is near the seed, the fungus reaches the vital crown region before the tissues become resistant. In the field most plants survive to produce the stunted, whitehead stage characteristic of the take-all disease. In cultivated soils the chance of the inoculum being concentrated near the seed is small. Sufficient time may ensue between infection and when hyphae reach the crown region for tissues to become more resistant. Garrett (1948) warned that if you expect to study the effects of plant nutrition on take-all of wheat, do not put the inoculum close to the seed. The fungus, when close to the seed, will destroy the host before expression of nutritional factors is possible.

The effects of plant nutrition on tissue maturation are difficult to assess, because plants with good nutrition produce more roots. If an equal number of severe infections occur, well-nourished plants will have more healthy roots. Garrett (1941) referred to this as an escape mechanism, demonstrating that the least loss occurs with normal, balanced nutrition, the most loss with phosphorus deficiency, and that any nutrient deficiency, either of an individual major nutrient or in total concentration, increases disease loss.

Deacon and Henry (1980) inoculated wheat and barley roots with *G. graminis* under septic and aseptic conditions. They grew the seedlings in perspex root-observation boxes with a sloping, transparent, removable side. Nineteen days after sowing inoculum of the pathogen was placed 5 cm below seed depth on a root tip (0 days of age), and on root tissues 5 and 15 days old. *G. graminis* attacked the 0-day-old seminal root tissue (root tip) most severely, the 5-day-old root tissue next, and the 15-day-old root tissue least severely. Results were the same, with or without the presence of normal microflora, so they reasoned that the reduced disease on the older root was not due to a more active microflora on the root tissue surface or in the rhizosphere.

Tissue Senescence

In maize, bred for high conversion of dry matter into grain, stalk tissues can become senescent early enough for *Gibberella zeae* and *Diplodia zeae* to cause severe losses. Even though the stalk may be infected early, these pathogens do not become aggressive until the time of pollination or shortly thereafter (Michaelson, 1957). Vegetative development of stalks of single-cross hybrids ceases at pollination, but in resistant hybrids dry matter in stalks increases for several weeks after pollination. Resistant hybrids maintain physiologic activity within the stalk at a higher level longer than susceptible hybrids (Wall and Mortimore, 1965). *Diplodia zeae* (Pappelis and Smith, 1963) and *Gibberella zeae* (Pappelis, 1965) spread mainly in dead cells within the pith. Craig and Hooker (1961) and Mortimore and Ward (1964) found higher levels of soluble sugars in stalks of resistant corn at physiologic maturity than in susceptible corn. The greater the carbohydrate sink is the greater the susceptibility of hybrids of the same genetic constitution (Dodd, 1980). These observations and those with *Macrophomina phaseolina* and Pythium root rot of spring barley stress the need to sustain metabolic balance between plant organs.

Cortical Sloughing

In contrast to charcoal rot of sorghum or Fusarium or Diplodia stalk rot of maize, which are favored by carbohydrate stress either during kernel formation (charcoal rot of sorghum) or after pollination and early kernel formation (the stalk rots of maize), Phymatotrichum root rot of cotton is not influenced by flowering. Symptoms appear at the same time on cotton from which the flowers were or were not removed (Rush et al., 1984a). A nonfruiting, sterile cotton is fully susceptible.

Phymatotrichum omnivorum causes a cortical rot of young cotton roots, but this rot causes no above-ground symptoms (Rush et al., 1984b). Death of cotton occurs 27 to 50 days after seedling emergence. At 18 to 25 days the root cortex is sloughed, even by healthy cotton roots. Lethal invasion of the vascular system occurs after the cortex dies. What accounts for the change from nondamaging "juvenile" attack to lethal attack after cortical death is not known, but flowering plays no role in the increased susceptibility of plants past the seedling stages.

Death of cotton from Phymatotrichum root rot occurs about 40 days after emergence when sclerotia are 5 cm deep in the soil and at about 61 days when sclerotia are 60 cm deep, as is common in the field. The studies of Rush et al., prove that the time when above-ground symptoms develop depends on soil temperature (it must be above 22°C at the depth of the sclerotia), soil moisture (moist soil favors disease), and the depth of the sclerotia in the soil, and that the lethal attack is initiated after the cortex is sloughed. These observations explain how cotton can appear healthy until the squares form, and then die suddenly. Many aspects of root health in dicots, including cortical death, are discussed in a comprehensive manner by Wilhelm (1959).

EFFECT OF ENVIRONMENTAL STRESS

Maize Seedlings at Low Temperature

Early pathologists and agronomists increased maize yields in the northern part of the North American maize belt by selecting lines adapted to survive and emerge in cold, wet soil (Bruehl, 1983). Modern seed-treatments have reduced the significance of their work, but research on maize at low temperature continues. Most North American and European maize cultivars are incapable of significant photosynthesis below 15°C. Eagles and Hardacre

(1979) tested the germination and survival of maize at a constant 10°C from hand-picked, hand-shelled, carefully dried, and fungicide-treated seed, so that pathogenic effects were minimized. Maternal effects on low-temperature germination were strong, with males influencing speed of emergence and total emergence. Hardacre and Eagles (1980) grew several corns at a sustained 13°C to determine their photosynthetic power. Survival at 104 days of a Confite Pueno hybrid and a hybrid of Huancavelicano, both from Peru, was 96%. Three United States maize belt hybrid controls did not sustain growth at 13°C.

Growth of ordinary maize at temperatures below 15°C is totally dependent on food reserves in the seed (Hardacre and Eagles, 1980). Seedlings develop slowly from 10 to 13°C but earlier workers demonstrated that even though green leaves were present at 15°C or below, very little if any photosynthesis occurred in standard maize. Thus if breeders reduce the temperature threshold for photosynthesis in this important plant, another increment of progress in early stand vigor is probable. With fungicides to minimize pathologic effects at the seedling stage, the ability to synthesize food at low temperature should increase resistance of the roots to *Pythium* spp.

Carbohydrate Reserve

Mechanisms of host resistance (hypersensitivity, phytoalexins, corking-off, or others) require energy. Energy is particularly important in the resistance of winter wheat to snow molds caused by species of *Typhula* and *Fusarium*. These pathogens are most destructive to winter wheat when deep snow persists for long periods on unfrozen soil, usually for 110 days or longer. These fungi grow at near 0°C in the dark in humid conditions, and deep snow on unfrozen soil maintains the soil surface near 0.5°C and no photosynthesis occurs. The ability of wheat to sustain the attack and to survive with vigor depends on its stage of development, which is governed by seeding date, autumn growing conditions, and the genetics of the cultivar.

In Douglas County, Washington, farmers learned by experience that large, robust plants were more resistant to snow molds than wheats of intermediate size. Wheat from very late seedings that barely emerge in autumn or emerge under the snow escape attack. These very young seedlings are seldom destroyed, because they have no senescent leaves and have reserve food in the seed, but they yield no more than spring wheat, losing the advantage of winter wheat. Plants of intermediate size with few tillers have lesser reserves in the crown, and the endosperm is exhausted. Large plants with 8 or more tillers have many senescent leaves that are susceptible, even to the least virulent of the snow mold pathogens (*T. incarnata*), but the large plants have carbohydrate reserves in the crown that sustains metabolic activity for a long period (Fig. 5.1).

Resistant wheats store more carbohydrates in the autumn (Kiyomoto and Bruehl, 1977; Amano and Osanai, 1983) than susceptible wheats.

Figure 5.1. (A) Snow mold, caused by *Typhla idahoensis*, destroys all the leaves of early-seeded winter wheats. (photographed 1 week after snow melt). (B) New growth from living crowns (recovery) is pronounced 5 weeks after snow melt in resistant wheats.

Amano and Osanai reported positive correlations between resistance to *T. ishikariensis* and *F. nivale* and plant dry weight, total carbohydrates, total sugars, nonreducing sugars, crude starch, and a high carbohydrate nitrogen ratio in plants hardened for winter. The snow-enduring wheats accumulated carbohydrates in greater concentration and faster in the autumn than the other wheats.

The chemical analyses of Amano and Osanai (1983) confirmed the hypothesis of earlier Japanese work (Tomiyama, 1955, and citations therein) that as food reserves of respiring wheat under snow at 0°C become exhausted, proteins are hydrolyzed, freeing simple nitrogen compounds potentially toxic to tissues, further reducing resistance. Proteinaceous nitrogen in susceptible wheats was 3.4% November 20 and 2.8% on April 20; a decline of 0.6%. Resistant wheats had 2.7% proteinaceous nitrogen November 20 and 2.4% on April 20; a decline of 0.3%. The greater carbohydrate reserves of the snow-enduring wheats partially protected the proteins against hydrolysis. In starvation, whether of plants or animals, breakdown of proteins results in products of potential toxicity. Animals excrete these products in urine. Plants probably have less effective ways to dispose of them.

Nutrient Imbalance

Lahaina, a "noble" cane (pure *Saccharum officinarum*), began to fail in localized spots in Hawaii in the 1920s and 1930s. The soils are volcanic in origin and they vary in nutrient content. C. W. Carpenter, 1919–1935, determined that part of the failure of Lahaina was due to Pythium root rot on soils low in phosphorus. Although phosphorus deficiency was the main predisposing factor in the field, Martin (1934) found that the disease was most severe when calcium was deficient, was severe with iron, phosphorus, sulphur, manganese, and potassium deficiency, and was unaffected by nitrogen deficiency.

Edgerton (1958) noted that the soils of Louisiana are alluvial, with deposits from various regions. Severe nutrient deficiencies of local distribution are rare or absent. Additions of either nitrogen or phosphorus do not alleviate losses from Pythium root rot in Louisiana. The disease is worst on poorly drained soils during seasons with cool, wet winters.

In June of 1917 in southern Saskatchewan, Canada, spring wheat turned yellowish to brownish in patches. Wheat that suffered most was on land previously summer fallowed. In 1926 Simmonds found Pythium root rot widespread on spring cereals in the Canadian prairies. The disease is favored by cool, wet springs followed by a sudden warm, dry period (Vanterpool and Ledingham, 1930). Several species of *Pythium* are involved, including *P. arrhenomanes, P. tardicrescens, P. graminicola,* and *P. aristosporum*. The pathogens occur in native grasses, and they have been isolated from wheat sown on virgin land. Oospores form rapidly in diseased roots, and they remain alive at least 4 years. The pathogen is disseminated in wind-blown dust, and it has been recovered from soil taken from rain gutters. Simmonds found the pathogens in every soil sampled in the Canadian prairies.

Authentic severe cases of Pythium root rot on wheat following wheat without summer fallow did not occur during a 20-year observation period (Vanterpool, 1940). Phosphate is the most deficient nutrient in root-rot soil, with nitrogen second. Application of phosphate reduces the losses, but the roots are still invaded by *Pythium* spp. Phosphates increases the number of roots. The number of diseased roots remain the same, but the number of healthy roots increased. If phosphate is added, nitrogen, too, can be added. The two nutrients must be in proper ratio to each other for best control of the disease. By 1946 the use of phosphate fertilizer was standard practice in Saskatchewan (Mitchell, 1946) and Pythium root rot has not been serious since.

Barley is more susceptible than wheat to Pythium root rot, and *P. arrhenomanes* affects the physiology of the host. Waller (1979) grew seedlings in water for 2 weeks or in 1/2 strength Hoagland's solution. Nutrients increased the growth of healthy seedlings by 84%, and of diseased seedlings by 34% relative to seedlings without the pathogen. Diseased plants are unable to utilize nutrients efficiently.

Most of us have thought of Pythium root rot of cereals in a physical sense, as a loss of roots (root-pruning), particularly of the fine rootlets of the seminal root system. Singleton and Ziv (1981) amputated portions of wheat seedling roots and destroyed the same amount of roots by inoculation with *P. arrhenomanes*. Infection by the fungus stops root elongation, but physical amputation interrupts it only temporarily. Seedlings remove materials from the seed more rapidly either when healthy or after amputation than after infection of roots by *P. arrhenomanes*. The fungus reduces the ability of the seedling to utilize the food reserves in the endosperm. The damage from pathogenesis is greater than that caused only by the physical loss of roots.

Climatic Adaptation

Pythium root rot was much reduced in Louisiana when interspecies hybrids replaced pure *S. officinarum* canes. *S. barberi* from India and *S. spontaneum* contributed to a more vigorous root system better adapted to low winter temperatures. Soil temperatures at 7.5 cm depth at Houma, Louisiana, near the Gulf of Mexico, are below 21°C from mid-October to early April, at least 5 months. Air temperatures in January average 13°C, so tropical sugar canes are stressed for a considerable period of time (Rands and Dopp, 1938). D-74 and other noble canes fail at a time when canes produced by hybridization with *S. barberi* or *S. spontaneum* survive (Edgerton et al., 1929).

Most of the barley produced in South Dakota has been of the "Manchurian" type, adapted to being seeded in a northern latitude in April and early May. Barley breeders attempted to reduce the drought hazard by developing cultivars that matured rapidly, in the hope that earliness would lead to drought escape. Earliness was achieved by crossing local barleys with barley native to North Africa, a Coast type barley. In North Africa barley emerges in November during a period of moderate-cool weather and gradually shortening days. The Coast barley responds to photoperiod and matures very early when grown from spring planting farther north.

P. arrhenomanes is severe on an infertile sandy clay soil in South Dakota, and Bruehl (1953, 1955) grew many spring barleys on this soil. Early-maturing varieties are most severely diseased, and moderately late varities yield best. Two mid-maturity varieties (Trebi and Odessa) outyielded two early South Dakota varieties (Plains, Feebar) on the root rot nursery soil in four successive seasons by 19 to 49% per year. Odessa (midmaturity) outyields Feebar (early maturity) on natural soil, but the reverse is true on soil treated with chloropicrin. Yields in the fumigated soil are three times that in natural soil. Water is not the limiting factor, disease is (Fig. 5.2).

Bruehl (1955) theorized that earliness results from a strong hormonal effect, favoring the inflorescence over the root system, and that later barleys maintain a better balance between the roots and reproductive structures. The soil sterilization experiment has proved the agronomists to be right, that early cultivars should be advantageous. However, when root rot is involved, earliness renders the plant more vulnerable. Climatic adaptation, the vegetative rhythm of the plant being in harmony with temperature and moisture regimes, cannot always be at the highest level for best yields.

Figure 5.2. Most fungal pathogens of cereal roots cause premature death, but Pythium root rot delays maturity. In the foreground Feebar spring barley on natural soil has not headed; in the background on chloropicrin-treated soil the heads have emerged. Tripp County, South Dakota, 1950.

COMPLEX STRESS

Charcoal Rot

Charcoal rot was studied in cotton, cowpea, jute, and peanut by Shaw in India in 1912; sweet potato in New Jersey by Taubenhaus in 1913; beans, lima beans and cowpeas in the southeastern states by various workers; maize and beans in California in 1932 by Mackie; and sorghum in India by Uppal. The names most frequently applied to the pathogen in literature are *Sclerotium bataticola, Rhizoctonia bataticola,* and *Macrophomina phaseolina.* No sexual state is known. Pycnidia are produced on some hosts by some strains in some environments, but usually sclerotia are the important propagules, and we will concern ourselves with them only. Diseases caused by *M. phaseolina* are common in warm areas on warm-season crops.

Although seedling blights occur, charcoal rot causes an internal rot of the stalk, extending down into the crown, of maize and sorghum. The parenchyma is destroyed, and small sclerotia develop in profusion on the rind and on the vascular strands within the stalk. These sclerotia become so abundant that charcoal appears to have been sprinkled within the stalk. Cook et al. (1973) scraped sclerotia from the inside of corn and sorghum stalks and obtained 0.4 g of sclerotia (about 560,000) from the base to the third nodal plate of a single corn stalk and 0.02 g (or 28,000) from each sorghum stalk.

Charcoal rot became serious during a series of dry seasons in the Midwest. Low soil moisture and high temperature favor the stalk rot (Livingston, 1945). Stalk rot of sorghum was worst in New Mexico on light soils under inadequate rainfall or inadequate irrigation and in seasons with favorable early growing conditions followed by drought Hsi (1956).

Edmunds (1964) found the balance between environmental conditions (temperature, moisture) and stage of host development critical for development of typical symptoms in the field. The pathogen is very active in plants subjected to near-lethal drought prior to inoculation, but high temperature without drought stress does not predispose the plants at any stage of development. He grew sorghums at 20°C and 80% available soil moisture until 4 to 7 days before inoculation. At 4 to 7 days prior to inoculation the soil temperature was adjusted to 35 or 40°C, and the soil moisture was decreased to the 25% available moisture level. This sudden, severe stress predisposes the plants to the pathogen. The moisture level was raised to 80% of available water immediately after inoculation, so that water stress occurred only prior to infection. No stalk rot developed in plants inoculated 0 to 12 days after anthesis; maximum rot developed when plants were inoculated 17 to 28 days after anthesis. Stalk rot diminished rapidly with later inoculations.

Disease progress in plants stressed prior to inoculation was rapid: stalks were internally rotted within 24 to 48 hr. Sclerotia were present at or shortly after death of the plant, 3 to 7 days after inoculation. The rapid and abundant development of sclerotia in diseased stalks led Edmunds to conclude that saprophytic activity in stalks long after host death was probably negligible. Lateral rot within the stalk is more devastating than longitudinal, and lateral rot progresses more rapidly at a soil temperature of 40 than at 35°C.

Odvody and Dunkle (1979) confirmed that seed production is associated with drought-stress-induced charcoal stalk and root rot of sorghum. A male sterile and a fertile sorghum were grown in a potting medium infested with sclerotia. At the time, the fertile plants were in the soft dough stage, and some of the plants were drought-stressed. Drought stress increased symptomless infections on the roots of sterile plants, and led to severe foot rot of fertile plants. No root infection occurred before imposition of drought stress. These authors postulated that drought stress increased root exudate that stimulated sclerotium germination.

Although dry weather can accentuate charcoal rot of soybeans, this disease is common in Missouri during all growing seasons, whether moist or dry, hot or cool (Wyllie and Calvert, 1969). They conducted experiments in which one set of soybeans was watered to near field capacity and another set was watered intermittantly so that it fluctuated from near the permanent wilting point to about 50% of field capacity. Inoculation was by soil infestation prior to planting. Nearly all the plants were infected, regardless of the temperature and water treatments. Blossoms were removed daily as they formed from half the plants. No sclerotia developed in these plants, but sclerotia formed in practically all the plants that produced beans. Wyllie and Calvert concluded that in soybeans producing seed is the greatest predisposing factor. They suggest that the association of warm, dry weather with charcoal rot in soybeans is due to earlier fruiting, providing a longer period for sclerotium formation in such seasons.

NEMATODE REPRODUCTION

Nematodes are unable to synthesize foods to the extent that many fungi and bacteria do, and their development is strongly influenced by the quality of host nutrition and they respond to the mineral nutrition of the host plant. Luedders et al. (1979) grew soybeans in soil naturally infested with *Heterodera glycines*. Because the soil was low in potassium, they added KCl or K_2SO_4 at four rates. The lowest rate of fertilization increased the number of cysts in the soil by 38% and the number of mature females on the roots by 89%. Higher rates resulted in declining reproduction until, at the highest rate, the number of new females on roots was the same as in the potassium-deficient plants. Davide and Triantaphyllou (1967b) grew a root-knot nematode on tomatoes with varying levels of nitrogen, phosphorus, and potassium. More larvae became females on well-nourished plants than on plants with nutrient deficiencies. The same authors (1967a) found that the age of the plant did not affect the ratio of males to females in *Meloidogyne incognita* or *M. javanica*. Under favorable conditions most larvae become females.

Defoliation also influences nematodes. Sustained egg production depends on continued feeding on intact (not defoliated) plants. Cook (1977) reared *Heterodera avenae* on spring barley defoliated on different dates, ranging from 28 days after inoculation to past maturity (112 days). No reproduction occurred on plants from which the tops were removed 28 days after inoculation, and no females survived to form cysts. Cysts per plant increased until the series defoliated at 65 days after inoculation, when the number of eggs per cyst (368) had reached maximum. Food within the females and within the roots was adequate for maximum fecundity by that stage of host development. Because of these observations, Cook thought that a very early-maturing cultivar might curtail egg production. This hypothesis is false. The nematodes delay host maturity, and maximum egg-production occurs even on the "earliest" barley.

The host has a finite carrying capacity. DiVito et al. (1980) grew maize in clay pots. Eggs of race 1, *Meloidogyne incognita*, were poured on the surface of the soil, and the corn was planted. After 75 days at 24 to 26°C (two generations) eggs at 0.5/g of soil resulted in 229 eggs/g of soil at termination, an increase of 458 times. The eggs were introduced in a geometric series, and the increase in egg production decreased in proportion with the initial inoculum level, until, when 1024 eggs/g constituted the initial inoculum, 1075 eggs/g were recovered. A point was reached at which the nematodes were just capable of maintaining the initial level of inoculum. Well-nourished nematodes reproduce at a high rate. As competition and deleterious effects on the host increase, the rate of reproduction declines proportionately. The carrying capacity of the host is finite and is reached midway in the inoculum-density series.

Cylindrocladium crotalariae, the anamorph of *Calonectria crotalariae*, cause Cylindrocladium black rot of peanut. Two important but contrasting nematodes, *Meloidogyne hapla* and *Macroposthonia (Criconemoides) ornata*, attack peanut. Diomande and Beute (1981) studied the interactions of these parasites on a black-rot-resistant and susceptible peanut. They used different inoculum densities of all three pathogens. Both nematodes increased black rot of the susceptible peanut, but only *M. hapla* increased black rot of the resistant cultivar. The failure of *Macroposthonia ornata* to increase black rot on the resistant peanut is puzzling, because the disease is increased by wounding the root tips with fine needles. Both nematodes increased most when on the host alone, and both decreased with increasing levels of fungal disease.

Nematodes, as obligate parasites, usually do less well when the root system suffers from another problem (Powell, 1971). Organisms accompanying the nematode may gain, but not the nematode.

TRAPPING MICROCONIDIA

Wilt-inducing fusaria penetrate roots of many plants, including nonsuscepts. Even though the cortex or extra-vascular tissues vary in susceptibility to invasion by these fungi, resistance lies mainly within the vascular system. Many inoculation techniques that distinguish between susceptible and resistant plants employ wounds that give direct access to the xylem, so we assume that real resistance or susceptibility in most instances depends on response to the fungus within xylem vessels.

One of the mechanisms by which wilts impair water movement is by occlusion of the vessels by gels and tyloses. These occlusions are important in vascular wilts, as is the increase in viscosity of the xylem stream if the pathogen produces polysaccharide slime or hydrolyses pectic substances. *F. oxysporum* f. sp. *lycopersici* produces pectinase enzymes resulting in pectinaceous plugging of vessels. The brilliant work of Beckman and his associates (1961) reveals an interesting, and at first contradictory, explanation of the role of vascular occlusions in the fusarial wilts. If occlusion occurs promptly, it is beneficial. If occlusion occurs late, it is harmful. Like blood clotting, it can save, and it can kill.

The end walls of xylem vessels disappear in some plants (simple perforation plate) or, in other plants, develop into elongate (scalariform) or circular (foraminate) openings through the end walls of the cells. These holes facilitate rapid movement of the xylem stream. The wilt-inducing fusaria produce microconidia in the xylem, and in banana and tomato roots, these spores are trapped by the perforation plates in xylem vessels. The microconidia germinate, and the germling grows through and sporulates above the site of physical entrapment. If the plant is resistant, gel forms rapidly about the germling prior to sporulation. In response to the presence of the pathogen, tyloses form in advance of the entrapment site. The gel can be hydrolysed, but the tyloses are permanent. Rapid gel formation followed by tyloses localizes the pathogen in the roots so that it does not become systemic.

In a susceptible plant host response is so slow that microconidia proceed from vessel to vessel in the root. Vascular occlusion occurs, but not until after extensive host invasion. Timing and vigor of the host response determines whether the plant is susceptible or resistant. In banana at low temperature, even a genetically susceptible plant reacts with a prompt, resistant reaction. The trapping sequence has been demonstrated in cotton, elm, hop, mimosa, and tomato, so it is considered a general phenomenon.

Beckman and Halmos (1962) introduced miscellaneous microorganisms not pathogenic to banana into xylem vessels, and if they grew, rapid vascular occlusion followed, localizing them. The only fungus they tested that did not elicit the prompt occlusion of xylem vessels of susceptible banana was *F. oxysporum* f. sp. *cubense*. Roots are wounded in nature by high winds that move the shoots, by cultivation, by nematodes, and by insects. Soil microorganisms are introduced repeatedly into plants, but only those compatible with the plant to such a degree that they do not initiate a strong host response can become systemic. Refer to Beckman and Talboys (1981) for additional details and literature.

One of the reasons there are no systemic fungal wilts of gymnosperms may be because they have tracheids but no vessels. All fungi that regularly produce systemic infections in the xylem have a *Cephalosporium*-type spore stage, a simple conidiophore with a small conidium adapted to dispersal in the xylem stream, regardless of the taxonomic position of the fungus (*Fusarium, Verticillium, Graphium, Dothiorella, Ophiostoma, Ceratocystis, Hymenula* (Bruehl, 1963). The bordered pits of tracheids do not permit free movement of these small spores.

TRAPPING BACTERIA

When rhizoplane microorganisms (mostly bacteria?) are washed from banana roots and introduced into xylem vessels of banana roots, within 24 hr the vessels become occluded (Beckman and Halmos, 1962). Bacteria pass through the openings of the perforation plates, so the physical trapping mechanism of limiting spread within the plant could be assumed to be less effective than for fungal pathogens. It is probable that bacteria are localized more by being bound to cell walls. Sequeira (1981) reviewed the literature on binding of bacteria to plant cell walls.

Evidence is accumulating that incompatible or avirulent bacteria in some host-pathogen relations are fixed to cell walls, thereby being localized and rendered incapable of causing significant disease. *Pseudomonas solanacearum*, the southern bacterial wilt pathogen, produces extracellular polysaccharide. Pathogenic isolates produce smooth colonies in culture, and nonpathogenic or weakly virulent isolates form rough colonies (Husain and Kelman, 1958). Smooth colonies result from abundant extracellular polysaccharide slime which contributes directly to the wilting phenomenon. Lack of the extracellular polysaccharide may also be important, in that cells without the slime appear to be recognized by cell walls as foreign bodies, leading to attachment and immobilization within the plant. The slime, produced by the virulent forms, apparently prevents recognition and effective response by the host. The outer bacterial membrane contains lipopolysaccharide responsible for host-specific recognition. The incompatible reaction results when the lipopolysaccharide of the outer bacterial membrane reacts with the host cell wall polymers i.e., the lectins. Sequeira and Graham (1977) found that isolates of *P. solanacearum* virulent to potato were weakly agglutinated by potato lectin, or not agglutinated at all, whereas all avirulent isolates agglutinated strongly.

TRANSLOCATION PATTERNS

Translocation patterns of elaborated foods (phloem transport) are an important factor in soilborne virus diseases of plants. According to Webb (1927,

1928) infection of winter wheat with wheat soilborne mosaic virus is favored by a soil temperature of about 16°C. This temperature is fairly warm for autumn-seeded wheat. In spite of this, symptoms are slow to develop in the field. In Kansas, even though autumn infection may be prevalent, symptoms are usually not observed for at least 7 weeks or until the following spring. Rao and Brakke (1970) inoculated 2-week-old wheat seedlings by root washings of diseased plants and incubated them at 17°C. One week after inoculation, half the plants were placed in the dark for 4 days. After 6 weeks 86% of the plants with the dark treatment had mosaic symptoms. Plants not placed in the dark had no symptoms. Rao and Brakke found that WSBMV moved slowly in the wheat plant and that either dark treatment or removing leaves after inoculation speeded virus movement. In contrast, Bockus and Niblett (1984) obtained 90% symptomatic plants 8 weeks after emergence at 15°C with a 12-hr day, which indicates that low temperature rather than a lack of translocation may govern the time when symptoms appear in the field. When many plant species are inoculated with tobacco necrosis virus, either mechanically or by using viruliferous zoospores of *Olpidium brassicae*, a group of 20 species are susceptible to *O. brassicae*, and the roots are infected by the virus, but the virus remains in the roots (Hiruki, 1967).

A classic experiment by Bennett (1937) with curly top virus is important, even though it is not a soilborne virus. Bennett partially split a sugar beet plant into three parts so that the three parts were united in the tap root. Viruliferous *Circulifer tenellus* were fed on one part of the beet, another part was kept in the dark for 5 days, and the third part was unshaded. In 11 days symptoms developed on the inoculated part, in 13 days symptoms appeared on the part darkened for 5 days, and 90 days later the unshaded, noninoculated part was still symptomless. He treated another beet the same way, except that instead of shading one part he removed its leaves and allowed new leaves to develop on that part. In this case symptoms developed in 12 days on the inoculated shoot, and a few days later on the defoliated shoot, and in 150 days no symptoms were present on the noninoculated, nondefoliated shoot. Bennett concluded that the curly top virus moved in the phloem and that the beet root is a strong sink with little upward translocation in phloem during the growing season.

COMMENTS

Many diseases of plants are affected by the degree of development of tissues, by their state of metabolism, and by energy levels and stress. Of great interest to me was what appears to be a real difference between the effects of host stress on unspecialized fungal pathogens and on nematodes (obligate parasites). Host stress favors many fungi but not nematodes.

Adaptation is a much-used word, but it has real significance. A plant in harmony with its environment, in harmony with seasonal sequences of temperature, light, and water, has many advantages. Adaptation reduces stress.

KEY REFERENCES

Beckman, C. H. and S. Halmos. 1962. Relation of vascular occluding reactions in banana roots to pathogenicity of root-invading fungi. *Phytopathology* **52**:893–897.
Davide, R. G. and A. C. Triantaphyllou. 1967b. Influence of environment on development and sex differentiation of root-knot nematodes. II. Effect on host nutrition. *Nematologica* **13**:111–117.
Edmunds, L. K. 1964. Combined relation of plant maturity, temperature, and soil moisture to charcoal stalk rot development in grain sorghum. *Phytopathology* **54**:514–517.
Halpin, J. E. and E. W. Hanson. 1958. Effect of age of seedlings of alfalfa, red clover, Ladino white clover, and sweet clover on susceptibility to Pythium. *Phytopathology* **48**:481–485.
Mortimore, C. G. and G. M. Ward. 1964. Root and stalk rot of corn in southwestern Ontario. III. Sugar levels as a measure of plant vigor and resistance. *Can. J. Plant Sci.* **44**:451–457.

REFERENCES

Amano, Y. and S.-I. Osanai. 1983. Winter wheat breeding for resistance to snow mold and cold hardiness. III. Varietal differences of ecological characteristics on cold acclimation and relationships of them to resistance. *Bull. Hokkaido Prefect. Agric. Expt. Sta.* **50**:83-97.
Beckman, C. H., M. E Mace, S. Halmos, ad M. W. McGahan. 1961. Physical barriers associated with resistance in Fusarium wilt of bananas. Phytopathology **51**:507–515.
Beckman, C. H. and P. W. Talboys. 1981. Anatomy of resistance. Pages 487–521 in *Fungal Wilt Diseases of Plants*. M. E. Mace, A. A. Bell, and C. H. Beckman, eds. Academic Press, New York.
Bennett, C. W. 1937. Correlation between movement of the curly top virus and translocation of food in tobacco and sugar beet. *J. Agric. Res.* **54**:479-502.
Bockus, W. W. and C. L. Niblett. 1984. A procedure to identify resistance to wheat soilborne mosaic in wheat seedlings. *Plant Dis.* **68**:123–124.
Bruehl, G. W. 1953. Pythium root rot of barley and wheat. *U. S. Dept. Agric. Tech. Bull.* 1084.
Bruehl, G. W. 1955. Barley adaptation in relation to Pythium root rot. *Phytopathology* **45**:97–102.
Bruehl, G. W. 1963. *Hymenula cerealis*, the sporodochial stage of *Cephalosporium gramineum*. *Phytopathology* **53**:205–208.
Bruehl, G. W. 1983. Nonspecific genetic resistance to soilborne fungi. *Phytopathology* **78**:948–951.
Cook, G. E., M. G. Boosalis, L. D. Dunkle, and G. N. Odvody. 1973. Survival of *Macrophomina phaseoli* in corn and sorghum stalk residue. *Plant Dis. Reptr.* **57**:873–875.
Cook, R. 1977. The relationship between feeding and fecundity of females of *Heterodera avenae*. *Nematologica* **23**:403–410.
Craig, J. and A. L. Hooker. 1961. Relation of sugar trends and pith density of Diplodia stalk rot in dent corn. *Phytopathology* **51**:276-282.

Davide, R. G. and A. C. Triantaphyllou. 1967a. Influence of the environment of development and sex differentiation of root-knot nematodes. I. Effect of infection density, age of host plant and soil temperature. *Nematologica* **13**:102–110.

Deacon, J. W. and C. M. Henry. 1980. Age of wheat and barley roots and infection by *Gaeumannomyces graminis* var. *tritici*. *Soil Biol. Biochem.* **12**:113–118.

Diomande, M. and M. K. Beute. 1981. Effects of *Meloidogyne hapla* and *Macroposthonia ornata* on cylindrocladium black rot of peanut. *Phytopathology* **71**:491–496.

DiVito, M., N. Vovlas, and R. N. Inserra. 1980. Influence of *Meloidogyne incognita* on growth of corn in pots. *Plant Dis.* **64**:1025–1026.

Dodd, J. L. 1980. Grain sink size and predisposition of *Zea mays* to stalk rot. *Phytopathology* **70**:534–535.

Eagles, H. A. and A. K. Hardacre. 1979. Genetic variation in maize (*Zea mays* L.) for germination and emergence at 10°C. *Euphytica* **28**:287–295.

Edgerton, C. W. 1958. Sugarcane and Its Diseases. State Univ. Press, Baton Rouge.

Edgerton, C. W., E. C. Tims, and P. J. Mills. 1929. Relation of species of Phythium to the root-rot disease of sugar cane. *Phytopathology* **19**:549–564.

Fellows, H. and C. H. Ficke. 1934. Effects on wheat plants of *Ophiobolus graminis* at different levels in the soil. *J. Agric. Res.* **49**:871–880.

Garrett, S. D. 1941. Soil conditions and the take-all disease of wheat. VI. The effect of plant nutrition upon disease resistance. *Ann. Appl. Biol.* **28**:14–18.

Garrett, S. D. 1948. Soil conditions and the take-all disease of wheat. IX. Interaction between host plant nutrition, disease escape, and disease resistance. *Ann. Appl. Biol.* **35**:14–17.

Hardacre, A. K. and H. A. Eagles. 1980. Comparisons among populations of maize for growth at 13°C. *Crop Sci.* **20**:780–784.

Hiruki, C. 1967. Host specificity in transmission of tobacco stunt virus by *Olpidium brassicae*. *Virology* **33**:131–136.

Hooker, A. L. and J. G. Dickson. 1952. Resistance to Pythium manifest by excises corn embryos at low temperatures. *Agron. J.* **44**:443–447.

Hsi, C. H. 1956. Stalk rots of sorghum in eastern New Mexico. *Plant Dis. Reptr.* **40**:369–371.

Husain, A. and A. Kelman. 1958. Relation of slime production to mechanism of wilting and pathogenicity of *Pseudomonas solanacearum*. *Phytopathology* **48**:155–165.

Kiyomoto, R. K. and G. W. Bruehl. 1977. Carbohydrate accumulation and depletion by winter cereals differing in resistance to *Typhyla idahoensis*. *Phytopathology* **67**:206–211.

Livingston, J. E. 1945. Charcoal rot of corn and sorghum. *Neb. Agric. Expt. Sta. Res. Bull.* 136.

Luedders, V. D., J. G. Shannon, and C. H. Baldwin, Jr. 1979. Influence of rate and source of potassium on soybean cyst nematode reproduction on soybean seedlings. *Plant Dis. Reptr.* **63**:558–560.

Martin, J. P. 1934. Symptoms of malnutrition manifested by the sugar cane plant when grown in culture solutions from which certain essential elements are omitted. *Hawaii Platn. Rec.* **38**:3–31.

Michaelson, M. E. 1957. Factors affecting development of stalk rot of corn caused by *Diplodia zeae* and *Gibberella zeae*. *Phytopathology* **47**:499–503.

Mitchell, J. 1946. The effect of phosphatic fertilizers on summer-fallow wheat crops in certain areas of Saskatchewan. *Sci. Agric.* **26**:566–577.

Odvody, G. N. and L. D. Dunkle. 1979. Charcoal rot of sorghum: Effect of environment on host-parasite relations. *Phytopathology* **69**:250–254.

Pappelis, A. J. 1965. Relationship of seasonal changes in pith condition ratings and density to *Gibberella* stalk rot of corn. *Phytopathology* **55**:623–626.

Pappelis, A. J. and F. G. Smith. 1963. Relationship of water content and living cells to spread of *Diplodia zeae* in corn stalks. *Phytopathology* **53**:1100–1105.

Powell, N. T. 1971. Interactions between nematodes and fungi in disease complexes. *Ann. Rev. Phytopathol.* **9**:253–274.

Rands, R. D. and E. Dopp. 1938. Pythium root rot of sugarcane. *U. S. Dept. Agric. Tech. Bull.* 666.

Rao, A. S. and M. K. Brake. 1970. Dark treatment of wheat inoculated with soilborne wheat mosaic and barley stripe mosaic viruses. *Phytopathology* **60**:714–716.

Robertson, H. T. 1932. Maturation of foot and root tissue in wheat plants in relation to penetration by *Ophiobolus graminis* Sacc. *Sci. Agric.* **12**:575–592.

Rush, C. M., T. J. Gerik, and S. D. Lyda. 1984a. Factors affecting symptom appearance and development of Phymatotrichum root rot of cotton. *Phytopathology* **74**:1466–1469.

Rush, C. M., S. D. Lyda and T. J. Gerik. 1984b. The relationship between time of cortical senescence and symptom development of Phymatotrichum root rot of cotton. *Phytopathology* **74**:1464–1466.

Sequeira, L. 1981. Induction of host physiological responses to bacterial infection: a recognition phenomenon. Pages 143–153 in *Plant Disease Control*. R. C. Staples and G. H. Toenniessen, eds. Wiley, New York.

Sequeira, L. and T. L. Graham. 1977. Agglutination of avirulent strains of *Pseudomonas solanacearum* by potato lectin. *Physiol. Plant Pathol.* **11**:43–54.

Singleton, L. L. and O. Ziv. 1981. Effects of *Pythium arrhenomanes* infection and root-tip amputation on wheat seedling development. *Phytopathology* **71**:316–319.

Tomiyama, K. 1955. Studies on the snow blight disease of winter cereals. *Hokkaido Natl. Agric. Expt. Sta., Report No. 47*, Sapporo, Japan.

Vanterpool, T. C. 1940. Present knowledge of browning root rot of wheat with special reference to its control. *Sci. Agric.* **20**:735–749.

Vanterpool, T. C. and G. A. Ledingham. 1930. Studies on "browning" root rot of cereals. I. The association of *Lagena radicicola* N. Gen., N. sp. with root injury of wheat. *Can. J. Res.* **2**:171–194.

Wall, R. E. and C. G. Mortimore. 1965. The growth pattern of corn in relation to root and stalk rot. *Can. J. Bot.* **43**:1277–1283.

Waller, J. M. 1979. Observations on Pythium root rot of wheat and barley. *Plant. Path.* **28**:17–24.

Webb, R. W. 1927. Soil factors influencing the developmeent of the mosaic disease in winter wheat. *J. Agric. Res.* **35**:587–614.

Webb, R. W. 1928. Further studies on the soil relationships of the mosaic disease of winter wheat. *J. Agric. Res.* **36**:53–75.

Wilhelm, S. 1959. Parasitism and pathogenesis of root disease fungi. Pages 356–366 in *Plant Pathology, Problems and Progess, 1908–1958*. C. S. Holton, G. W. Fischer, R. W. Fulton, H. Hart, and S. A. E. McCallan, eds. Univ. of Wisc. Press, Madison.

Wyllie, T. D. and O. H. Calvert. 1969. Effect of flower removal and pod set on formation of sclerotia and infection of *Glycine max* by *Macrophomina phaseoli*. *Phytopathology* **59**:1243-1245.

Chapter 6

Fungistasis

CONCEPT OF SOILBORNE

What is a soilborne plant pathogen? Soilborne implies no more than the term seedborne. Spores of *Tilletia caries* are borne externally on the surface of the wheat kernel. Hyphae of *Ustilago nuda* are within the embryo in intimate association with host tissue. Both are seedborne. Wallace (1978) considered an organism to be soilborne if any part of the life cycle is subterranean. Soilborne pathogens range from those whose propagules contaminate the soil and function there in some way to those, such as ectoparasitic nematodes, that exist entirely within the soil. This is an inclusive concept of soilborne. Organisms whose existence in soil is casual are not stressed, and pathogens that do not exist within the soil itself, but spread from plant to plant by root grafting, are excluded.

Three Cereal Smuts

Mathieu Tillet (1755, translated in 1937) put the black dust (teliospores) of bunted wheat kernels (*Tilletia caries*) on winter wheat seed and planted it in southern France. On November 10, 1751 he scattered bunt dust over the surface of soil prepared for planting the next day. Tillet proved that *T. caries* could be both seed- and soilborne, even though he did not know the dust was alive.

Under natural conditions in most of the world *T. caries* is not soilborne. Tillet spread fresh spores preserved dry until just before planting in late autumn. Teliospores of this species lack endogenous dormancy and the dormancy induced by fungistasis necessary to make it soilborne in most climates. In most of the world wind-borne teliospores deposited on soil germinate before winter wheat is planted. Benedict Prevost (1807, translated in 1939) saw germinated spores in drops of water, in soil, and on the surface of wheat seedlings.

For many years common bunt was epidemic in Idaho, Washington, and Oregon, in spite of use of the most effective seed treatments available. Heald and George (1918) explained the failure of seed treatments successful in the rest of the world. Killing spores on the seed is not enough. The summers and early autumn are so dry the spores from spore showers during harvest survive ungerminated on the surface of summer-fallowed land. Harrowing and seeding mix the viable spores in the surface soils. Young seedlings emerge through the infested soil, and infection through the coleoptile occurs. It was not until fungicides that protect both seed and coleoptiles were available that common bunt was controlled in this region. Teliospores of *T. caries* do not require light for germination. They are adapted to germination within the soil at or near seed depth.

Tilletia controversa, the cause of dwarf bunt, is so closely related to *T. caries* that the two can hybridize (Silbernagel, 1964). The teliospores of the two species are difficult to distinguish, yet *T. controversa* is completely soilborne. Dusting seed with teliospores is not an effective means of inoculation. The teliospores of this species persist for several years in soil, particularly within unbroken bunt balls (Hoffmann, 1982). Spores germinate on the soil surface, not restrained by fungistasis, and germination is stimulated by low temperature and light. Germination is prolonged, extending over a period of months (Trione, 1982). Young tillers are infected near the soil line. Late and shallow seeding favors infection. Infection occurs mostly in January and February, favored by snow cover. Common bunt is a seedling-infecting smut; dwarf bunt is a shoot-infecting smut. Physiologic differences make the epidemiology of two morphologically similar fungi different. By practical concept, *T. caries* is soilborne only under special circumstances, and *T. controversa* is strictly soilborne.

Ustilago maydis is soilborne, but by a different mechanism than *T. controversa*. Spores of this species are widely disseminated in the Corn Belt by airborne spores from sori ruptured during mechanical harvest. The airborne spores contaminate the soil. They respond to fungistasis (Jackson, 1958), so they do not germinate on or in soil. Spores are deposited by wind in leaf whorls of young maize plants along with dust. They germinate in water within the leaf whorl, and meristematic tissues are infected. Though the spores germinate in water, nutrients that leak into the water stimulate germination. The soil infestation is so general and aerial spread is so effective in the Corn Belt that neither Christensen (1963) nor Dickson (1956) has recommended crop rotation in that region. *U. maydis* has one critical attribute of many soilborne plant pathogens, i.e., response to fungistasis.

CONCEPT OF FUNGISTASIS

In attempting to describe what constitutes being soilborne, the only special term used is the word *fungistasis*. Spores of *Tilletia caries* and *T. controversa* do not respond to fungistasis; those of *Ustilage maydis* do. Fungistasis refers to the property of natural soil that inhibits the germination of otherwise germinable propagules within or in contact with the soil. Response to fungistasis, or failure to germinate within or upon soil in the absence of sugars, amino acids, or other stimulants leaking from potential host organs (seeds, roots, etc.), is an essential attribute of propagules of many soilborne fungi. Fungi have devised ways to restrict germination of propagules under conditions unfavorable for sustained growth. Fungi are heterotrophs, dependent on nutrients produced by other organisms, and germination in the absence of potential food would lead to death.

The general inhibition of germination of fungal propagules by natural soil was called fungistasis by Dobbs and Hinson (1953). Lockwood (1964) stated that Simmonds, Sallans and Ledingham in 1950 recovered ungerminated spores of *Bipolaris sorokiniana* from soil that germinated readily on clean moist filter paper, and that Chinn in 1953 found that spores of several fungi did not germinate on glass slides buried in soil, but adding nutrients enabled them to germinate. Dobbs and Hinson realized that the inhibition of germination of many fungal propagules by natural soil is important, that it serves to restrict germination of propagules at an inappropriate time. Fungal propagules of many species, lacking endogenous dormancy, would germinate when temperature, moisture, and aeration are favorable without respect to potential food if it were not for fungistasis. Jackson (1957) found that root exudates in the rhizosphere overcame fungistasis. These observations established fungistasis as one of the major factors in survival of most soilborne fungi.

The term *mycostasis* has been used in place of fungistasis by Dobbs and others, but I agree with Watson and Ford (1972), that fungistasis is preferred. Stasis means to arrest development, to maintain inactivity, to make static, but not to kill (fungicidal).

Natural soil free of undecomposed organic debris contains limited nutrition for fungi. Propagules that germinate in the absence of adequate food eventually exhaust themselves, even with recycling, and die. Soilborne fungi are masters of the waiting game (Park, 1960). Most of them are immobile within soil, waiting for a rootlet or other structure to come within range. The exudate from the host structure overcomes fungistasis, the spore germinates, and the germ tube responds chemotropically by growing toward the potential source of food. The propagule does not waste itself by untimely germination.

A three-stage sequence—induction, maintenance, and release from fungistasis—was proposed by Watson and Ford (1972). Fungistasis occurs essentially on contact of the propagule with natural, microbiologically active soil. Watson and Ford visualized maintenance and release from fungistasis as being governed by the balance between germination inhibitors (fungistasis factors) and germination stimulators. The latter are generally sources of organic carbon or specific stimulators. As the concentration of inhibitors is increased, so must the concentration of stimulators required to release the spore from fungistasis. The soil contains varied inhibitors influenced by the soil, the season, and other factors, and propagules of different species could be affected differentially by these inhibitors. Inhibitors and stimulators are ephemeral, thereby not persisting in soil. Sugars and amino acids would not accumulate, and presumably inhibitors, whatever they are, would be metabolized by something or be dissipated in some way. The stimulators would establish concentration gradients. Watson and Ford urged that research be conducted on the levels of stimulants required to counteract fungistasis. The inhibitors, largely unknown, could not be directly measured, but based on the balance hypothesis, their concentration could be evaluated by the amount of stimulators required to overcome them. Inhibitors resulting from biologic activity would be produced in greatest amount in a soil in an active phase, but their degradation in such soils should also be at a higher level.

Dobbs and Bywater (1959) discovered a residual fungistasis not biological in origin and not destroyed by heat. This finding indicates that fungistasis can have two components, with the main one being of biological origin, the other being abiotic. Watson and Ford cautioned that spore germination, or rather failure to germinate, in distilled water has little biologic significance. Duggar in 1901 found that spores of some fungi did not germinate in distilled water but that they germinated when minerals were added to the water.

Griffin and Roth (1979), in a later review of fungistasis, discussed carbon- or energy-deficient spores in contrast to carbon-independent spores. Energy-independent propagules germinate in mineral salt solutions or in distilled water, whereas few energy-dependent spores do so. Results of germination trials can differ greatly if high or low spore concentrations are tested. Such findings probably reflect the presence of germination inhibitors. Fungistasis and its causes have been reviewed (Lockwood, 1964, 1977, 1984; Watson and Ford, 1972; Griffin and Roth, 1979), and thorough treatment of this concept will not be attempted by me.

Microbiostasis

Ho and Ko (1982) believe that the term fungistasis is too narrow, that the phenomenon of nongermination or cell division of germinable propagules on natural, unamended soil applies to bacteria and actinomycetes as well. Russell and Hutchinson found in 1909 that bacteria failed to multiply in natural soil, and Katznelson in 1940 observed this for actinomycetes and fungi. Ho and KO added *Agrobacterium radiobacter, Streptomyces scabies,* and *Penicillium funiculosum* to natural soil and no multiplication occurred. They used the words bacteriostasis, actinostasis, and fungistasis and proposed the broader term, *microbiostasis,* to apply to a general phenomenon by which these microorganisms are rendered static in soil.

In an excellent review of soil microbiostasis, Ko and Ho (1984) found no evidence that antibiotic-producing organisms are more effective in establishing microbiostasis than nonantibiotic producers, that germination inhibitors of microbial origin are not important, and that absence of sufficient food is the real cause. They found that fungistasis is stronger than bacteriostasis or actinostasis.

FOOD AS A TIMING FACTOR

This chapter marks a turning point in the book. The first chapters (soil, water, temperature, aeration) dealt primarily with physical factors of the environment. The chapter on host physiology dealt with stresses on the host, whether imposed by physical factors or by internal stresses within the plant. Fungistasis leads into another general subject, the obtaining and conserving of food by many soilborne plant pathogens.

Food is a driving force in selection. *Claviceps purpurea*, an ergot fungus, produces sclerotia in the inflorescences of its hosts. The sclerotia fall to the ground, over winter, germinate to form ascocarps, and produce airborne ascospores at the time grass flowers so that the cycle can repeat. The sclerotia respond to the environment (temperature, moisture, and time) in such a way as to eject ascospores (once a year) at the right time to ensure the presence of food (young ovaries to be infected). Examination of most life cycles reveals that timing is mostly oriented toward obtaining food.

Oospores of *Pythium debaryanum* remain dormant because of fungistasis within soil until food (exudates from germinating seeds) initiates germination. Food stimulates germination. Germination of sclerotia of *C. purpurea* is not timed directly by food but by cold, heat, light and time. The airborne propagules of *C. purpurea* "had to seek food" by dispersal. The oospore of *P. debaryanum* remains fixed within the soil, and it germinates when food comes to it. In the aerial-disseminated examples, physical factors determine timing. In the soil-trapped propagule, food times germination. The ultimate goal of both fungi is food; only the timing mechanisms differ. Release from fungistasis is a major timing mechanism.

INDUCTION OF FUNGISTASIS

Bacteria

Epstein and Lockwood (1984) wetted air-dry sandy clay loam, incubated it moist for 16 to 24 hr, then flooded it and shook the mixture for 1 hr. The suspension was allowed to settle, centrifuged, and finally, filtered. The filtrate, containing 0.3% soil, was mixed into sand to produce a dilution of 0.3% soil in sand. Other treatments, such as finer filtration, more severe centrifugation, or dilution, were used to reduce bacterial numbers within the soil suspensions. Using conidia of *Bipolaris victoriae*, fungistasis declined as the numbers of bacteria in the preparations declined. Germination was 0% on raw soil, 98% on sterile soil, 82% on sterile sand, and 9% on the sand + 0.3% soil. Fungistasis was lost when bacteria were removed from the soil suspension. Chemicals added to soil further implicated bacteria as the cause of fungistasis. Antifungal substances, (PCNB, carboxin, or dicloran) did not reduce fungistasis. Antibacterial substances (penicillin G, methicillin, vancomycin, chloramphenicol and rifampin) annulled fungistasis. Tetracycline was ineffective, but most anti-bacterial agents not toxic to conidia of *H. victoriae* suppressed fungistasis. Most soil bacteria and actinomycetes, when tested individually, induced fungistasis. Epstein and Lockwood concluded that soil fungistasis is of microbial origin and that the fungistatic action is reduced by reduction in numbers of bacteria. They doubt that accumulation of microbial wastes (staling products), as postulated by Robinson and Park (1966), is important. Epstein and Lockwood found no evidence of germination inhibitors, either in the soil solution or soil atmosphere.

Fradkin and Patrick (1985) enclosed conidia of *Bipolaris sorokiniana* within inert membranes with openings 0.2 μm in diameter and incubated the conidia in soil. Even though bacteria were excluded by the membranes, the conidia did not germinate in soil. They isolated bacteria that colonized the surface of the conidia in other treatments, grew them in liquid culture, and the bacteria-free filtrates of several of them induced fungistasis, providing evidence for the presence of fungistatic factors of bacterial origin within the soil solution.

Iron Deficiency

Water-soluble fluorescent pigments produced by *Pseudomonas fluorescens* are fungistatic (Misaghi et al., 1982). The fungistasis is overcome by adding soluble iron in excess of the chelating power of the pigments. This fungistasis is attributed to iron-deficiency caused by the soluble pigment, similar to the action of siderophores. This phenomenon should not be confused with general soil fungistasis that is counteracted by sugars and amino acids. Iron, however, may play a major role in the distribution of several fungi, and this may establish the significance of bacterial siderophores in the rhizosphere.

Elad and Baker (personal communication, R. R. Baker, Colorado State University) found that chlamydospores of *Fusarium oxysporum* are subject to iron-deficiency-induced fungistasis and that chlamydospores of *F. solani* are not. Their studies included *F. oxysporum* f. sp. *vasinfectum* race 1, which is most damaging in acid soils of the Southeast, where iron should be relatively abundant. It would be of great value if chlamydospores of race 4, from India, would also be evaluated for iron dependency. This race thrives in alkaline soils in which iron should be relatively unavailable. Elad and Baker believe the chlamydospores of *F. solani*, being larger and containing more iron, are not affected by bacterial siderophores that compete for iron. *F. solani* is adapted to a wide range of soils, including those of neutral or alkaline pH.

Volatiles

Soluble fungistatic substances diffuse in soil moisture so that fungistasis can be detected by placing propagules directly on the surface of moist soil, cellophane membranes, filter paper, or on agar disks on soil (allow a period for diffusion before placing the spores on the agar disks). Hora and Baker (1974) studied volatile fungistatic substances. Disks of purified water agar were suspended above moistened soils to absorb any volatile substances emitted from the moist soils. Conidia were then added to the disks. Little inhibition resulted from unlimed soil, but liming increased inhibition of conidium germination. In general, the more alkaline the soil, the greater the inhibition. These volatile inhibitors were detected in sterile soil as well as in natural soil. They concluded that elevated pH released volatile inhibitors from soil and soil solutions. Soil treated with hydrogen peroxide or with 400°C for 12 hr, to reduce or destroy organic matter, still retained a trace of germination inhibitors in aqueous extracts, leading Ko and Hora (1974) to conclude that inorganic substances are involved. Fungistatic volatiles produced by actinomycetes are treated in Chapter 9.

Energy Dependence

Lockwood (1984) supported the belief that spores of many species are born energy-independent but that, on contact with natural soil, nutrients are at first rapidly lost from the spores, rendering them energy-dependent, a form of induced fungistasis. Many spores originally will germinate in balanced salt solutions containing no organic compounds, but after exposure to the soil they quickly become energy-dependent, remaining dormant until exudates in adequate quantity counteract the induced dormancy. Many lines of evidence support this hypothesis. Heavy soils enforce fungistasis more strongly than light soils, and clay adsorbs more nutrients than course-textured soil particles (Filonow and Lockwood, 1983a, b). They found no substances in soil extracts that inhibited germination. The rate of loss of nutrients in soil by conidia of *Bipolaris victoriae* and by sclerotia of *Macrophomina phaseoli* is correlated with the degree of fungistasis (Arora et al., 1983). Leaching nutrients from spores originally energy-independent renders them energy-dependent. Ko and Lockwood (1967) stated, "There seems to be no necessity for postulating the existence of fungistatic substances to account for the failure of fungal spores to germinate in natural soil."

Fradkin and Patrick (1985) found that conidia of *B. sorokiniana* buried within membranes with holes 10 μm, large enough for ready bacterial penetration, develop energy-dependence much more rapidly than when buried in membranes that excluded bacteria. The presence of bacteria on the spores leads to rapid development of a requirement for an exogenous energy source to germinate. In nature bacteria would have direct access to spores within soil, so that the requirement for exogenous food for germination would develop rapidly.

STRENGTH OF FUNGISTASIS

Soils Differ

Fungistasis is at its highest level in soils with high microbial activity. It is stronger in topsoil than in subsoil. Chinn (1967) studied 11 soils within a 140-mile radius of Saskatoon, Canada. The soil was partially dried, then used undiluted or diluted with varying amounts of silica sand by weight. Silica sand at 8% water and undiluted soil at optimum water content were controls. Spores of *Bipolaris sorokiniana* were spread on slides and buried in the soils at 24°C. Chinn placed the soils in four groups: (1) sandy, 2.3% organic matter, 19,400 microbes/g; (2) clay, 4.8% organic matter, 20,800 microbes/g; (3) loam, 8.4% organic matter, 25,000 microbes/g; and (4) peat, 40.3% organic matter, 50,800 microbes/g. Nine isolates of *B. sorokiniana* were tested. Germination of conidia on moist blotting paper and on silica sand was 86 and 84%, respectively. In general, fungistasis was strongest in peat soil, intermediate in the clay and loam, and weakest in the sandy soil. "Weak" does not really mean weak in that when each soil was undiluted, most conidia of all isolates did not germinate.

Propagules Differ

Griffin and Roth (1979) reviewed studies with conidia of *Verticillium dahliae* and *V. albo-atrum*. These fungi produce very small spores and, according to Steiner and Lockwood (1969), the smaller the spore the greater the need for exogenous food to germinate in natural soil; i.e., smaller spores should have the strongest responses to fungistasis. Griffin and Roth presented conflicting data and interpretations that I believe require further evaluation in terms of the question, "What is the function of the propagule?" Conidia of *Verticillium* and *Cephalosporium* spp. and microconidia of *Fusarium oxysporum* have special roles. These small conidia are disseminated in the xylem stream to establish systemic infections. How important are they as survival propagules within soil? In *Cephalosporium gramineum* they are the only known spore stage in the soil, along with the even smaller spores produced by its *Hymenula cerealis* (sporodochial) stage. These conidia are responsible for infection, but they do not live long in soil. They are not specialized for survival like chlamydospores of *F. oxysporum*. Theoretically, microconidia of *F. oxysporum* should be least sensitive to fungistasis, macroconidia intermediate in sensitivity, and chlamydospores should have the strongest response to fungistasis to fulfill their role as survival spores. In soil, macroconidia can transform into chlamydospores or germinate and then produce chlamydospores, transformations that could not occur during dormancy.

Isolates of *B. sorokiniana* differ in germinability in natural soil (Chinn and Tinline 1964). Those that germinate readily or are insensitive to fungistasis die quickly in natural soil. Chinn (1967) tested *Bipolaris sorokiniana* spores on undiluted soil; germination of a single isolate was 35% on sand, 17% on clay, 19% on loam, and 12% on peat. This isolate was atypical, because three other isolates did not germinate when soil exceeded 2% of the sand-soil mix. Not only did soils differ in fungistatic power, but isolates of the same fungal species differed in response to the fungistatic factors.

RELEASE FROM FUNGISTASIS

Fungistasis in Action

Agricultural soils contain sufficient germinable propagules of pathogenic *Pythium* spp. to cause seed rot or seedling blight under proper conditions, but they remain static in fallow soil (Fig. 6.1). On introduction of seeds into moist soil, nutrients escape from the seeds during imbibition. Fungistasis is overcome, and propagules exposed to sufficient exudate germinate. Only those propagules close to the seed respond. Response to fungistasis enables propagules near the seed to germinate and conserves propagules at a distance too great to establish a parasitic relationship with the host.

Wilt-inducing fusaria of flax and cabbage are capable of invading young root tips without wounds. Griffin (1969) studied the influence of root-tip exudates of peanut on germination of chlamydospores of *F. oxysporum* in soil. Germinated spores were first detected at the root surface 7 mm behind the advancing root tip. At 13 mm behind the root tip germination was 90% at the root surface and 27% at 0.9 mm to the side of the root surface (Fig. 6.2) Exudates from the advancing root tip initiated germination. The time sequence of this response is less rapid than for *Pythium* spp., but the slower response of *F. oxysporum* is appropriate for its type of attack (Fig. 6.3). Both examples illustrate conservation of propagules. Those too distant from the target do not germinate, and the timing of germination is "calibrated" to efficiently establish the parasitic relationship.

Death without Fungistasis

The sugar pine is a valuable timber tree in California, and reforestation was pursued vigorously in parts of the Sierra Nevada mountain range during the 1950s and 1960s. Sugar pine was grown in nurseries and transplanted to the

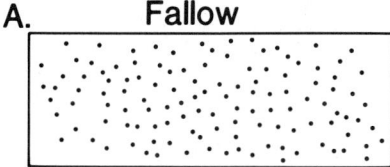

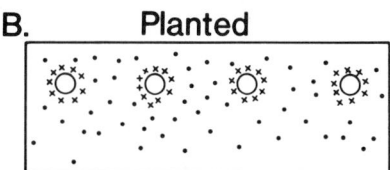

·· Dormant oospore
×× Germinated oospore
○ Pea seed

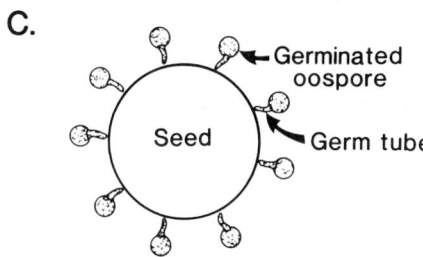

Figure 6.1. In fallow soil (A) the 100 or so oospores of *Pythium ultimum* per gram of soil are dormant. After seeds of peas are seeded in cool, moist soil (B), oospores within a millimeter of the seed germinate. Those distant from the seed, the majority of them, remain dormant, held so by fungistasis. The oospores germinated by means of germ tubes that grew toward the source of exudate; they responded by positive chemotropism (C).

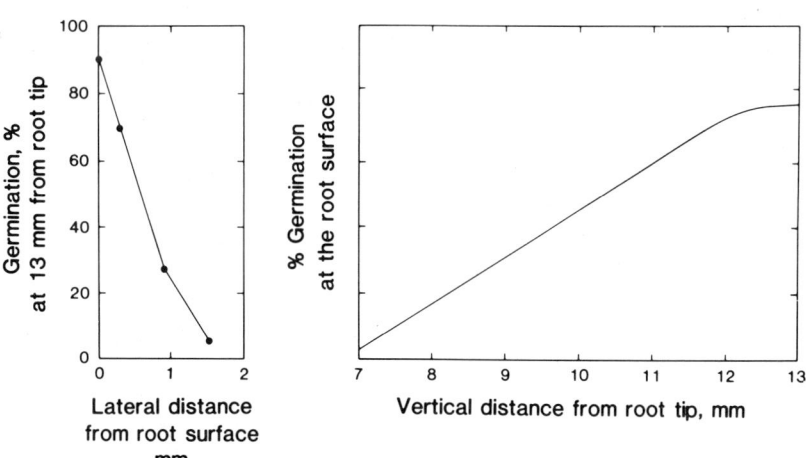

Figure 6.2. The percentage of germination of chlamydospores of *Fusarium oxysporum* is high near the root surface at a distance of 13 mm from the root tip (left). Germination had begun at the root surface by the time the peanut root tip had grown 7 mm beyond the chlamydospore (right). By the time the root tip had grown 12 to 13 mm past the spores, germination had reached maximum (after Griffin, 1969).

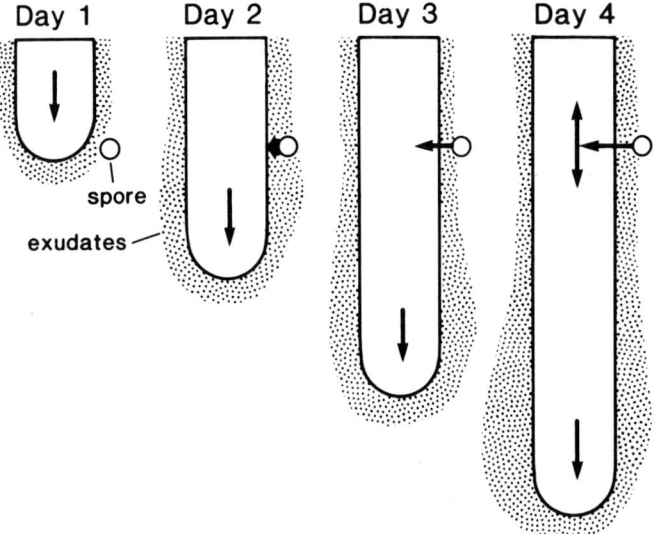

Figure 6.3. Exudate from a peanut root tip stimulates germination of a chlamydospore of *Fusarium oxysporum*. The germ tube grew toward the source of the exudate and penetration occurred behind the root tip. After Griffin, 1969.

forest. *Fusarium oxysporum* was a serious pathogen in the nursery, and Smith (1967) was concerned that infected transplants could lead to problems in the forest. He found that transplanted seedlings ranged in infection from 0 to 90% but that seedlings of the same species produced in the forest were not infected by *F. oxysporum*. *F. oxysporum* was present in the transplanted seedlings in the autumn after transplanting, but 4 years later *F. oxysporum* was no longer present. *F. oxysporum* is practically ubiquitous in cultivated soil, but it disappears in pine-covered sites in the Sierra Nevada. In contrast, *F. oxysporum* and *F. roseum* are present in forest areas where brush and grass grew (Toussoun et al., 1969). The soil and climate were the same as for the pine-covered sites, except for the absence of the pine needle litter.

Extracts of pine needles (litter), partially decomposed needles (duff) and decomposed needles (humus), stimulate germination of *F. oxysporum* and *F. solani* macroconidia and chlamydospores (Toussoun et al., 1969). When pine litter extract was added to a bean field soil containing chlamydospores of *F. oxysporum*, 85% of the chlamydospores germinated. A 1% solution of glucose germinated 10% of the chlamydospores, a 2.5% solution of asparagine germinated 7%, and 3% germinated in distilled water. Over 93% of the chlamydospores of *F. solani* f. sp. *phaseoli* germinated in a pine soil, and none germinated in a bean field soil. In forest soil germination occurred and the germ tubes lysed without forming replacement chlamydospores. Pine litter extract was so stimulatory that the *Fusarium* propagules germinated in the absence of suitable substrate and died.

COMMENTS

The existence and importance of fungistasis is unquestioned; only the mechanisms present problems. It is hard for me to accept the concept that many spores subject to fungistasis are born or rendered nutrient deficient. If nutrient responses are paramount, why are pine needle extracts so powerful in overcoming fungistasis of chlamydospores of *Fusarium* spp.?

Cochrane et al. (1963) produced macroconidia of *Fusarium solani* f. sp. *phaseoli* in liquid shake cultures, then removed and washed the macroconidia. They added 10 mg of fresh, fully hydrated but washed conidia to 5 ml of liquid to study germination. The amount of carbon required for complete germination, supplied either as glucose or ethanol, exceeded the dry weight of the conidia. Reducing the spore concentration by 25% reduced the carbon requirement to 20% of the above. They concluded that the endogenous reserves of the conidia cannot support germination. In nature, macroconidia of *F. solani* and *F. oxysporum* are washed by rain into the soil where, even in the absence of exogenous food, they convert into chlamydospores. This conversion probably requires as much energy as germination.

Cochrane et al. (1963) reported the spore concentration as 10 mg of spores in 5 ml of liquid, not by number of spores per ml. If we assume that a spore is about as dense as water and has a volume of about 2×10^3 μm^3, a cubic centimeter of spores would have about 5×10^8 spores. Cochrane's spore suspension would therefore approach 10^6 spores per ml. Chlamydospore populations in natural soil seldom exceed $5 + 10^3$ /g of air dry soil.

To continue this reasoning, assume that the 1 million spores, with 100% conversion to chlamydospores, would establish a high concentration (5×10^3 spores/g) of chlamydospores in 200 g of dry soil. If the soil contained 35% water (very wet for most soils), 70 g of water would be added to the soil, or the spore suspension would be diluted 70 times. In other words, even allowing for gross errors in the above approximations, the original spore suspension was unrealistically dense, and if germination inhibitors were present, the high requirement for exogenous carbon may have been necessary to counteract germination inhibitors. If we are to have real knowledge, we will have to keep our experiments as close to nature as possible.

In the original article by Ko and Lockwood (1967), conidia of *Bipolaris sorokiniana*, macroconidia of *Fusarium solani* f. sp. *phaseoli* and f. sp. *pisi*, and teliospores of *Ustilago maydis* did not germinate in water and would be considered energy-dependent. These spores surely contain enough food to support germination and germ-tube growth. Conidia of some powdery mildews, which contain as much as 70% water (Yarwood, 1950), can germinate and infect on a dry leaf from which little if any exogenous food could be available. Single zoospores of *Aphanomyces euteiches* (Cunningham and Hagedorn, 1962) and *Pythium* spp. (Spencer and Cooper, 1967) have sufficient energy to swim, encyst, and infect. These zoospores are attracted by exudates and gain energy from them, but they are small propagules high in water content that must, even with food gained from exudates, have much less total energy than a conidium of *B. sorokiniana*.

Oospores of *Pythium debaryanum* respond to food in the soil solution. Surely they are not energy-dependent because when no or very little food is present, they can germinate indirectly, producing several zoospores, each capable of infecting.

"Energy-dependent" propagules may have a sophisticated mechanism preventing germination in the absence of exogenous foods. Insufficient endogenous energy for germination seems unlikely. Energy- or exudate-sensitive may be a better concept than energy-dependent. The loss of nutrients from originally energy-independent propagules, as discussed under induction of fungistasis, may trigger an induced dormancy overcome by exogenous nutrients.

It is logical to assume that many substances, both volatile and nonvolatile, contribute to fungistasis and that microorganisms, particularly bacteria, induce fungistasis. Furthermore, nonspecific sugars and amino acids that exude from fresh organic matter overcome most fungistasis and signal the presence of potential substrate for further development. There may be little point in trying to define the precise "cause" of fungistasis. There are probably many causes, some more important to one organism, in one soil, climate, or season, than to another organism.

It sometimes seems to me that the more is written about a subject, the less we really know.

KEY REFERENCES

Dobbs, C. G. and W. H. Hinson. 1953. A widespread fungistasis in soils. *Nature* **172**:197–199.

Elad, Y. and R. Baker. 1985. The role of competition for iron and carbon in suppression of chlamydospore germination by *Pseudomonas* spp. *Phytopathology* **75**:1053–1059.

Epstein, L. and J. L. Lockwood. 1984. Effect of soil microbiota on germination of *Bipolaris victoriae* conidia. *Trans. Brit. Mycol. Soc.* **82**:63–69.

Fradkin, A. and Z. A. Patrick. 1985. Interactions between conidia of *Cochliobolus sativus* and soil bacteria as affected by physical contact and exogenous nutrients. *Can. J. Plant Pathol.* **7**:7–18.

Griffin, G. J. 1969. *Fusarium oxysporum* and *Aspergillus flavus* spore germination in the rhizosphere of peanut. *Phytopathology* **59**:1214–1218.

Lockwood, J. L. 1977. Fungistasis in soils. *Biol. Rev.* **52**:1–43.

Watson, A. G. and E. J. Ford. 1972. Soil fungistasis — a reappraisal. *Ann. Rev. Phytopathol.* **10**:327–348.

REFERENCES

Arora, D. K., A. B. Filonow, and J. L. Lockwood. 1983. Exudation from ^{14}C-labeled fungal propagules in the presence of specific microorganisms. *Can. J. Microbiol.* **29**:1487–1492.

Chinn, S. H. F. 1967. Differences in fungistasis in some Saskatchewan soils with special reference to *Cochliobolus sativus*. *Phytopathology* **57**:224–226.

Chinn, S. H. F. and R. D. Tinline. 1964. Inherent germinability and survival of spores of *Cochliobolus sativus*. *Phytopathology* **54**:349–352.

Christensen, J. J. 1963. Corn smut caused by *Ustilago maydis*. Monograph No. 2. Am. Phytopath. Soc., St. Paul, Minn.

Cochrane, J. C., V. W. Cochrane, F. G. Simon, and J. Spaeth. 1963. Spore germination and carbon metabolism in *Fusarium solani*. I. Requirements for spore germination. *Phytopathology* **53**:1155–1160.

Cunningham, J. L. and D. J. Hagedorn. 1962. Penetration and infection of pea roots by zoospores of *Aphanomyces euteiches*. *Phytopathology* **52**:827–834.

Dickson, J. G. 1956. Diseases of Field Crops. McGraw-Hill, New York.

Dobbs, C. G. and J. Bywater. 1959. Studies in soil mycology. II. *Brit. Forest. Comm., Rept. Forest Res.* **1958**:98–104

Filonow, A. B. and J. L. Lockwood. 1983a. Mycostasis in relation to the microbial nutrient sinks of five soils. *Soil Biol. Biochem.* **15**:557–565.

Filonow, A. B. and J. L. Lockwood. 1983b. Loss of nutrient-independence for germination by propagules incubated on soils or on a model system imposing diffusive stress. *Soil Biol. Biochem.* **15**:567–573.

Griffin, G. J. and D. A. Roth. 1979. Nutritional aspects of soil mycostasis. Pages 79–96 in *Soilborne Plant Pathogens*. B. Schippers and W. Gams, eds. Academic Press, New York.

Heald, F. D. and D. C. George. 1918. The wind dissemination of the spores of wheat. *Wash. Agric. Expt. Sta. Bull.* 151.

Ho, W. C. and W. W. Ko. 1982. Characteristics of soil microbiostasis. *Soil Biol. Biochem.* **14**:589–593.

Hoffmann, J. A. 1982. Bunt of wheat. *Plant Dis.* **66**:979–986.

Hora, T. S. and R. Baker. 1974. Abiotic generation of a volatile fungistatic factor in soil by liming. *Phytopathology* **64**:624–629.

Jackson, R. M. 1957. Fungistasis as a factor in the rhizosphere phenomenon. *Nature* **180**:96–97.

Jackson, R. M. 1958. An investigation of fungistasis in Nigerian soils. *J. Gen. Microbiol.* **18**:248–258.

Ko, W. H. and W. C. Ho. 1984. Soil Microbiostasis. Pages 175–184 in *Soilborne Crop Diseases in Asia*. Food and Fertilizer Technology Center for the Asian and Pacific Region, Taipei, Taiwan.

Ko, W. H. and F. K. Hora. 1974. Factors affecting the activity of a volatile fungistatic substance in certain alkaline soils. *Phytopathology* **64**:1042–1043.

Ko, W. H. and J. L. Lockwood. 1967. Soil fungistasis: Relation to fungal spore nutrition. *Phytopathology* **57**:894–901.

Lockwood, J. L. 1964. Soil fungistasis. *Ann. Rev. Phytopathol.* **2**:341–362.

Lockwood, J. L. 1984. Soil fungistasis:Mechanisms and relation to biological control of soil-borne plant pathogens. Pages 159–174 in *Soilborne Crop Diseases in Asia*. Food and Fertilizer Technology Center for the Asian and Pacific Region, Taipei, Taiwan.

Misaghi, I. J., L. J. Stowell, R. C. Grogan, and L. C. Spearman. 1982. Fungistatic activity of water-soluble fluorescent pigments by fluorescent pseudomonads. *Phytopathology* **72**:33–36.

Park, D. 1960. Antagonism—the background of soil fungi. Pages 148–159 in *The*

Ecology of Soil Fungi. D. Parkinson and J. S. Waid, eds. Liverpool Univ. Press, Liverpool, UK.

Prevost, I. B. 1939. Memoir on the immediate cause of bunt or smut of wheat, and of several other diseases of plants, and on preventives of bunt. 80 pp. Paris. 1807. Translated from the French by G. W. Keitt. *Phytopath. Classics* **6**:1-95. Am. Phytopath. Soc., St. Paul, Minn.

Robinson, P. M. and D. Park. 1966. Volatile inhibitors of spore germination produced by fungi. *Trans. Brit. Mycol. Soc.* **49**:639-649.

Silbernagel, M. J. 1964. Compatibility between *Tilletia caries* and *T. controversa*. *Phytopathology* **54**:1117-1120.

Smith, R. S., Jr. 1967. Decline of *Fusarium oxysporum* in the roots of *Pinus lambertiana* seedlings transplanted into forest soils. *Phytopathology* **57**:1265.

Spencer, J. A. and W. E. Cooper. 1967. Pathogenesis of cotton (*Gossypium hirsutum*) by *Pythium* species: zoospore and mycelium attraction and infectivity. *Phytopathology* **57**:1332-1338.

Steiner, G. W. and J. L. Lockwood. 1969. Soil fungistasis: Sensitivity of spores in relation to germination time and size. *Phytopathology* **59**:1084-1092.

Tillet, M. 1937. Dissertation on the cause of the corruption and smutting of the kernels of wheat in the head, and on the means of preventing these untoward circumstances. 150 pp. Bordeaux. 1755. Translated from the French by H. B. Humphrey. *Phytopath. Classics* **5**:1-191. Am. Phytopath. Soc., St. Paul, Minn.

Toussoun, T. A., W. Menzinger, and R. S. Smith, Jr. 1969. Role of conifer litter in ecology of *Fusarium:* stimulation of germination in soil. *Phytopathology* **59**:1396-1399.

Trione, E. J. 1982. Dwarf bunt of wheat and its importance in international wheat trade. *Plant Dis.* **66**:1083-1088.

Wallace, H. R. 1978. Dispersal in time and space. Pages 181-202 in *Plant Disease, an Advanced Treatise,* Vol. II. J. G. Horsfall and E. B. Cowling, eds. Academic Press, New York.

Yarwood, C. E. 1950. Water content of fungus spores. *Am. J. Bot.* **37**:636-639.

Chapter 7

The Rhizosphere

Many organic compounds escape or are excreted from roots (into the rhizosphere), from stems in contact with soil (laimosphere), seeds (spermosphere), peanut fruits (geocarposphere), tubers, etc. (Curl, 1982). Most exudate contains sugars and amino acids, and these substances, when present in sufficient quantity, overcome fungistasis of propagules not requiring specific stimulants for germination or hatching. Some plant pathogens, such as eggs of some cyst-nematodes and sclerotia of *Sclerotium cepivorum*, are stimulated by chemicals from particular plants, usually hosts. These special hatching or germination factors are not amino acids and sugars common to all plants. Diffusion gradients guide propagules to infection courts. Germ tubes grow toward and motile propagules move toward the source of attractive substances, increasing the efficiency of the inoculum. The concentration of exudates is low, but their use by organisms in the rhizosphere accelerates the release of exudates by increasing the diffusion gradient (Burstrom, 1965).

CONCEPT: RHIZOSPHERE

Hiltner in 1904 viewed the rhizosphere as that portion of the soil directly influenced by substances issuing from roots into the soil solution, favoring certain bacteria. He envisioned harmful organisms around roots of unthrifty plants and beneficial organisms around roots of healthy plants (Curl, 1982). "The primary biological fact of the rhizosphere or zone of root influence is the greater number and activity of soil microorganisms in this region than in root-free soil. Between these two zones is an area of transition in which the root influence diminishes with distance. It is generally accepted that the term rhizosphere soil refers to the thin layer adhering to a root after the loose soil and clumps have been removed by shaking," (Katznelson, 1965).

The term *rhizoplane* was used by Clark (1942) to denote the root surface itself as a special habitat or site of microbial activity. The rhizoplane, or root surface, supports relatively high biologic activity and reflects more sensitively than the rhizosphere the effect of the root on soil microflora and microfauna. Katznelson (1965) stated that use of rhizosphere soil enables quantitative comparison (on an oven-dry basis) with bulk soil that is not possible with rhizoplane counts.

Rhizosphere bacterial counts are higher than bulk soil counts. Ratios 10:1 or 20:1, rhizosphere:bulk soil, are not uncommon. Rhizosphere soil, in comparison with bulk soil, also has higher numbers of nematodes, actinomycetes, and fungi and slightly higher populations of protozoa and algae. The rhizosphere of legume roots is usually richer in nutrition than that of nonlegumes. Although increases in numbers of bacteria are greater than for other groups of microbes in rhizospheres, Katznelson (1965) cautioned thusly: What is the biologic equivalency of nematodes versus bacteria? How many active bacteria equal one active nematode?

In general, the rhizospheres of most plants favor gram-negative, nonsporulating, rod-shaped bacteria, with the selective action greatest at the rhizoplane. *Pseudomonas* species usually predominate on wheat, with *Arthrobacter* second. *Bacillus* species are not normally abundant on roots, but occasionally *B. polymyxa* may be abundant on wheat roots and *B. megaterium* on cotton roots. Bacteria that require amino acids are more abundant on the rhizoplane and in the rhizosphere than in the bulk soil. Antibiotic-producing antinomycetes are more abundant in rhizosphere than in root-free soil.

Garrett (in Katznelson, 1965) commented that fresh organic residues are colonized rapidly by fungi that neither produce nor tolerate antibiotics. They exploit the ephemeral residue rapidly. In contrast, the rhizosphere is relatively stable environment with a limited, continuous source of food. This condition apparently favors antibiotic producers and organisms tolerant to antibiotics. Katznelson lists rapid growth rate and simple nutritional requirements in addition to production of or tolerance to (or both) antibiotics as important attributes among the most successful rhizosphere organisms.

The rhizosphere effect strengthens in proportion to the plant's vegetative activity, then declines with senescence and maturity. The death or decline of a root leads to the changing of its associated microflora. As active and declining roots intermingle, a complex rhizosphere arises. Riviere

(1959) believed that in wheat the most intense rhizosphere activity was during tillering. Other workers placed the peak at flowering (Rovira, 1965).

Another factor of importance in comparing the relationships among rhizospheres (R) and bulk soils (S) is the background population in nonrhizosphere (bulk) soils. In soils high in organic matter, the bulk soil is rich in microbial numbers. In sands, microbial populations are generally low. If the rhizosphere is equally rich in a loam and sand, the R:S ratio will be greater in the sandy soil, indicating a stronger rhizosphere effect. Katznelson cited an example of rhizosphere counts around roots of yellow birch seedlings in which the R:S ratio was 15:1 in the humus layer, and 60:1 in the underlying sand layer. The actual counts in the rhizospheres were the same; the ratios reflected differing background counts.

Clark (1940) and Timonin (1940) both observed increased rhizosphere counts in drier soil (Clark for wheat and soybeans; Timonin for flax) than in wetter soil. Temperature and moisture have direct effects (on the microorganisms) and indirect effects (on the plant), and indirect effects appeared to be more important (Katznelson, 1965). Bacterial counts around wheat roots are highest in cool soil, around soybean roots they are higher in warm soil. Chemicals applied to leaves often increase or decrease the activities of portions of the microflora in rhizospheres.

Few organisms are truly adapted to the rhizosphere. Rovira (1963) inoculated maize seeds individually with *Azotobacter chroococcum, Bacillus polymyxa, Clostridium pasteurianum,* and *Pseudomonas fluorescens. Azotobacter chroococcum* failed to colonize maize roots, even with no competition. *Bacillus polymyxa* and *Clostridium* reached modest populations. Only *Pseudomonas fluorescens* reached substantial numbers.

EXUDATES

Quantity

Using radioactive $^{14}CO_2$, Martin (1977) determined that 17.3% of total carbon assimilates from wheat roots is lost to the soil. Sugars accounted for less of the carbon than amino acids. Martin believed much of the carbon loss results from autolysis of cortical cells rather than from direct exudation. Cortical cells die in the absence of microorganisms (Henry and Deacon, 1981), so this degeneration is a natural process. Exudation in general is greater in natural than in sterile conditions (Barber and Martin, 1976), amounting to 5 to 10% of photosynthate in sterile conditions and 12 to 18% under nonsterile conditions. Total carbon includes exudates, mucigel, cell wall residues, and sloughed plant cells.

The quantity of assimilate lost by roots leads to the question, Why? Has evolution been unsuccessful in curtailing loss, or is some of this loss not a loss but a function of normal, efficient roots? Olsen et al. (1981) emphasized that quantitative relationships between most higher plants and minor elements are narrow. Boron is required at about 0.04 ppm in the soil solution and at 1 ppm is injurious to many plants. Heavy metals, harmless at low concentration, can be toxic in excess. Plants are exposed to soils that vary in soil chemistry. Plants apparently influence the solubility of certain minor elements by active processes involving exudates and expenditure of energy.

Ferric iron (Fe^{3+}) is too insoluble at pH 7 to meet the needs of higher plants (by about 10^9 times). A plant undergoing iron deficiency stress acidifies the soil solution at the rhizoplane from about pH 7 to about 3.7 within hours. This pH change increases the solubility of iron by about 10^9 times. Root exudates in this case contain reductants that convert Fe^{3+} to Fe^{2+}, which is more soluble and is taken up by the plant (Olsen et al., 1981).

Iron chelated by microbial siderophores exists in significant quantities in the soil solution. Plants compete with microorganisms for this chelated Fe^{3+} by changing the rhizoplane pH and absorbing Fe^{2+} (Olsen et al., 1981). They do not say so, but rhizoplane and rhizosphere microflora often include significant numbers of fluorescent pseudomonads that produce siderophores, and root exudates could have indirect value to the plant by supporting these bacteria.

Root exudates, at least sometimes, represent an adaptation to some soil problems. Iron deficiency, associated with aerobic conditions (Fe^{3+}) in most soils, may change to iron excess under water-logging or reducing conditions (Fe^{2+}). This conversion may produce toxic results. Mn^{2+} is affected similarly. Rice protects itself from iron toxicity (excess soluble Fe^{2+}) by oxidizing the iron to Fe^{3+} at the root surface and rendering it essentially insoluble. The quantity of exudate reported may be misleading, because exudate is not liberated uniformly over the root surface, and higher local concentrations must have occurred.

Severe freezing greatly increased exudation from young wheat roots, and Bailey et al. (1982) believed this increased exudation favors infection of roots by *Cephalosporium gramineum*.

Gradients

Papavizas and Davey (1961) studied the gradient of total bacteria, streptomycetes, and fungi from the rhizoplane to 80 mm laterally from the tap root of lupin plants. Bacteria and streptomycetes were most abundant at the rhizoplane, and they declined gradually to 80 mm from the tap root. In contrast, total *Penicillium* spp. and *P. piscarium* declined abruptly from the rhizoplane outward, and *P. lilacinum* showed little response to either the rhizoplane or the rhizosphere.

Most growth of microorganisms on the root surface (rhizoplane) occurs in the grooves between epidermal cells (Bowen and Foster, 1978). The highest amounts of exudation probably occur along the junctions of cells.

Rovira in 1956 observed this multiplication of bacteria in the grooves, and Bowen and Rovira (1976) found that even under favorable conditions, only 8 to 10% of the root surface is colonized by bacteria. Bowen (1979) doubted that this small degree of overall occupancy can deter fast-growing, virulent fungal pathogens.

Quality

Smith and Peterson (1966) collected exudates from the entire root system of aseptically grown red clover seedlings up to 12 days old. Glucose, fructose, ribose, galactose, and xylose were exuded. Macroconidia of three *Fusarium* spp. were exposed in natural soil to sugars solutions calculated to be five times the average exudate concentration of the whole root rhizosphere. Only glucose and fructose stimulated germination of macroconidia. Chlamydospores responded most strongly to glucose. When the fungi were grown in liquid cultures with nitrate nitrogen, ribose yielded the greatest weight of mycelium, followed by xylose. Glucose yielded the least mycelial dry weight. The authors concluded that some sugars are effective in stimulating germination, but other sugars may sustain the fungus after germination has occurred.

When asparagine and other amino acids are added to soils containing 10 ppm nitrogen, germination of chlamydospores of *F. solani* f. sp. *phaseoli* is three times that when sugars alone are added. Inorganic nitrogen, particularly ammonium nitrogen, added with sugars increases germination to that near amino acids alone (Cook and Schroth, 1965). In low nitrogen soils, amino acids are more effective than sugars in stimulating germination. In high nitrogen soils, sugars alone are effective. Bean hypocotyls exude primarily sugars, so nitrogen content of the soil influences the number of chlamydospores that germinate.

Qualitative differences in exudate from the root tip and from the root-hair zone of sunflower plants exist. Frenzel (1960), using specific nutrient-requiring mutants of *Neurospora*, found threonine and asparagine more abundant near the root tip, and leucine, valine, glutamic acid and phenylalanine more abundant in the root hair zone.

Root Age

In young roots exudates and mucigel provide the main source of nutrition for rhizosphere microorganisms. As the roots age, moribund root hairs and epidermal and cortical cells augment sloughed root cap cells. Exudates are important some distance back from the root tip, but less so with greater distance. Pearson and Parkinson (1961) (with broad bean) and Schroth and Snyder (1961) (with beans) demonstrated the highest concentration of root exudates near the root tip.

Cortical and epidermal cells of wheat roots die as they age (Deacon and Henry, 1978). Van Vuurde and Schipper (1980) confirmed this by growing wheat seedlings in soil in a container with a glass side tilted so roots grew against the glass. The growth made by each root each day was marked on the glass, and at 10 days the container was dismantled and the roots studied in daily increments. Roots not influenced by other roots were selected for study. Van Vuurde and Schippers found that the greatest numbers of rhizoplane bacteria correlated with death of host cells. Cell death increased from day 1 to day 4 or 5, paused during days 4 to 6, and increased again during days 6 to 8. The pattern of host cell death paralleled bacterial numbers. The pause in cell death (days 4 to 6) correlated with the time during which secondary roots emerged, as observed by Deacon and Henry.

The outer cortical cell layers of rootlets are expendable, and soil microflora respond to the concomitant release of nutrients. On the older part of the root (9 to 10 days) actinomycetes increase at the expense of bacteria. Nuclei persist longest in the innermost layer of the cortex, next to the endodermis (Holden, 1975, 1976). Deacon (1981) concluded that root cortical death is a normal process and that it proceeds in the absence of pathogens.

Hamlen et al. (1972) studied alfalfa under sterile conditions and found 151 mg of glucose per gram of dry root weight in exudates at 4 weeks of age, 86 mg at 6 weeks, 4 mg at 8 weeks, and less than 1 mg thereafter. In older plants an increase in exudation of four of five carbohydrates occurred at flowering, but this rise was low (3 mg/g or less) compared to exudation from young roots. They stressed the significance of exudation from young roots when roots are susceptible to many pathogens.

Host Genetics

Rescue spring wheat is very susceptible to *Bipolaris sorokiniana*; Cadet is resistant (Larson and Atkinson, 1970). Resistance is determined primarily by genes on chromosome 5B. If 5B from the resistant wheat is substituted into the susceptible wheat, the susceptible wheat becomes resistant, and substitution of 5B from the susceptible wheat into the resistant wheat renders the resistant wheat susceptible. Dead cortical cells were more abundant in Rescue (susceptible) than in Cadet (resistant) (Deacon and Lewis, 1982). Substitution of chromosome 5B from one cultivar to the other alters the extent of cortical cell death, with more live cells when 5B from Cadet is present. Small increases in cortical cell senescence increase susceptibility to *B. sorokiniana*. Rhizosphere bacteria are more abundant around roots of Rescue (susceptible) than around those of Cadet (resistant). Deacon and Lewis did not attribute resistance to rhizosphere bacteria. Deacon and Henry (1980) found the same relationship with *Gaeumannomyces graminis*. General vigor of cortical cells influences susceptibility to several parasites of roots, especially those favored by predisposition. Part of this retention or loss of cell vigor is genetically controlled.

Oxygen and Minerals

Oxygen consumption by microorganisms in the root region exceeds that consumed by the roots (Trolldenier, 1979). Bacterial numbers and oxygen consumption are consistently highest on roots of plants supplied with excessive nitrogen and insufficient potassium. When Trolldenier grew spring wheat supplied with varying amounts of potassium, the number of bacteria decreased as potassium increased. Doubling the amount of potassium reduced the amount of exudates from bean roots by half. Other authors report that phosphorus and calcium deficiencies increase exudation, and nitrogen deficiency decreases exudation.

pH It is probably improper to deal with inorganic substances in a chapter on exudates, but the form of nitrogen absorbed by roots influences the rhizosphere pH. *Gaeumannomyces graminis* var. *tritici* is favored by alkaline soils. Gould et al. (1966) observed a reduction of Ophiobolus patch of turf following the use of ammonium sulfate. In this case, large amounts of the fertilizer were used, and the pH of the bulk soil was decreased.

Huber and others reported a striking difference in take-all severity of winter wheat supplied with ammonium (reduced disease) or nitrate (increased disease). In this case, bulk soil pH could not have been altered appreciably (see Chapter 22 for details). Smiley and Cook (1973) and Smiley (1974, 1975) investigated this phenomenon. The rhizosphere pH of roots fed NH_4^+ nitrogen is lower than the bulk soil pH, and the rhizosphere pH of wheat given NO_3^- nitrogen is usually increased. When NO_3^- is absorbed, some OH^- or HCO_3^- will be excreted, raising the rhizosphere pH. This result is strongest in plants that assimilate NO_3^- in the roots (peas, beans, radishes, apples). In contrast, when NH_4^+ nitrogen is assimilated in roots, H^+ is excreted, causing a decrease in rhizosphere pH. The pH changes in the rhizosphere resulting from NO_3^- versus NH_4^+ nutrition are so great that the rhizosphere pH can deviate from bulk soil pH by as much as 1.2 units. These differences significantly altered the severity of take-all of wheat (Smiley, 1975).

Hatching Factors

Baunacke (1922) found that eggs of *Heterodera schachtii*, the sugar beet cyst-nematode, hatched well for a while when placed in water, but that hatching soon declined. If the eggs were placed in water containing beet roots or roots of several members of the Chenopodiaceae or Cruciferae, hatching was rapid and sustained for a long time. If these roots were placed in water, removed, and the eggs then placed in the water, hatch proceeded as if the roots were still present. Roots of many other plants did not stimulate hatching. Baunacke concluded that something exuded from the roots of some plants stimulated hatching, and thus the existence of "hatching factors" was discovered. The hatching factors, still largely unknown chemically, are not the common sugars and amino acids present in the root exudates of all plants. The eggs of different species of cyst-nematodes vary in their dependence on hatching factors.

Hatching factors, reviewed by Clarke and Perry (1977), are specific chemicals characteristic of certain plants that stimulate eggs to hatch. Most plants that exude hatching factors effective for a particular cyst-nematode are hosts, but not all. *Globodera rostochiensis*, the golden potato nematode, responds to hatching factors. Oxygen consumption (respiration) is increased within 24 hr and hydration of larvae is increased within 48 hr of exposure to hatching factors. At least 3 days are required for vigorous movement of larvae within eggs after exposure to hatching factors. The larvae of *G. rostochiensis* almost completely fill the egg, so their movements are restricted, and they do not become fully hydrated until emergence from the egg. The action of hatching factors is still not clearly understood, but they are believed to increase permeability of the egg shell. Eggs of *G. pallida* exposed to potato root diffusated for 5 min and transferred to distilled water are stimulated. The hatching factor is quite stable in soil because hatching is stimulated by a soil extract obtained 1 month after potato roots had been removed.

There is too little difference in hatching stimulation between *Solanum vernei* × *S. tuberosum* hybrids (resistant) and susceptible potatoes to support a role of exudates in resistance (Turner and Stone, 1981). Williams (in Turner and Stone) had previously found that exudates of *S. tuberosum* spp. *andigena* (resistant) were as stimulatory as those of potatoes susceptible to *G. rostochiensis*. Price and Steele (in Turner and Stone) tested the eggs of the sugar beet cyst-nematode, *Heterodera schachtii,* against root diffusates from resistant and susceptible breeding lines of sugar beets. Diffusates from resistant beets stimulated hatching as well as diffusates from susceptible beets, so resistance is not the result of lack of hatching factors. The eggs of *H. major* and *H. glycines* hatch about as well in water as in the presence of root diffusates.

Shepherd (1970) reviewed the influence of root exudates on the activities of some parasitic nematodes. The cyst-nematodes with wide host ranges respond to hatching factors from many plants, some of which are not hosts. The cyst-nematodes with narrow host ranges respond to exudates of a narrow range of plants, mostly those of hosts. Some hatching factors are effective at concentrations of 1×10^{-7}.

Root exudates of a general nature (amino acids, sugars) guide the motile forms of nematodes to host structures through chemotaxis and affect molting of some species. *Paratylenchus dianthus* persists in soil as dormant fourth stage juveniles for long periods. Rhoades and Linford (1959) found that few of these preadults molted when in water (1 to 2%), but in root exudates 85% transformed into adults. *P. projectus* responded similarly. Some stages in the life cycle of some nematodes, other than hatching, respond to signals received from host exudates.

TOXIC OR REPELLENT SUBSTANCES

Longidorus elongatus, an important vector of soilborne viruses of raspberry, declines on raspberry. Raspberry is toxic to this nematode (Taylor and Murant, 1966). Adult females placed in aqueous extracts of raspberry roots die. The rapid decline of *L. elongatus* following incorporation of raspberry residue into soil results from toxic action rather than from complex biological interactions with predaceous nematodes, fungi, or bacteria parasitic to nematodes.

Root-knot of eggplant is a serious problem in Rajasthan, India. Nematicides are not always practical or safe under some farm situations, and Varma et al. (1978) interplanted eggplant (susceptible) with marigolds and with sesame. Marigold and sesame roots are toxic or repellant to *M. incognita* and *M. javanica*. The sesame gave slightly better yields of eggplant than did marigolds, and sesame has market value. The yield response and nematode control from sesame was comparable to that obtained after incorporating aldicarb nematicide within the top 15 cm of soil.

COMMENTS

A series of chapters ("Fungistasis,""Rhizosphere,""Spermosphere,""Volatiles") emphasizes the significance of chemicals in the soil, regardless of their origin. It is logical that microorganisms within soil respond to these chemicals, either positively or negatively, depending on their nature. These substances are sources of information to the microorganisms, evidence of the existence of harmful or useful substances in the environment.

The root system is dynamic, with the quantity and quality of exudates being affected by the condition and genetics of the host. The quantity is so great one cannot help wondering if Olsen et al. (1981) are correct, that exudation is not all waste, that it may have positive aspects that more than offset the effects of exudates on plant pathogens.

KEY REFERENCES

Bailey, J. E., J. L. Lockwood, and M. V. Wiese. 1982. Infection of wheat by *Cephalosporium gramineum* as influenced by freezing of roots. *Phytopathology* **72:**1324–1328.

Clarke, A. J. and R. N. Perry. 1977. Hatching of cyst-nematodes. *Nematologica* **23:**350–368.

Katznelson, H. 1965. Nature and importance of the rhizosphere. Pages 187–209 in *Ecology of Soilborne Plant Pathogens*. W. C. Snyder and K. F. Baker, eds. Univ. of Calif. Press, Berkeley.

Olsen, R. A., R. B. Clark, and J. H. Bennett. 1981. The enhancement of soil fertility by plant roots. *Am. Sci.* **69**:378–384.

Rovira, A. D. 1965. Plant root exudates andheir influence upon soil microorganisms. Pages 170–186 in *Ecology of Soilborne Plant Pathogens*. Univ. of Calif. Press, Berkeley.

REFERENCES

Barber, D. A. and J. K. Martin. 1976. The release of organic substances by cereal roots into soil. *New Phytol.* **76**:69–80.

Baunacke, W. E. 1922. Untersuchungen zur Biologie and Bekampfung des Rubennematoden, *Heterodera schachtii* Schmidt. *Arb. Biol. Reichsanst. Landund Forstwirt.* **11**:185–288.

Bowen, G. D. 1979. Integrated and experimental approaches to the study of growth of organisms around roots. Pages 209–227 in *Soil-Borne Plant Pathogens*. B. Schippers and W. Gams, eds. Academic Press, New York.

Bowen, G. D. and R. C. Foster. 1978. Pages 231–256 in *Soil Factors in Crop Production in a Semi-arid Environment*. J. S. Russell and E. L. Greacen, eds. Queensland Univ. Press, Australia.

Bowen, G. D. and A. D. Rovira. 1976. Microbial colonization of plant roots. *Ann. Rev. Phytopathol.* **14**:121–144.

Burstrom, H. G. 1965. The physiology of roots. Pages 154–166 in *Ecology of Soil-Borne Plant Pathogens, Prelude to Biological Control*. K. F. Baker and W. C. Snyder, eds. Univ. of Calif. Press, Berkeley.

Clark, F. E. 1940. Notes on types of bacteria associated with plant roots. *Trans. Kan. Acad. Sci.* **43**:75–84.

Clark, F. E. 1942. Experiments toward the control of the take-all disease of wheat and the *Phymatotrichum* root rot of cotton. *U. S. Dept. Agric. Tech. Bull.* **835**:1–27.

Cook, R. J. and M. N. Schroth. 1965. Carbon and nitrogen compounds and germination of chlamydospores of *Fusarium solani* f. *phaseoli*. *Phytopathology* **55**:254–256.

Curl, E. A. 1982. The rhizosphere: relation to pathogen behavior and root disease. *Plant Dis.* **66**:624–630.

Deacon, J. W. 1981. Ecological relationships with other fungi: competitors and hyperparasites. Pages 75–101 in *Biology and Control of Take-all*. M. J. C. Asher and P. J. Shipton, eds. Academic Press, New York.

Deacon, J. W. and C. M. Henry. 1978. Death of the cereal root cortex: its relevance to biological control of take-all. *Ann. Appl. Biol.* **89**:100 (Abstract).

Deacon, J. W. and C. M. Henry. 1980. Age of wheat and barley roots and infection by *Gaeumannomyces graminis* var. *tritici*. *Soil Biol. Biochem.* **12**:113–118.

Deacon, J. W. and S. J. Lewis. 1982. Natural senescence of the root cortex of spring wheat in relation to common root rot (*Cochliobolus sativus*) and growth of a free-living bacterium. *Plant Soil* **66**:13–20.

Frenzel, B. 1960. Zur Atiologie der Anreicherung von Aminosaurem und Amiden im

Wurzelraum von *Helianthus annuus* L.: ein Beitrag zur Klarung der Probleme der Rhizophare. *Planta* **55**:156-207.

Gould, C. J., R. L. Goss, and V. L. Miller. 1966. Effects of fungicides and other materials on control of Ophiobolus patch disease on bent grass. *J. Sports Turf Res. Inst.* **42**:41-48.

Holden, J. 1975. Use of nuclear staining to access rates of cell death in cortices of cereal roots. *Soil Biol. Biochem.* **7**:333-334.

Holden, J. 1976. Infection of wheat seminal roots by varieties of *Phialophora radicicola* and *Gaeumannomyces graminis*. *Soil Biol. Biochem.* **8**:109-119.

Larson, R. I. and T. G. Atkinson. 1970. A cytogenetic analysis of reaction to common root rot in some hard red spring wheats. *Can. J. Bot.* **48**:2059-2067.

Martin, J. K. 1977. Factors influencing the loss of organic carbon from wheat roots. *Soil Biol. Biochem.* **9**:1-7.

Papavizas, G. C. and C. B. Davey. 1961. Extent and nature of the rhizosphere of *Lupinus*. *Plant Soil* **14**:215-236.

Pearson, R. and D. Parkinson. 1961. The sites of excretion of ninhydrin positive substances by broad bean seedlings. *Plant Soil* **13**:391-396.

Rhoades, H. L. and M. B. Linford. 1959. Molting of preadult nematodes of the genus *Pratylenchus* stimulated by root diffusates. *Science* **130**:1476-1477.

Riviere, J. 1959. Contribution to the study of the wheat rhizosphere. *Ann. Agron.* **45**:93-337.

Rovira, A. D. 1959. Root excretions in relation to the rhizosphere effect. IV. Influence of plant species, age of plant, light, temperature and calcium nutrition on exudation. *Plant Soil* **11**:53-64.

Rovira, A. D. 1963. Microbial inoculation of plants. I. Establishment of free living nitrogen fixing bacteria in the rhizosphere and their effects on maize, tomato and wheat. *Plant Soil* **19**:304-314.

Schroth, M. N. and W. C. Snyder. 1961. Effect of host exudates on chlamydospore germination of the bean root rot fungus, *Fusarium solani* f. *phaseoli*. *Phytopathology* **51**:389-393.

Shepherd, A. M. 1970. The influence of root exudates on the activity of some plant-parasitic nematodes. Pages 134-137 in *Root Diseases and Soil-Borne Pathogens*. T. A. Toussoun, R. V. Bega, and P. E. Nelson, eds. Univ. of Calif. Press, Berkeley.

Smiley, R. W. 1974. Take-all of wheat as influenced by organic amendments and nitrogen fertilizers. *Phytopathology* **64**:822-825.

Smiley, R. W. 1975. Forms of nitrogen and the pH in the root zone and their importance to root infections. Pages 55-62 in *Biology and Control of Soilborne Plant Pathogens*. G. W. Bruehl, ed. Am. Phytopath. Soc., St. Paul, Minn.

Smiley, R. W. and R. J. Cook. 1973. Relationship between take-all of wheat and rhizosphere pH in soils fertilized with ammonium vs. nitrate-nitrogen. *Phytopathology* **63**:882-890.

Smith, W. H. and J. L. Peterson. 1966. The influence of the carbohydrate fraction of the root exudate of red clover, *Trifolium pratense* L., on *Fusarium* spp. isolated from the clover root and rhizosphere. *Plant Soil* **25**:413-424.

Taylor, C. E. and A. F. Murant. 1966. Nematicidal activity of aqueous extracts from raspberry canes and roots. *Nematologica* **12**:488-494.

Timonin, M. I. 1940. The interaction of higher plants and soil microorganisms. II.

Study of the microbial populations of the rhizosphere in relation to resistance of plants to soil-borne diseases. *Can. J. Res., Sec. B.* **18**:444–456.

Trolldenier, G. 1979. Effects of mineral nutrition of plants and soil oxygen on rhizosphere organisms. Pages 235–240 in *Soilborne Plant Pathogens.* B. Schippers and W. Gams, eds. Academic Press, New York.

Turner, S. J. and A. R. Stone. 1981. Hatching of potato cyst-nematodes (*Globodera rostochiensis* and *G. pallida*) in root exudates of *Solanum vernei* hybrids. *Nematologica* **27**:315–318.

Van Vuurde, J. W. L. and B. Schippers. 1980. Bacterial colonization of seminal wheat roots. *Soil Biol. Biochem.* **12**:559–565.

Varma, M. K., H. C. Sharma and V. N. Pathak. 1978. Efficacy of *Tagetes patula* and *Sesamum orientale* against root-knot of eggplant. *Plant Dis. Reptr.* **62**:274–275.

Chapter 8

Spermosphere

Bacteria and fungi are more numerous near germinating seeds, and actinomycetes are less numerous. Slykhuis (1947) was impressed by changes in the microflora near seeds, and he coined the name *spermatosphere* to emphasize the special environment near the seed. The zone of influence of the seed has subsequently been referred to as the spermosphere, a term parallel with rhizosphere.

One reason *Pythium* spp. are so successful as seed-rotters is that *Pythium* propagules respond rapidly to exudates, and invasion of host tissue occurs rapidly, before significant antagonism in the spermosphere develops. The richness in stored foods of seeds results in more intense microbial activity about a seed than about a root. *Fusarium solani,* which attacks more slowly, is less important in this region. It is proficient in attacking hypocotyls of beans and epicotyls of peas, because the laimosphere is less rich in exudates. Each underground structure influences soil microorganisms in slightly different ways.

The spermosphere is different from the rhizosphere. The nature of the seed, its proneness to threshing injuries, the effects of maturation and age, and other factors, make it a special subject, even though the same sugars, amino acids, and other substances that exude from roots are involved. Gross anatomy and structure of seeds affect susceptibility to mechanical injury during harvest and handling. Hulled grains (barley, oats, rice) are protected during threshing. Webster and Dexter (in Roberts, 1972a) ranked wheat, maize, and beans as being susceptible to physical damage in order of their increasing susceptibility to harvest injury.

In grasses the "seed" is a mature, dried fruit, with the true seed surrounded by the pericarp. The pericarp (pure maternal tissue), the endosperm, and the embryo differ genetically, making for complex relationships. Efforts to lengthen the growing season of maize by seeding it early in cool, wet soil has led to much study of resistance to *Pythium* spp. Pinnell (1949) believed that the female parent of hybrids had a strong influence on germination in cold soil, which he attributed to differences in the endosperm. Hooker and Dickson (1952) excised embryos and found part of the resistance to *P. debaryanum* in the embryo, with both parents contributing to resistance. Maternal effects could include differential proneness of the pericarp to injury. Early evidence implicates the pericarp (maternal), embryo, and endosperm each as important to germination and emergence in cold soil.

SEED FACTORS

Maturity of the Seed When Harvested

Maize seed does not attain maximum resistance to seed rot until it reaches full maturity, when membranes and tissues are fully formed. Seed harvested in the milk stage imbibes water and germinates more rapidly than mature seed, but seedlings from mature seed are more vigorous and weigh more (Dungan, 1924). Koehler et al. (1934) collected maize seed in the milk stage that contained 74% water (Koehler et al., 1934). Seed in the early dent stage contained 59% water, and fully mature dented seed contained 29% water. At husking time in late November, the water content of seed had decreased to 22%. The ears from each maturity class were husked and placed on wires in a well-ventilated room to dry. Germination was tested in mid-winter, and only those ears that gave 100% germination in the laboratory trial were used. High-grade seed is typically seeded in central Illinois 1–10 May to obtain highest yields. Planting was delayed to mid-May (soil temperature was 61°F) to favor weaker seeds. Even with the stated precautions, field stands and yields reflect seed maturity. Without seed treatment, the best stand and highest yield are obtained from mature seed. Both immaturity and overmaturity reduce stands and yield. Under adversity, best emergence in cold, wet soil is attained when maize at harvest contains about 28% water (Koehler et al., 1934, and Rush and Neal, 1951). Laboratory germination tests often fail to predict seedling emergence under adverse conditions (Table 8.1).

Age of Seed

The effect of length of storage of maize seed has been tested by incubating seed with sound pericarps in moist soil for 12 days at 50°F, after which in-

Table 8.1. Effect of seed maturity on germination of corn in a laboratory trial and in cold, wet soil

	Harvest dates, 1948					
	30 Aug	9 Sept	19 Sept	29 Sept	9 Oct	19 Oct
Moisture, % at harvest	72	53	40	35	28	26
Laboratory germination, %	88	96	97	98	98	98
Cold soil germination, %	1	7	36	65	80	57

Source: Rush and Neal, 1951.

cubation was at 70 to 75°F (Koehler, 1954). The stands from untreated seeds stored for 0.5, 1.5 and 2.5 years were 83, 55, and 40% respectively. Stands were 100, 99, and 98% when seeds were treated with thiram. Increased age predisposed the seed to Pythium seed rot and seedling blight. Abu-Shakra and Ching (in Roberts, 1972b) compared soybean seeds stored 3 months with seed stored 3 years at room temperature. Even though the percentage of germination of both lots was high, the seedlings from the older seed were weaker.

Roberts (1972b) defines viability as the ability to produce vigorous seedlings, not just germination. He concluded that degenerative changes in ultrastructure result from a general degeneration of the lipoprotein membranes of the cells. Degeneration is usually first detected in mitochondria. Chromosome damage is also common in aged seeds, and it, too, is associated with reduced viability.

Injuries

Koehler (1954) harvested 5–10 September when about 90% of the kernels were denting and the grain was at 42% moisture, wetter than at physiologic maturity. The ears were picked by hand, dried slowly, and the kernels shelled by hand. In spite of careful handling, immature seed was damaged by *Pythium* in cold, moist soil. Wrinkling and folding of the pericarp occurs during the drying of immature seed, making natural wounds.

Nature endowed the mature maize kernel with good protection, but modern farm technology reduces this protection. Well-matured, hand-picked, slowly dried, hand-shelled kernels produce stands and yields equal to those produced by commercially harvested and dried kernels treated with fungicides. (Koehler, 1954). Pericarp fractures can be present on 81% of commercially prepared seed. The fungicides compensate for wounds resulting from commerical farm practice. Physical injuries increase the rate of imbibition, and rapid water-uptake injures membranes.

The seed coat of white-seeded snap beans generally adheres less tightly than those of dark-seeded beans, and the seed coat of white beans is more frequently injured (Atkins, in Moore, 1972). Moore observed that white seed coats are thin, adhere loosely to the cotyledons, and are more susceptible to rapid water uptake. Most commercial flax produced for linseed oil in North America has dark brown seeds, in spite of higher oil content and quality in light-seeded flaxes (Culbertson and Kommedahl, 1956). Split seed coats occur on 5% of brown seeds and on 53% of light colored seed.

Scarified soybeans absorb water much faster than intact seed, reaching a high water content in a few hours (Schlub and Schmitthenner, 1978). This can lead to anaerobiosis within the seed by filling the cavity between the cotyledons with water, and improper hydration may result in membrane damage.

The water content of large seeds at harvest affects the type and extent of harvest injuries. Moore (1972) reported that the water content of soybeans at 10 A.M. was 13.5% and that at 1 P.M. of the same sunny day it was 12.2%. When the combine cylinder was set at 700 rpm, 4% of the seed coats were broken at 13.5% water and 5% at 12.2% water. At a combine cylinder speed of 1155 rpm, 12% were injured at 13.5% water and 48% at 12.2% water. Damp seed tends to bruise, and very dry seed tends to crack during harvest. Bruising is not as injurious as cracking if the damp seed is dried properly and promptly after harvest. Seeds at optimum water content when impacted are dry enough to prevent cell rupturing but not dry enough to promote fractures.

QUANTITY OF EXUDATES

The quantity of exudates from three bean cultivars is correlated with their susceptibility to preemergence damping-off caused by *Rhizoctonia solani* and *Pythium ultimum* (Schroth and Cook, 1964). Cracked seed coats result in greater exudation and greater susceptibility. The data of Scroth and Cook suggest that the increased preemergence damping-off resulting from cracked seeds is not due to increased exudation. Intact and cracked seed resulted in 30% and 31% germination of chlamydospores, respectively, but disease incidence increased from 53 to 88% (Table 8.2). Exudation by uninjured seed was sufficient to stimulate germination of the chlamydospores.

Schlub and Schmitthenner (1978) added 12.7 mg of sucrose to individual soybean seed. When sound seed, seed with fine cracks, or wounded seed received the added sucrose, seedling quality was not affected by the added sugar in the presence of 130 *Pythium ultimum* propagules per gram of dry soil. They concluded that even sound soybean seeds exude enough food to initiate and sustain attack by *P. ultimum*. Exudation levels of sound, cracked, and wounded soybeans seeds correlate with the rate of water absorption. Schlub and Schmitthenner also germinated wounded and sound

Table 8.2. Effect of sound and cracked seed coats on germination of *Fusarium solani* f. sp. *phaseoli* chlamydospores and on preemergence damping-off of three bean varieties

	Intact		Craked	
Variety	Chlamydospore germination	Preemergence damping-off	Chlamydospore germination	Preemergence damping-off
Pinto	30%	53%	31%	88%
Small white	25	25	30	60
N203	18	18	33	41

Source: Schroth and Cook, 1964.

soybean seed in saturated soil at 15°C in natural and autoclaved soil. The seedlings of wounded seed are weaker than those of sound seed, whether in the presence of pathogens or not. Wounds reduce the vigor of the seedlings. In contrast to the results of Schlub and Schmitthenner, Keeling (1974) found that coating soybean seed with glucose reduced stands. *P. ultimum* and *P. debaryanum* were the main pathogens. Keeling concluded that seed rot and seedling blight are directly rated to the amount of carbohydrate exuded from the seeds.

High soil moisture supports rapid diffusion of exudates in soil (Fig. 8.1) and rapid hyphal growth (up to 300 μm per hour) by *P. ultimum*. At saturation, the spermosphere about bean seed, as judged by germination of sporangia, extends 10 mm from the seed within 24 hr (Stanghellini and Hancock, 1971). Within 6 hr mycelial growth is abundant within 2 to 4 mm of the seed, and water-soaked lesions are present on the cotyledons within 24 hr.

EFFECT OF SOIL OXYGEN AND TEMPERATURE

Periods of low oxygen (4 to 8%) in the soil atmosphere, for as few as 24 hr, predispose soybean seeds to rot by *P. ultimum* (Brown and Kennedy, 1966). Sugar loss from soybeans is greater at 4 to 8% than at higher O_2 concentrations, and bacterial contamination of the seeds increases sugar loss. Exudation is most abundant near the hilum, and the radicle emerges in this region. At 4 to 8% O_2 concentrations, the radicles are destroyed. At 11.5% O_2 seedlings are attacked later, commonly in the neck region of the hypocotyl. Grinyova (in Brown and Kennedy) hypothesized that curtailment of aerobic respiration shifts metabolism so that some products are not utilized properly, increasing unused hydrolysed materials. In addition, low O_2 increases permeability and stimulates exosmosis.

Mung bean seeds exude heavily at 42°C (Kraft and Erwin, 1967). They found 13 amino acids in the 42°C-exudate and 9 in the 12°C-exudate. Unfavorable temperature alters cell permeability and favors exudation, as

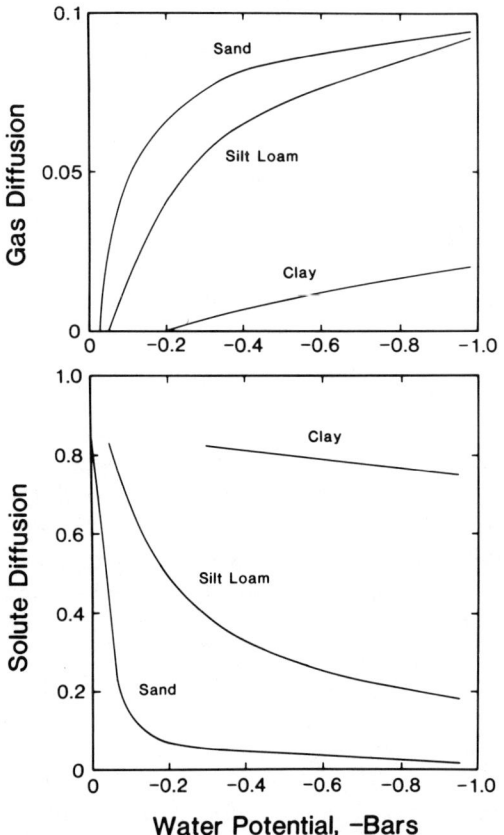

Figure 8.1. Gas diffusion (oxygen, volatiles) is favored by coarse-textured soil and reduced water content (upper), and solute diffusion is favored by heavy soils that contain greater quantities of water over a wider range of soil matric water potentials (lower). After Papendick and Campbell, 1981.

suggested by Grineva (in Kraft and Erwin). The rate of sugar loss at 42°C is about 12 times faster than at 24°C, and rate of nitrogen loss is even more accelerated. Seeds exposed to 42°C grow normally when transferred to 24°C, so germinability is not destroyed.

COMMENTS

Controversy exists, but wounds, in my opinion, have their greatest effect in reducing host resistance rather than by increasing exudation. Seeds of most plants are greatly dehydrated, and hydration following planting is a critical stage. Collins et al. (1984) determined the rate of imbibition by unwounded maize seeds, and they calculated that the unwounded membranes in the seed regulate the rate of imbibition. In initial stages, imbibition by unwounded seed is slow. Imbibition accelerates after the initial few hours of imbibition. Initial slow acquisition of water must favor rehydration of membranes and organelles. In addition, evidence indicates that rapid entrance of water leads to oxygen deficiency within the seed. Anyone who has inoculated growing plants with a hypodermic needle is impressed with how well a growing plant withstands mutilation. Wounding a dried seed has far more serious consequences.

KEY REFERENCES

Brown, G. C. and B. W. Kennedy. 1966. Effect of oxygen concentration on Pythium seed rot of soybean. *Phytopathology* **56**:407–411.

Koehler, B. 1954. Some conditions influencing the results from corn seed treatment tests. *Phytopathology* **44**:575–583.

Schroth, M. N. and R. J. Cook. 1964. Seed exudation and its influence on pre-emergence damping-off bean. *Phytopathology* **54**:670–673.

REFERENCES

Collins, G., E. Stibbe and B. Kroesbergen. 1984. Influence of soil moisture stress and soil bulk density on inbibition of corn seeds in sandy soil. *Soil Tillage Res.* **4**:361–370.

Culbertson, J. O. and T. Kommedahl. 1956. The effect of seed coat color upon agronomic and chemical characters and seed injury in flax. *Agron. J.* **48**:25–28.

Dungan, G. H. 1924. Some factors affecting the water absorption and germination of seed corn. *J. Am. Soc. Agron.* **16**:473–481.

Hooker, A. L. and J. G. Dickson, 1952. Resistance to Pythium manifest by excised corn embryos at low temperatures. *Agron. J.* **44**:443–447.

Keeling, B. L. 1974. Soybean seed rot and the relation of seed exudate to host susceptibility. *Phytopathology* **64**:1445–1447.

Koehler, B., G. H. Dungan and W. L. Burleson. 1934. Maturity of seed corn in relation to yielding ability and disease infection. *J. Am. Soc. Agron.* **26**:262–274.

Kraft, J. M. and D. C. Erwin. 1967. Stimulation of *Pythium aphanidermatum* by exudates from mung bean seeds *Phytopathology* **57**:866–868.

Moore, R. P. 1972. Effect of mechanical injuries on viability. Pages 93–113 in *Viability of Seeds*. E. H. Roberts, ed. Chapman and Hall, London.

Papendick, R. I. and G. S. Campbell. 1981. Theory and measurement of water potential. Pages 1–22 in *Water Potential Relations in Soil Microbiology*. J. F. Parr, W. R. Gardner, and L. F. Elliott, eds. Soil Science Society of America.

Pinnell, E. L. 1949. Genetic and environmental factors affecting corn seed germination at low temperatures. *Agron. J.* **41**:562–568.

Roberts, E. H. 1972a. *Viability of Seeds*. Chapman and Hall, London.

Roberts, E. H. 1972b. Cytological genetical and metabolic changes associated with loss of viability. Pages 253–306 in *Viability of Seeds*. E. H. Roberts, ed. Chapman and Hall, London.

Rush, G. E. and N. P. Neal. 1951. The effect of maturity and other factors on stands of corn at low temperatures. *Agron. J.* **43**:112–116.

Schlub, R. L. and A. F. Schmitthenner. 1978. Effects of soybean seed coat cracks on seed exudation and seed quality in soil infested wih *Phythium ultimum*. *Phytopathology* **68**:1186–1191.

Slykhius, J. T. 1947. Studies on *Fusarium culmorum* blight of crested wheat and brome grass seedlings. *Can. J. Res.* **25**:155–180.

Stanghellini, M. E. and J. G. Hancock. 1971. Radial extent of the bean spermosphere and its relation to the behavior of *Pythium ultimum*. *Phytopathology* **61**:165–168.

Chapter 9

Volatiles

Organic substances are released into soil from living or dead plant and animal substrates. Soluble substances (solutes) within the soil solution are referred to as *exudates*, and those within the soil atmosphere as gases are named *volatiles*. Volatiles have significant affects on the microbiology of soils. The review of Linderman and Gilbert (1975) is a highly recommended treatment of this topic.

TOXIC VOLATILES

Before the days of modern chemical technology, between 1907 and 1918, Schreiner, Shorey and Skinner (in Rands and Dopp, 1938) extracted organic compounds from poorly drained soils. They identified vanillin, salicylic aldehyde, and several organic acids. These compounds are injurious to plants in low concentrations. Rands and Dopp investigated root rot of sugarcane, caused mainly by *Pythium arrhenomanes,* on poorly drained Louisiana soils. In pot trials salicylic aldehyde alone at 20 ppm did not reduce the growth of sugarcane. *P. arrhenomanes* alone reduced top growth by 22%. When salicylic aldehyde and *Pythium* were both present, growth was reduced 51%. They concluded that aldehydes increase root rot by predisposing the host. Hydrogen sulfide, in parallel experiments at 10 ppm, reduced growth of cane but did not increase severity of Pythium root rot, indicating that generalizations cannot be made about the effects of volatiles on diseases. Graham and Greenberg (1939) and Bruehl (1953) confirmed the effect of salicylic aldehyde in increasing the severity of Pythium root rot on wheat and barley, respectively.

Flooding under warm conditions, especially when the soil contains fresh organic matter, usually results in the death of many propagules within the soil. Okazaki (1985) found that, after flooding soil with a 3% glucose solution, the gases produced in the soil were more lethal to chlamydospores of *Fusarium oxysporum* f. sp. *raphani* than was liquid centrifuged from the soil. When confined to gases within the head space above incubation chambers, fewer than 1% of the chlamydospores were germinable after 2 days. The nature of the toxic volatiles is not known.

Ammonia

Calcium oxide controlled a seedling problem in peas when Lewis and Lumsden (1984) mixed several calcium and magnesium salts in soil infested with *Pythium ultimum*. The sporangia of *P. ultimum* and oospores of *P. aphanidermatum* were killed by NH_3 vapors and by volatiles emitted by soil heavily amended by calcium oxide (CaO). Lewis and Lumsden found that germinable propagules of *P. ultimum* and *P. aphanidermatum* decline rapidly in soil amended with 0.2% CaO. They believe that CaO reacts with inorganic nitrogen in the soil, and liberates NH_3 and that NH_3 kills enough propagules to protect the pea seeds. CaO and Calcium hydroxide [$Ca(OH)_2$] are effective but magnesium oxide, hydrated lime, calcium carbonate, calcium sulfate, magnesium carbonate, and rock phosphate have little or no effect.

A 0.2% application is 4400 kg/ha (2 tons/acre) on a broadcast basis. Evidence that the reaction of CaO and $Ca(OH)_2$ liberates NH_3 and that the NH_3 has fungicidal properties is important. The repeated addition of calcium would increase the soil pH. An in-the-seed-row treatment at 25 g/m of row, with rows 20 cm apart (= 1200 kg/ha, or 0.5 ton/acre), applied at planting time improved stands (Lewis and Lumsden, 1984). This rate of application could be applied more frequently. Whether or not CaO has significance as a practical control measure, the fungicidal effects of NH_3 to *Pythium* spp. propagules is demonstrated.

INHIBITORY VOLATILES

Volatiles from *Fusarium oxysporum* temporarily inhibit germination of sporangiospores of *Rhizopus stolonifer*. Robinson and Park (1966) found that cultures of *F. oxysporum* emitted the fruity aroma of acetaldehyde. It escapes rapidly from unsealed cultures. Robinson and Park proposed the name *sporostasis* to define the inhibition of germination of fungal spores by inhibitors produced by the parent mycelium or by other organisms. Sporostasis is a widespread phenomenon; the mycelium of one species of fungus often inhibits germination of spores of many other species. The term *sporostasis* is useful in studies of pure cultures, but the term *fungistasis* is preferred for studies in soil. The origin of the germination-inhibiting substances in soil is less certain; thus a more general term can be used with greater confidence.

Volatiles of general significance in soil microbiology have been investigated by Hora and Baker (1972). They taped a sterile glass slide to the inside of a Petri dish lid, put a block of sterile purified water agar on the slide, and then put the lid over either moist soil or a culture of an organism. The agar was exposed to volatile substances for 24 hr. The slide with agar was removed, spore suspensions were added to the agar, and observations were made of spore germination. Germination of spores of several fungi was inhibited, and they concluded that *Streptomyces* spp. were responsible for most of the inhibition. Sterile soils colonized by *Streptomyces* spp. that produce an earthy smell inhibit germination of conidia of *Trichoderma viride, Zygorhynchus vuilleminii*, and *Gonatobotrys simplex*. *Streptomyces* that do not produce the earthy smell do not produce fungistatic volatiles. Neutral and alkaline soils inoculated with actinomycetes produce the strongest inhibition. Miscellaneous soil bacteria and fungi do not produce effective volatile inhibitors. When chitin is added to natural soil, volatile fungistatic factors are strongly produced. According to Alexander (1961), chitin is attacked primarily by actinomycetes, including *Streptomyces*.

Boletus variegatus forms mycorrhizae with *Pinus sylvestris* roots. Krupa and Fries (1971) identified volatiles produced by *B. variegatus* in pure culture, by *P. sylvestris* roots alone, and by mycorrhizal roots. Mycorrhizal roots produce several times more volatiles than nonmycorrhizal roots. The volatiles are primarily terpines that are not formed by the fungus alone. Terpines exhibit fungistatic effects by reducing the growth of *B. variegatus* (Melin and Krupa, 1971), *Phytophthora cinnamomi*, and *Fomes annosus* (Krupa and Nylund, 1972) in culture. Krupa and Fries (1971) hypothesized that volatiles from mycorrhizae may aid in maintaining the symbiotic balance between the fungus and the host, and these volatiles may alter the microbial composition of the rhizosphere, aiding particularly in defense against pathogens.

STIMULATORY VOLATILES

General Stimulation

Moist alfalfa hay emits volatiles that stimulate growth and respiration of many fungi and bacteria within soil (Menzies and Gilbert, 1967). The response exceeds the energy in the volatiles, being more stimulatory than sustaining. Hyphae emerge downward from natural soil toward moist, nonsterile hay separated from the soil by a 5-cm air space. No growth occurs in the absence of the moist hay. Dried corn leaves, wheat straw, bluegrass clippings, tea, and tobacco leaves are also stimulatory; thus the effect is not a special effect of alfalfa hay. Linderman and Gilbert (1969) found that an exposure of 15 min to volatiles results in as much mycelial growth as exposure for 5 days, evidence for a stimulatory rather than a nutritional function.

When microsclerotia of *Verticillium dahliae* or sclerotia of *Sclerotium rolfsii* are exposed to stimulatory volatiles, germination usually results. Gilbert and Griebel (1969) found that exposure of sclerotia of *V. dahliae* to proper concentrations of alfalfa hay distillates increases colony counts of *V. dahliae* several fold in soil dilution plates on selective media. Farley et al. (1971) showed the increase to be due to sporulation subsequent to the germination of the microsclerotia. Germination of the microsclerotia in the absence of a suitable substrate reduce the vigor and longevity of the microsclerotia. Exposure of sclerotia of *S. rolfsii* to volatiles leads to germination and death of many sclerotia that did not form secondary sclerotia. Many of the weakened sclerotia are killed by *Chaetomium, Trichoderma* and by *Fusarium solani* (Linderman and Gilbert, 1969).

When eruptively germinable *S. rolfsii* sclerotia are exposed to volatiles from moistened alfalfa hay, the hyphae grow both more vigorously and directionally toward the source of the volatiles (Punja and Grogan, 1981). Infection from eruptively germinated sclerotia occurs on moist field soil from a distance of 3.5 cm, and from 6 cm in the presence of volatiles. Punja and Grogan speculated that this increased effective range is due to directional growth or to use of the volatile as substrate plus directional growth (Fig. 9.1). In their illustration the mycelium was more dense when volatiles were present.

Most evidence supports a stimulatory rather than sustaining role for volatiles in relation to germination of and growth from sclerotia. In contrast, evidence with fungal spores supports both stimulatory and nutritional effects. The mass of volatiles in relation to the mass of sclerotia or spores differs greatly. What could be a trivial food source to a sclerotium could be substantial to a conidium. Griebel and Owens (1972) exposed macroconidia of *F. solani* f. sp. *phaseoli* to radioactive acetaldehyde or ethanol. Most of the increased respiration is at the expense of the volatile (70 to 90%), with little use of endogenous reserves. Griebel and Owens concluded that stimulation by ethanol or acetaldehyde does not deplete endogenous reserves.

Specific Stimulation

Sclerotia of *Sclerotium cepivorum* live at least 10 years in field soil, and even though *S. cepivorum* attacks only *Allium* species, control by rotation has been impractical. Coley-Smith (1979), as early as 1960, obtained evidence that volatiles from onion seedlings stimulate germination of sclerotia of *S.*

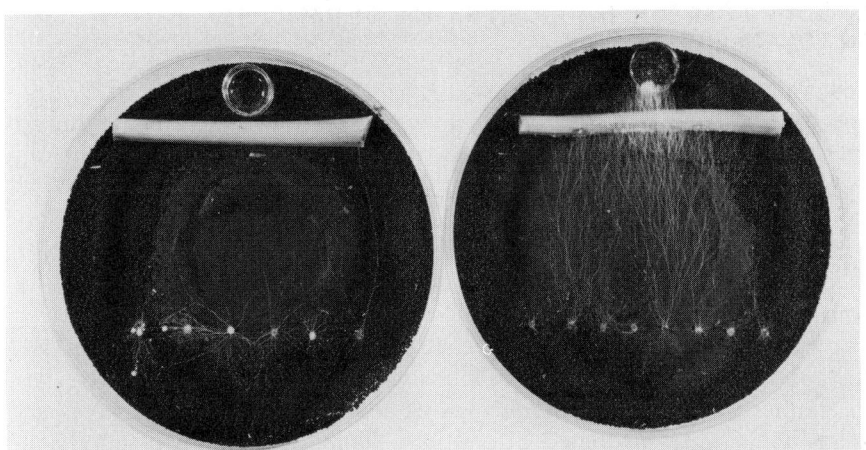

Figure 9.1. Effect of stimulatory volatiles on the germination of dried sclerotia of *Sclerotium rolfsii* on natural soil. Left, nondirectional growth with formation of replacement (secondary) sclerotia. No volatiles present. Right, directional, abundant hyphal growth toward the volatiles without formation of replacement (secondary) sclerotia. No volatiles present. After Punja and Grogan, 1981.

cepivorum and that these sclerotia in soil are stimulated only by exudates or volatiles from *Allium* spp. *A. aflatunense* and *A. cristophii*, which do not stimulate germination of sclerotia and normally escape attack, are attacked when germination stimulants are added to the soil (Coley-Smith et al., 1981). The sclerotia can germinate in sterile soil without the need for an external stimulant.

Merriman et al. (1980) applied commercial onion oil to infested soil 4 weeks before onions were planted. Onion oil is no more expensive than many soil fumigants. Application was by commercial equipment used for soil fumigation. Yield increases, disease reduction, and sclerotium mortality result when sclerotia are stimulated to germinate in the absence of a host. Onion oil at a concentration of 5% in water injected 10 cm deep is most effective. The numbers of sclerotia in the 0 to 10 and 10 to 20 cm depth layers were reduced by 77 and 91%, respectively. Disease was reduced from 57 to 13% and the yield increased by 103%.

The successful use of onion oil was in soil containing an average of 110 sclerotia/kg of a red brown earth. Results were unsatisfactory in a black clay loam with 180 sclerotia/kg of soil. Only 50% of the sclerotia were killed in the black clay loam. Adsorption of the oil onto clay minerals may have reduced the effectiveness of the treatment in the heavy soil. The active ingredients disperse more readily downward than upward, and proper soil preparation, temperature, and moisture effects need further study.

EFFECT OF CARBON: NITROGEN RATIOS OF RESIDUES ON VOLATILES

Lewis and Papavizas (1974) exposed *Rhizoctonia solani* on potato dextrose agar to volatiles eminating from various crop residues. Most immature residues with low C:N ratios increase growth of *R. solani;* volatiles from mature residues with high C:N ratios do not. Residues with C:N ratios below 15:1 increase melaninization of vegetative hyphae. Residues with C:N ratios greater than 15:1 have no effect on melaninization. They believed that NH_3 or a low-molecular-weight volatile amine, or both, stimulate the pigmentation.

Volatiles from decomposing cabbage reduce colonization of mature buckwheat stem pieces by *R. solani* in soil from 53% in the controls to 14%. In another experiment *R. solani* was established in buckwheat stems and these stems were incubated in soil with vapors of decomposing residue passed through the soil. The fungus was retrieved from 85% of the controls and from 37% of the soils exposed to volatiles from decomposing immature cabbage (Lewis and Papavizas, 1974).

Ethylene

Witchweeds (*Striga* spp.), parasitic plants in the family Scrophulariaceae, are particularly important in Africa. *S. asiatica* was first found in the United States in North Carolina in 1956, and it is now present on 153,960 ha in southeastern North Carolina and northeastern South Carolina (Eplee, 1981). A witchweed plant produces up to 500,000 microscopic seeds, and the seeds remain viable up to 20 years in soil. The seeds germinate when stimulated by strigol exuded from potential hosts.

If strigol could be produced economically, it could be used to stimulate germination in the absence of a host. Egley and Dale (1970) discovered that ethylene stimulates germination of seeds physiologically capable of germination (preconditioned). Ethylene injected 15–20 cm deep stimulates germination of at least 90% of the germinable seeds. If done 3 consecutive years, along with eradication of escapes, it is thought that *S. asiatica* can be eradicated. Ethylene diffuses 1 m from the point of injection in soil. A more complete account is given by Cook and Baker (1983, pp. 243–244).

Ethylene, produced in soil by some fungi and bacteria, affects many biological processes and has growth regulatory effects on higher plants. It was thought (Cook, 1975) that ethylene played a prime role in regulating the balance between aerobes and anaerobes in soil and that it is an important cause of fungistasis. Studies of ethylene in soil are summarized by Cook and Baker (1983, 241–244). Smith (1978) studied the consumption of O_2 and the production of CO_2 by cultivated, meadow, and woodland soil supplied with an atmosphere containing about 50 ppm of ethylene. He observed no effect of ethylene on soil respiration under aerobic conditions, and he concluded that ethylene has little effect on regulating the activities of soil microorganisms.

COMPLEX EXPERIMENTS WITH VOLATILES

When sclerotia of *Sclerotium rolfsii* produced on oat grains on the surface of a natural field soil and on potato dextrose agar are incubated on moist soil, 7% of the soil-sclerotia and 60% of the PDA-sclerotia germinated (Beute and Rodriguez-Kabana, 1979a). Soaking the sclerotia for 1 hr increases germination of the soil-produced sclerotia to 18% and the PDA-sclerotia to 80%. Both types of sclerotia have 100% germination on a selective medium. Sclerotia produced in the presence of natural soil are more subject to fungistasis.

Volatiles from dried peanut leaves, in quantities similar to those found under peanut plants, stimulate germination of the sclerotia, and leaves of five different cultivars are equally stimulatory regardless of their resistance or susceptibility to southern stem rot. The sclerotia are incubated on soil in desiccators. If a vessel of sodium hydroxide solution is placed within the desiccator to remove CO_2, germination of sclerotia by volatiles from peanut hay increases from 53% in the presence of CO_2 to 87% with CO_2 removed. Placing barium oxide in the desiccator to remove CO_2 and release O_2 gives a further but insignificant increase in germination (91%). When potassium chromate, sulfuric acid, and activated charcoal are used to absorb the volatiles, germination is 37%. This experiment demonstrates the aerobic nature of this fungus and the role of volatiles in stimulating germination of the sclerotia.

Seventy-one volatile organic compounds from peanut residues have been identified. Methanol is a major component, and methanol alone is equivalent in stimulatory power to all the volatiles from peanut hay. When tomato plants are transplanted into inoculated soil, 6 days after the sclerotia and plants within plastic bags are exposed to methanol, 100% of the tomato plants are dead, compared with 33% without the methanol treatment. Germination of sclerotia is 72% in the presence of methanol and 28% in the controls.

When fresh, green leaves of peanut are used, no stimulation occurs nor do *S. rolfsii* sclerotia respond to volatiles from partially decomposed peanut plants. The presence of vigorous, live peanut tissue does not stimulate infection. Autocatalysis of fresh, dead tissue results in enzymatic release of stimulatory volatiles. The response of propagules to tissues in various states of decomposition could provide clues as to the degree of their saprophytic characteristics in nature.

COMMENTS

How important are volatiles? In the case of *Sclerotium cepivorum* and *Striga* spp., they are crucial. The sclerotia or seeds, respectively, remain dormant in soil for years, to germinate only when stimulated by specific chemicals emitted by host plants. This survival mechanism, a long-lived structure responsive to specific host emissions, is ideal for an organism of limited host range.

The response of sclerotia of *Sclerotium rolfsii* to volatiles from fresh, dried plant debris has fundamental implications. Hyphae from eruptively germinating sclerotia (see Chap. 13) grow directionally toward moistened hay (Punja and Grogan, 1981). Sclerotia that have not been conditioned (dried) to germinate eruptively respond similarly but more weakly. Dried sclerotia are more sensitive to volatiles than undried sclerotia. Freshly fallen, dried leaves, by autolysis, release the volatiles which stimulate germination and direct hyphae toward the source of the gases. The distance from which infection can occur and the number of sclerotia that germinate are increased, both responses leading to increased disease (Fig. 9.1).

The failure of rotting materials to release stimulatory volatiles is evidence of discrimination. *S. rolfsii* colonizes fresh host debris and gains inoculum potential from saprophytism. It does not compete with soil saprophytes for substrates already colonized.

Gases diffuse faster than solutes, and Linderman and Gilbert (1975) stressed that volatiles are often lost during the soil manipulations and extraction procedures used to study soil chemistry. Another problem in evaluating the results of experiments with products of organic decomposition on soilborne plant pathogens and diseases is that many experiments involve unrealistic quantities of soil amendments and or unusual conditions of incubation. The same criticism has been made about some studies with antibiotics. But like antibiotics, small amounts may have important effects in microsites within soil.

The studies of Hora and Baker (1972) with *Streptomyces* spp. that produce the earthy smell should be given careful thought. Many studies in Chap. 6 suggest that no chemicals within soil enforce fungistasis, that only a shortage of food is involved. Yet volatiles from soil diffuse through air and reduce germination.

KEY REFERENCES

Beute, M. K. and R. Rodriguez-Kabana. 1979. Effect of wetting and the presence of peanut tissues on germination of sclerotia of *Sclerotium rolfsii* produced in soil. *Phytopathology* **69**:869–872.

Coley-Smith, J. R. 1979. Survival of plant pathogenic fungi in soil in the absence of host plants. Pages 39–57 in *Soilborne Plant Pathogens*. B. Schippers and W. Gams, eds. Academic Press, New York.

Epplee, R. E. 1981. Striga's status as a plant parasite in the United States. *Plant Dis.* **65**:951-954.

Hora, T. S. and R. Baker. 1972. Soil fungistasis: Microflora producing a volatile inhibitor. *Trans. Brit. Mycol. Soc.* **59**:491-500.

Linderman, R. G. and R. G. Gilbert. 1975. Influence of volatiles of plant origin on soil-borne plant pathogens. Pages 90-99 in *Biology and Control of Soil-Borne Plant Pathogens.* G.W. Bruehl, ed. Am. Phytopathol. Soc., St. Paul, Minn.

Menzies, J. D. and R. G. Gilbert. 1967. Responses of the soil microflora to volatile components in plant residues. *Proc. Soil Sci. Am.* **31**:495-496.

REFERENCES

Alexander, M. 1961. *Introduction to Soil Microbiology.* Wiley, New York.

Beute, M. K. and R. Rodriguez-Kabana. 1979. Effect of volatile compounds from remoistened plant tissues on growth and germination of sclerotia of *Sclerotium rolfsii. Phytopathology* **69**:802-805.

Bruehl, G. W. 1953. Pythium root rot of barley and wheat. *U.S. Dept. Agric. Tech. Bull.* 1084.

Coley-Smith, J. R. 1960. Studies of the biology of *Sclerotium cepivorum* Berk. IV. Germination of sclerotia. *Ann. Appl. Biol.* **48**:8-18.

Coley-Smith, J. R., G. A. Esler, and C. M. New. 1981. Possibilities for biological and integrated control of white rot disease of *Allium.* Pages 459-466. *Proc. British Crop Prot. Conf.—Pests and Diseases.*

Cook, R. J. 1975. Foreward. In *Biology and Control of Soil-Borne Plant Pathogens.* G. W. Bruehl, ed. Am. Phytopathol. Soc., St. Paul, Minn.

Cook, R. J. and K. F. Baker. 1983. *The Nature and Practice of Biological Control of Plant Pathogens.* Am. Phytopathol. Soc., St. Paul, Minn.

Egley, G. H. and J. E. Dale. 1970. Ethylene, 2-chloroethylphosphonic acid and witchweed germination. *Weed Sci.* **18**:586-589.

Farley, J. D., S. Wilhelm, and W. C. Snyder. 1971. Repeated germination and sporulation of microsclerotia of *Verticillium albo-atrum* in soil. *Phytopathology* **61**:260-264.

Gilbert, R. G. and G. E. Griebel. 1969. The influence of volatile substances from alfalfa on *Verticillium dahliae* in soil. *Phytopathology* **59**:1400-1403.

Graham, V. E. and L. Greenberg. 1939. The effect of salicyclic aldehyde on the infection of wheat by *Pythium arrhenomanes* Drechsler, and the destruction of the aldehyde by *Actinomyces erythropolis* and *Penicillium* sp. *Can. J. Res.* **17**:52-56.

Griebel, G. E. and L. D. Owens. 1972. Nature of the transient activation of soil microorganisms by ethanol or acetaldehyde. *Soil Biol. Biochem.* **4**:1-8.

Krupa, S. and N. Fries. 1971. Studies on ectomycorrhizae of pine. I Production of volatile organic compounds. *Can. J. Bot.* **49**:1425-1431.

Krupa, S. and J. Nylund. 1972. Studies on ectomycorrhizae of pine. III. Growth inhibition of two root pathogenic fungi by organic volatile organic constituents of ectomycorrhizae root systems of *Pinus sylvestris* L. *Eur. J. For. Pathol.* **2**:88-94.

Lewis, J. A. And R. D. Lumsden. 1984. Reduction of pre-emergence damping off of peas caused by *Pythium ultimum* with calcium oxide. *Can. J. Plant Pathol.* **6**:227–232.

Lewis, J. A. and G. C. Papvizas. 1974. Effect of volatiles from decomposing plant tissues on pigmentation, growth, and survival of *Rhizoctonia solani*. *Soil Sci.* **118**:156–163.

Linderman, R. G. and R. G. Gilbert. 1969. Stimulation of *Sclerotium rolfsii* in soil by volatile components of alfalfa hay. *Phytopathology* **59**:1366–1372.

Melin, E. and S. Krupa. 1971. Studies on ectomycorrhizae of pine. II. Growth inhibition of mycorrizal fungi by volatile organic constituents of *Pinus sylvestris* (Scots Pine) roots. *Physiol. Plant* **25**:337–340.

Merriman, P. R., S. Isaacs, R. R. MacGregor and G. B. Towers. 1980. Control of white rot on dry bulb onions with artificial onion oil. *Ann. Appl. Biol.* **96**:163–168.

Okazaki, H. 1985. Volatile(s) from glucose-amended flooded soil influencing survival of *Fusarium oxysporum* f. sp. *raphani*. *Ann. Phytopath. Soc. Japan* **51**:264–271.

Punja, Z. K. and R. G. Grogan. 1981. Mycelial growth and infection without a food base by eruptively germinating sclerotia of *Sclerotium rolfsii*. *Phytopathology* **71**:1099–1103.

Rands, R. D. and E. Dopp. 1938. Influence of certain harmful soil constituents on severity of Pythium root rot of sugarcane. *J. Agr. Res.* **57**:53–67.

Robinson, P. M. and D. Park. 1966. Volatile inhibitors of spore germination produced by fungi. *Trans. Brit. Mycol. Soc.* **49**:639–649.

Smith, K. A. 1978. Ineffectiveness of ethylene as a regulator of soil microbial activity. *Soil Biol. Biochem.* **10**:269–272.

Chapter 10

Antibiotics

Many of us owe our lives to the proper use of antibiotics. They are powerful biologic agents of long use in human, animal, and plant pathology, yet their importance within the soil is far from understood. Some workers question whether they play a significant role. To me, the safest assumption is that they are important to those organisms that produce them, even though real evidence to support their importance in nature is scant. Organisms that produce strong antibiotics are most commonly found in soil. Many are slow-growing, obscure organisms of seemingly minor importance. Actinomycetes, true bacteria, and fungi, particularly *Penicillium* and *Cephalosporium* spp., are among the best-known producers of antibiotics. Most plant pathogens either do not produce them or produce them in small quantities. Antibiotics are active at low concentration. "Staling" products, changes in substrate pH, or other such effects that may reduce the growth of other organisms are not considered antibiotics.

When antibiotics were first appreciated, several biologists assumed they were of major importance in the soil because most organisms that produced powerful antibiotics were soil inhabitants. Efforts to demonstrate the significance of antibiotics in bulk soil failed. Antibiotics could not be extracted in significant quantities from soil unless it had been greatly enriched with fresh organic matter. When antibiotics were added to soil, their concentrations quickly decreased to undetectable levels. Siminoff and Gottlieb (1951) found that streptomycin was adsorbed to clay and organic matter and Martin and Gottlieb (1952) that illite and bentonite adsorbed terramycin and aureomycin. Gottlieb and others concluded that antibiotics were of little importance in natural soil. They either were not produced in sufficient quantity, were degraded chemically, or were physically adsorbed or inactivated.

Wright (1956a,b) presented an opposing viewpoint on the importance of antibiotics in soil. Instead of working with the soil in bulk, bits of straw inoculated with *Trichoderma viride* were incubated in soil and extracted for antibiotics. Gliotoxin was present in greater concentration in the straw than in the surrounding soil. Wright also inoculated viable seeds with spores of fungi that produced antibiotics, then sowed them in natural soil, and on recovery found frequentin, gladiolic acid, and gliotoxin in the seed coats. Wright shifted our focus from the bulk soil to microsites. Microorganisms are most active in and on food sources and on bits of undigested organic matter, rather than in bulk soil.

Brian (1957) reviewed the literature on the fate of antibiotics in soil, and the following concepts were taken from his paper. Basic antibiotics are strongly adsorbed on clays and organic materials, but neutral or acidic antibiotics are not. Antibiotics, even when adsorbed, may influence the soil microflora. Heat-labile antibiotics are usually degraded quickly in soil. A high proportion of organisms isolated from organic debris in soil produce antibiotics. Some pathogens produce antibiotics in host tissue before death of the host, and in some cases the antibiotics are phytotoxic. In general, antibiotic producers tend to be tolerant to antibiotics.

Brian asked, Why have so many root parasites failed to develop tolerance to antibiotics? In my opinion this is logical. Most root parasites increase while within the host and most are dormant within the soil, resuming activity when fungistasis is overcome by exudates. With little or no true saprophytic activity within soil, susceptibility to antibiotics would be of little consequence. If antibiotics produced by plant pathogens were highly phytotoxic, the host tissue would be killed quickly, before the pathogen reproduced to high levels, or a hypersensitive host reaction could quickly curtail pathogenic activity.

Sustained life depends on recycled elements, and the rapid decomposition of organic matter in soil is evidence that the natural system works. If strong antibiotic producers prevail in nature, the recycling of nutrients would be less effective. Nature has not favored organisms that effectively prevent other organisms from completing the cycling of nutrients. Antibiotic producers are a small part of the soil microflora.

CONDITIONS FAVORING ANTIBIOTIC PRODUCTION

Antibiotics are formed by most organisms after they have attained full vegetative development or as conditions become less favorable for further development. Some workers think of antibiotics as waste products that happen to be toxic to other organisms. It is probable that the excretion of secondary metabolites is more fundamental than just an excretion process, that production of secondary metabolites permit an alternative metabolic pathway, prolonging activity after it otherwise would cease. From the standpoint of survival during adversity, the coincidence of increased antibiotic production with stress would be a positive adaptation. When the producer is weak, it becomes most harmful to susceptible organisms in its vicinity.

Isolates of *Cephalosporium gramineum* that produce the highest amounts of antibiotic tend to grow more slowly than isolates that produce little antibiotic (Bruehl et al., 1969). The quantity of antibiotic produced is so small it surely does not tax the food supply. It is more probable that the metabolic pathways favoring antibiotic production are less efficient, resulting in less growth compensated for by the advantage of antagonism toward competitors.

C. gramineum grows well at high water potentials, decreasing to almost no growth at −98 bars (Bruehl et al., 1972). The greatest inhibition zones are in wet media, but the width of inhibition zones declines less rapidly than growth by *C. gramineum*. When the ratios of the diameter of the inhibition zone and of the *C. gramineum* colony are calculated, the ratios of inhibition relative to growth are greatest between about −27 bars and −60 bars, all well below the optimum water potential for growth of *C. gramineum*. Relative antibiotic production is greatest when *C. gramineum* is still growing but is under moderate-to-severe stress (Fig. 10.1).

Wong and Griffin (1974) studied the effect of water stress on the production of antibiotic by a *Streptomyces* sp. Growth of *Streptomyces* declines linearly with decreasing water potential, down to −100 to −150

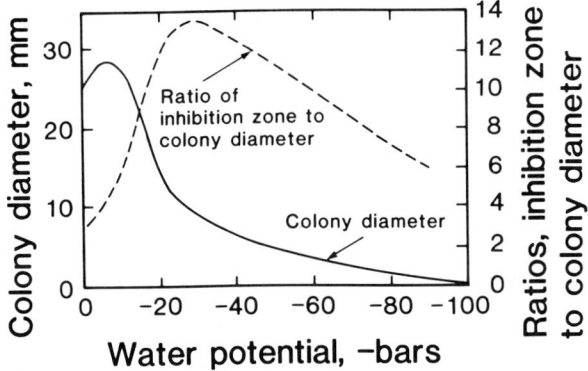

Figure 10.1. *Cephalosporium gramineum* grows most rapidly on high osmotic water potential medium, but production of antibiotics relative to mycelial growth is greater, judged by inhibition of *Rhodotorula rubra*, when it encounters water stress. After Bruehl et al. 1972.

bars. The culture filtrate of cultures grown at -20 to -30 bars is more inhibitory to *Pythium ultimum* than filtrates from other water potential ranges. *P. ultimum* is tolerant at high water potentials and increasingly sensitive as water potentials are reduced with salt mixtures. When water potential is reduced with sucrose, *P. ultimum* is highly tolerant. The results of Wong and Griffin not only confirm the production of antibiotic in greater concentration when the producer experiences stress, but they also illustrate the ability of *P. ultimum* to tolerate the antibiotic under conditions of high water potential and abundant energy. ". . .inhibitory substances have their greatest effect when the energy of growth is small,. . . an increase in the amounts of available nutrients may enable an organism to overcome a positive inhibitor" (W. Brown, 1922, in Park 1960) (recall similar concepts in reference to fungistasis, Chapter 6).

ANTIBIOTICS IN BULK SOIL

Pythium root rot of sugar cane in Louisiana is severe on certain soils and of minimal constraint to yields on other soils. Cooper and Chilton (1950) found the rhizosphere and rhizoplane of sugar cane roots from soils with the least root rot richest in actinomycetes antagonistic to *P. arrhenomanes*. They divised an *antibiotic value* based on the average inhibition zones in vitro and total numbers of antagonistic actinomycetes isolated. The antibiotic values are highest in the most disease-free soils. The most acid soils (pH 5.25) have the lowest antibiotic value. Antagonism by individual isolates from the bulk soil is slightly higher than that of those from the rhizosphere and rhizoplane, but total number of antagonists is much higher on or near the roots. Cooper and Chilton did not claim that higher yields and less root rot resulted from antibiosis.

Johnson (1954) inoculated sterilized soil with both the pathogen and antagonists, planted corn in the soil 1 week later, and measured the resulting degree of disease control produced in the presence of each antagonist. In this manner 58 fungi (mostly *Penicillium, Spicaria* and *Aspergillus* spp.), 55 bacteria, and 58 actinomycetes were tested. Greatest disease control is achieved by actinomycetes, and the correlation between antagonism in vitro and disease reduction is high. True bacteria and the fungi are relatively ineffective as antagonists of *P. arrhenomanes*. Most of the actinomycetes in the Louisiana sugar cane soils are *Streptomyces* spp. The ability to produce antibiotics does not guarantee disease reduction, but it contributes to it. Both *Pythium* and the antagonists grow in sterile soil. Johnson concluded that antibiotics are produced in the soil; that they are not immediately adsorbed, inactivated, or decomposed within the soil; and that the antibiotic effects are of some duration.

This experiment, using sterile soil, does not prove that actinomycetes that produce antibiotics are responsible for healthier roots in one soil than in another, but it would be hard to deny that they play some role in competitive relationships. The growth of several of the ineffective fungi in sterile soil surely exceeded that of the actinomycetes, so exhaustion of nutrients or benign competition is not as effective as antibiotic production in suppressing *P. arrhenomanes*.

ANTIBIOTICS IN THE RHIZOPLANE

Suppression of take-all of wheat in some soils has been associated with *Pseudomonas* spp. on the rhizoplane by some workers. *Gaeumannomyces graminis* advances along the root surface by means of dark runner hyphae prior to penetration. The runner hyphae, external to the host, are fully exposed to possible competition or antagonism from rhizoplane organisms. *G. graminis* is favored by neutrality or alkalinity, and the rhizosphere-rhizoplane is made more acid by $(NH_4)_2SO_4$ and more alkaline by $Ca(NO_3)_2$ uptake.

Smiley (1979) studied *Pseudomonas* spp. obtained from the rhizoplane of wheat roots growing in take-all suppressive and conducive soil, with wheat fertilized with $(NH_4)_2SO_4$ or with $Ca(NO_3)_2$. Bacteria antagonistic to *G. graminis* in culture are abundant in the suppressive soil and rare in the nonsuppressive soil. The higher the antagonism ratings in vitro, the lower the disease severity in the suppressive soil. The two fertilizers do not influence the total numbers of bacteria isolated from rhizoplanes, but more antagonists are isolated when the soil is fertilized with $(NH_4)_2SO_4$ than with $Ca(NO_3)_2$, and more antagonists are isolated on media at pH 5.5 than at pH 7.0. Acidity in some way favors these antagonists. D. Weller (USDA-ARS, Pullman, Wash., personal communication) tested many bacteria as seed treatments (see Chapter 16), and most of those effective in reducing take-all produced antibiotic. Evidence is accumulating to support a role of antibiotics in the ecology of *G. graminis*.

STALING, ANTIBIOTICS, AND FUNGISTASIS

When microorganisms grow on ordinary culture media, the media are generally rendered less favorable for growth of other organisms. When reduction in growth is relatively slight, in contrast to marked or complete reduction in growth as caused by antibiotics, the medium is said to be staled. How much staling is due to reduction of nutrients within the medium and to accumulation of miscellaneous materials deleterious to other organisms is not known. Organisms change the substrate on which they grow by partial depletion of nutrients and by production of waste products. Park (1960) suggested that staling in cultures has some features in common with fungistasis in natural soil; i.e., that low levels of deleterious metabolites restrict development. Fungistasis may be caused by the presence in soil of many deleterious substances, each in low concentration and each, if alone, would not be markedly fungistatic. Together, however, these substances may restrict development.

Griffin (1962) attempted to determine whether production of antibiotics is essential to fungistasis. He tested several fungi, *Streptomyces griseus, Bacillus subtilis,* and *B. megaterium,* for the production of antibiotics effective against *Pythium ultimum* and *Gliocladium fimbriatum* in vitro, then inoculated sterile soil with these organisms. Griffin then assayed these soils for their ability to prevent germination (fungistasis) of conidia of *G. fimbriatum.* He found little correlation between antibiotic production in vitro and strength of fungistasis in soil. Fungistasis is weakest following inoculation with *Bacillus subtilis, B. megaterium* and *R. solani,* but all organisms produced some fungistatic effect. Griffin concluded that antibiotics are not essential to fungistasis and that mildly toxic fungal and bacterial metabolites, such as staling products in laboratory cultures, are involved.

The degree of staling acts in a selective manner. Dwivedi and Garrett (1968) inoculated the surface of an agar medium with air-dried soil. After 0, 1, 2, 3, and 4 days of staling the agar was inverted, and bits of air-dried soil were placed at four spots on the exposed agar surface. On the 0-time dishes four species of fungi emerged; on 1-day dishes *Trichoderma koningi* was identified; on 2-day dishes, *Fusarium culmorum* and two other fungi; on 3-day agar, *Trichoderma harzianum* and *Penicillium janthinellum;* and on 4-day agar, three species of *Penicillium* and *Trichoderma viride* emerged. Competitive relationships were altered by the degree of staling prior to the second inoculation. Because staling is such a common phenomenon, it is tempting to assume that it must occur in soil as well as in laboratory media and that it contributes to fungistasis and to competitive relations among organisms capable of saprophytic growth. Staling is discussed briefly by Griffin (1981).

TOLERANCE IN PLANT PATHOGENIC BACTERIA

Patrick (1954) studied 28 bacterial plant pathogens and potential antagonists isolated from nine soils. In general, *Xanthomonas* species are most sensitive to antibiotics, *Pseudomonas* spp. are intermediate, and *Erwinia* spp. are most tolerant to miscellaneous antibiotics. Although *E. amylovora, E. carotovora,* and *E. stewartii* are relatively tolerant to antibiotics, *E. tracheiphila* is sensitive to many. *E. tracheiphila* overwinters in the bodies of adult cucumber beetles, being completely dependent on the insect vector for survival (Walker, 1950). Patrick's study presented little evidence that tolerance to antibiotics plays a significant role in survival of bacterial plant pathogens in soil. See *Pseudomonas solanacearum,* Chap. 14.

BACTERIOCINS

Antibiotics produced by one organism are generally toxic to unrelated organisms. The antibiotic penicillin, produced by a fungus, inhibits growth of many bacteria. In contrast, bacteriocins are toxic to closely related organisms, usually members of the same species (Vidaver, 1976). The producer is always resistant. In biologic control, bacteriocins are so selective they attack the target pathogen with essentially no effect on nontarget organisms.

The first commercial application of organisms producing bacteriocins was the use of *Agrobacterium tumefaciens* isolate 84 for the control of crown gall. United States bacteriologists refer to *A. tumefaciens* as the cause of crown gall and to *A. radiobacter* as a nonpathogenic, closely related, soil bacterium. Pathogenicity is governed by a tumor-inducing plasmid (Ti plasmid) transferrable from the pathogen to the nonpathogen. On infection of the plant host, part of the genome of the Ti plasmid is incorporated into the genome of host cells, rendering them tumerogenic. Kerr (1980) selected a nonpathogenic strain (no. 84) of *A. tumefaciens* capable of growth in host wounds that produced a bacteriocin that killed the pathogen. Susceptibility to the bacteriocin is governed by the Ti plasmid. Susceptibility to the bacteriocin and virulence are essentially linked, with only pathogenic cells destroyed.

In Kerr's initial experiments peach seeds were dipped in a suspension of living strain 84, seeded in pathogen-infested soil, and the resulting seedlings were root-dipped in a suspension of strain 84 and transplanted. Control of crown gall was nearly 100%. This method is now widely used in many countries. As might be expected, all pathogenic strains of *A. tumefaciens* are not susceptible, and pathogenic bacteria that produce a bacteriocin that attacks strain 84 have been found. But the potential use of such bacteria as protectants is great. The control agent is adapted to the same environment as the pathogen, making it ideal for use wherever the pathogen is found.

COMMENTS

Studies of saprophytic survival, rhizosphere organisms, and microbialization provide evidence that antibiotics, staling, and bacteriocins affect the lives of soil organisms. A general staling may contribute to fungistasis, but it is unlikely that antibiotics do.

Weller's observation (see "Antibiotics in the Rhizoplane," this chapter) that antibiotic production is important in the action of *Pseudomonas fluorescens* applied to wheat seed as a protectant against *Gaeumannomyces Graminis* var. *tritici* has been confirmed. The antibiotic is a dimer of phenazine-1-carboxylic acid (Gurusiddaiah et. al, 1986). It is highly active against *G. graminis* var. *tritici, Pythium aristosporum, P. heterothallicum, P. volutum,* and *Rhizoctonia solani.* Linda Thomashow (Root Disease and Biological Control Research Unit, Pullman, Washington 99164) removed the gene for antibiotic production from *P. fluorescens* strain 2-79 by genetic engineering, and it was no longer effective as a protectant.

This chapter completes the phase of microbial biology concerning the influence of organic compounds in the soil solution or in the soil atmosphere (host exudates, volatiles, and antibiotics) and provides a logical transition to the next general subject, survival of plant pathogens in infested residues. Chemicals are important to soil organisms, not only as food but as "intelligence." Nematodes and zoospores exhibit remarkable ability to respond to chemical attractants or repellents. Chemical stimuli are equally important to propagules stationary within the soil. Not germinating or hatching (in the case of nematode eggs) in the absence of potential food is essential to survival.

The chapter on exudates neglected nematodes, but a clear, readily available paper on chemotaxis among nematodes is that of Viglierchio (1961). Zoospore chemotaxis is widely treated, but the paper of Hickman (1970) is especially recommended. The review by Vidaver (1976) is recommended for bacteriocins.

KEY REFERENCES

Kerr, A. 1980. Biological control of crown gall through production of agrocin 84. *Plant Dis.* **64**:15–30.

Smiley, R. W. 1979. Wheat-rhizoplane pseudomonads as antagonists of *Gaeumannomyces graminis*. *Soil Biol. Biochem.* **11**:371–376.

Wong, P. T. W. and D. M. Griffin. 1974. Effect of osmotic potential on Streptomycete growth, antibiotic production and antagonism to fungi. *Soil Biol. Biochem.* **6**:319–325.

REFERENCES

Brian, P. W. 1957. The ecological significance of antibiotic production. Pages 168–188 in *Microbial Ecology*. R. E. O. Williams and C. C. Spicer, eds. Symposium Soc. Gen. Microbial, Cambridge Univ. Press, London.

Bruehl, G. W., B. Cunfer and M. Toivianen. 1972. Influence of water potential on growth, antibiotic production, and survival of *Cephalosporium gramineum*. *Can. J. Plant Sci.* **52**:417–423.

Bruehl, G. W., R. L. Millar and B. Cunfer. 1969. Significance of antibiotic production by *Cephalosporium gramineum* to its saprophytic survival. *Can. J. Plant Sci.* **49**:235–246.

Cooper, W. E. and S. J. P. Chilton. 1950. Studies on antibiotic soil organisms. I. Actinomycetes antibiotic to *Pythium arrhenomanes* in sugarcane soils of Louisiana. *Phytopathology* **40**:544–552.

Dwivedi, R. S. and S. D. Garrett. 1968. Fungal competition in agar plate colonization from soil inocula. *Trans. Brit. Mycol. Soc.* **51**:95–101.

Griffin, D. H. 1981. *Fungal Physiology*. Wiley, New York.

Griffin, G. J. 1962. Production of a fungistatic effect by soil microflora in autoclaved soil. *Phytopathology* **52**:90–91.

Gurusiddaiah, S., D. M. Weller, A. Sarkar, and R. J. Cook. 1986. Characterization of an antibiotic produced by a strain of *Pseudomonas fluorescens* inhibitory to

Gaeumannomyces graminis var. *tritici* and *Pythium* spp. *Antimicrob. Agents and Chemother.* **29**:488-495.

Hickman, C. J. 1970. Biology of *Phytophthora* zoospores. *Phytopathology* **60**:1128-1135.

Johnson, L. F. 1954. Antibiosis in relation to pythium root rot of sugarcane and corn. *Phytopathology* **44**:69-73.

Martin, N. and D. Gottlieb. 1952. The production and role of antibiotics in the soil. III. Terramycin and aureomycin. *Phytopathology* **42**:294-296.

Park, D. 1960. Antagonism - the background of soil fungi. Pages 148-159 in *The Ecology of Soil Fungi.* D. Parkinson and J. S. Waid, eds. Liverpool Univ. Press, Liverpool, UK.

Patrick, Z. A. 1954. The antibiotic activity of soil microorganisms as related to bacterial plant pathogens. *Can. J. Bot.* **32**:705-735.

Siminoff, P. and D. Gottlieb. 1951. The production and role of antibiotics in the soil. I. The fate of streptomycin. *Phytopathology* **41**:420-430.

Vidaver, A. K. 1976. Prospects for control of phytopathogenic bacteria by bacteriophages and bacteriocins. *Ann. Rev. Phytopathol.* **14**:451-465.

Viglierchio. D. R. 1961. Attraction of parasitic nematodes by plant root emanations. *Phytopathology* **51**:136-142.

Walker, J. C. 1950. *Plant Pathology.* McGraw-Hill, New York.

Wright, J. M. 1956a. The production of antibiotics in soil. III. Production of gliotoxin in wheat straw buried in soil. *Ann. Appl. Biol.* **44**:461-466.

Wright, J. M. 1956b. The production of antibiotics in soil. IV. Production of antibiotics in coats of seeds sown in soil. *Ann. Appl. Biol.* **44**:561-566.

Chapter 11

Saprophytic Survival

Parasites have their greatest advantage while within the living host, where they increase with little competition from other microorganisms. Nematodes, most phytopathogenic bacteria, and many parasitic fungi obtain essentially all elaborated substance from the living host, with saprophytism being of little or no importance. Obligate parasites obtain all their nutrition from the living host. The only useful use of the term obligate is in reference to conditions in nature. *Puccinia graminis* and other obligate parasites may eventually all be grown in culture, as can smut fungi, but if they obtain little or no food saprophytically in nature, they are obligate parasites. As knowledge and techniques progress, more and more obligate parasites will be cultured, but this should not diminish the concept of obligate parasitism.

Bacillus cereus, the cause of frenching of tobacco, and *Periconia circinata,* the cause of milo disease of sorghum, are primarily saprophytes. Pathogens of this type are not considered here. *Erwinia carotovora* is a saprophyte at the cellular or tissue level, because it multiplies between dead cells, and the cells die in advance of the bacteria. It is parasitic at the organismal level, because the dead cells or tissues, at least for a while, are part of a living host. Some types of worms in the stomach or intestines of a mammal are considered parasites even though they may live on contents of the digestive tract without directly attacking tissues or sucking blood. In the same way, microconidia of wilt fusaria may grow for awhile within xylem vessels, or dead host cells, and yet we consider this part of the parasitic phase. It occurs within a living organism.

There are two main ways by which organisms within soil possess or control their substrates. One, called *dormant survival* by most biologists, is the storage of nutrients in dormant structures (eggs, oospores, sclerotia, etc.). This mode of survival is passive possession of the substrate, because the substance obtained from the living host is possessed with little or no expenditure of energy while the pathogen is dormant. The other main method is called *saprophytic survival,* survival by active metabolism within colonized organic debris. This type of survival is active possession of substrate, because the possessor must maintain a critical level of metabolism to remain alive and in control of the substrate.

Most soilborne parasites of herbaceous, short-lived plants survive passively in soil by means of dormant structures. Many soilborne fungi that attack trees survive as active hyphae in woody tissues, but even among them a degree of passive or dormant survival occurs. Garrett (1970, pp. 164–174) described the survival of fungi in wood. Some of the basidiomycetes form zone-lines (darkened borders) around portions of wood occupied by the fungus in question. Campbell in 1933 and 1934 likened the dark zone to the rind of a sclerotium and called the tissues within the zone-lines pseudosclerotia. Readers interested in the survival of fungi in woody tissues are advised to read Garrett's lucid narrative.

PRIOR COLONIZATION OF SUBSTRATE

Prior colonization refers to the parasitic colonization of a host, establishment of the parasite in the tissues before the host dies. Being well established in the tissues through parasitism, essentially in pure culture, is comparable to establishing wheat in clean soil and removing all weeds until the wheat is well developed. It then has strong possession of the substrate and competes well with would-be colonists (weeds). The importance of being the first to colonize the substrate is evident in the recovery of *Fusarium culmorum* and *Penicillium* sp. from inoculated straw in practically pure culture after several weeks in soil (Walker, 1941). Barton (1960) established species of *Pythium* in little wooden blocks impregnated with sugar prior to burial in soil. *P. mamillatum,* an aggressive competitor, did not colonize the blocks already colonized by another *Pythium* spp. Bruehl and Lai (1966) established several fungi in sterilized wheat straws and buried them in fertile, moist silt loam. After 6 to 13 weeks at 15°C the straws were assayed for fungal content, and the only fungus emerging from 208 pieces of straw inoculated with *Trichoderma viride* was *T. viride*. It maintained complete possession of the substrate during the test period. *T. album* emerged from 194 of 208 inoculated straws, and *Fusarium culmorum* from 185 of 208 straws. Possession is nine-tenths of the law (Table 11.1).

Table 11.1. Recovery of fungi previously established in 208 straw segments and of other fungi after burial in moist soil at 15°C for 6–13 weeks

Initial colonists	Recovery of	
	Test fungus	Other fungi
Trichoderma viride	208	0
Penicillium sp.	208	10
Trichoderma album	206	14
Fusarium culmorum	208	23
Cephalosporium gramineum	208	46
Penicillium aculeatum	196	79
P. funiculosum	193	115
Fusarium nivale	178	156
None	0	208

Source: Bruehl and Lai, 1966.

Establishment within the substrate in competitive relationships is important, and possession of substrate is a result of initial, overwhelming inoculum potential (Garrett, 1956). The ability of *Pseudocercosporella herpotrichoides* to possess straw is "through prior and continuing utilization of free nutrients" (Macer, 1961). Several factors probably contribute to maintenance of saprophytic possession of substrate. If the substrate is occupied but not nutritionally exhausted, the occupying fungus has a great advantage. Being well established enables it to compete better.

The attributes of an ideal active possessor of organic substrates (Bruehl, 1975) are

1. To gain entry as a parasite ahead of would-be saprophytic competitors and become adequately established throughout the tissues (Fig. 11.1).
2. To maintain minimal metabolism to conserve the substrate, produce no excess hydrolysis of reserve nutrients, and allow no leakage of sugars and amino acids into the soil solution.
3. To produce a minimum but adequate, constant level of staling products or antibiotics to make the substrate unattractive and to counter the attraction by any nutrients that may escape into the surrounding environment.
4. To be both flexible and stable in relation to environmental factors, Because its survival depends upon maintenance of a certain minimal metabolism for a sustained period of time, it should not exhaust its substrate quickly under favorable conditions and it must maintain some activity under relatively adverse environmental conditions. The latter is impossible, and is the reason why fungi dependent upon this type of survival are vulnerable to the fluctuating conditions of surface soils, especially in temperate climates. Under warm, dry conditions *Penicillium* spp. are strong

competitors for substrates, and under more humid, acid conditions *Trichoderma* spp. are among the strongest of competitors.

Some relatively successful active possessors have the first three characteristics. All fail in the fourth, theoretical requirement for active possession of substrate.

Active possession of substrate is also influenced by the physical nature of the substrate. A leaf decomposes rapidly in or on the surface of moist soil. A log or tree root is a substantial substrate that, if dominated (possessed) by the pathogen, may sustain the organism for long periods of time. The significant portion of substrate is that which is subject to digestion by the enzymes produced by the pathogen. When digestible elements are exhausted, the substrate will no longer sustain the pathogen, and the normal succession of organisms will proceed.

SURVIVAL AIDED BY ANTIBIOTIC PRODUCTION

Among the pioneer discoveries of Nisikado et al. (1934) on Cephalosporium stripe of wheat are that the pathogen becomes systemic in the vascular system and that conidia are produced in xylem vessels of growing wheat. It is aerobic. If no stubbles or straw are present, little fungus survives through the summer in soil under paddy rice. Because the fungus is soilborne, they recommended crop rotation (rice in the summer, nongrass plants in the winter) and pulling the diseased plants and burning them for fuel. The specific

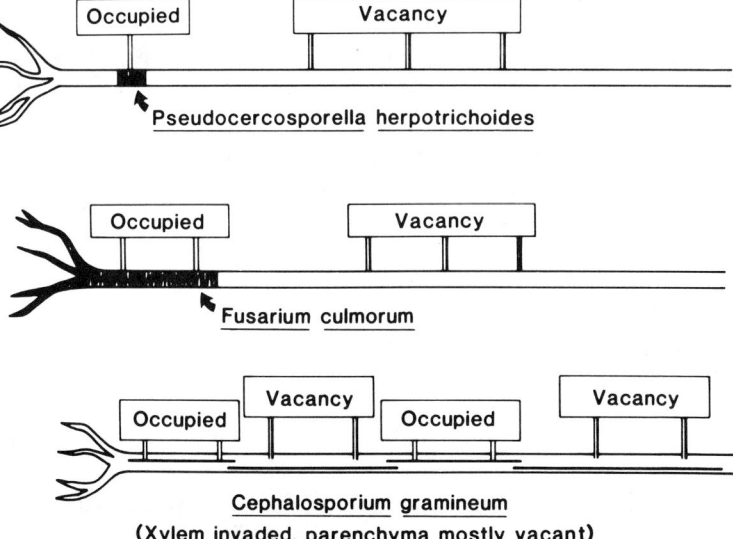

Figure 11.1. The amount of substrate occupied in the parasitic phase influences the amount of host tissue available to pathogenic fungi in the saprophytic phase. Most parasitic fungi compete poorly for dead substrate.

epithet, *gramineum*, indicates that the fungus attacks several grasses. Ellis and Everhart in 1894 described what Bruehl (1963) believed to be the sporodochial stage, *Hymenula cerealis*, of the stripe fungus on wheat straw from West Virginia. Light is not required for formation of the *Hymenula* stage because it is abundant on straws buried in soil as well as on the soil surface (Bruehl and Lai, 1966). The *Hymenula* stage develops on the surface of dead stubbles and straw, particularly near nodes, in cool, wet periods of late autumn (November) into early spring. The *Cephalosporium* stage is primarily restricted, while the plant is alive, to the xylem in wheat stems. Hyphae ramify the surrounding parenchyma when the host declines, or shortly thereafter. The extent of parenchyma invasion is probably highly variable but important because it determines the amount of host tissue "possessed" or controlled by the pathogen and accessible for its saprophytic survival.

Wiese and Ravenscroft (1975) studied the numbers of propagules in soil in Michigan after the harvest of diseased crops. The population begins to increase in September-October, is at its highest, sometimes over 100,000 propagules/gram of moist soil, from November through February or March, and then declines rapidly. Many of these propagules are conidia from the *Hymenula* stage. The sharp decline in the spring indicates that the conidia are relatively short-lived in soil. The half-life of conidia in soil at 23°C is 0.5 to 2.5 weeks, whether the soil is moist or dry. At 7°C conidia survive longer in moist than in dry soil, with a half-life near 12 to 14 weeks in moist soil. These studies prove that a new crop of spores must be produced each season if the disease is to develop significantly and, circumstantially, that *C. gramineum* lacks an effective long-lived resting propagule. No one has found a special resting stage for this fungus. In Michigan the fungus can survive in straw on the soil surface for at least 3 years, but it is mostly destroyed within 1 year if the refuse is plowed under. The drier summers of Oregon, Washington, and Idaho lead to slower decomposition of straw, and more than 1 year is required for its decomposition, even when plowed under.

All isolates of *C. gramineum* from nature produce antibiotics (Bruehl et al., 1969). Isolates that produce relatively little antibiotic can be obtained by single-sporing cultures that have been maintained for long periods on potato dextrose agar. We assume that such mutants occur in nature but that they are soon eliminated. Antibiotic-producing isolates survive longer and more vigorously than nonproducers in our experiments, and the producers reduce invasion of infested straws by other fungi. The antibiotic is most effective in acid media and survival in and possession of the substrate was greatest in acid soils. These observations indicate that the antibiotics of *C. gramineum* are important to its saprophytic survival.

Kobayashi and Ui (1979) isolated Graminin A from cultures of *C. gramineum*. Graminin A reduces the germination of spores of several fungi and the growth rate of *Bacillus subtilis*, but it had no effect on *Pseudomonas fluorescens*. Graminin A is toxic to wheat, and Kobayashi and Ui consider it important in pathogenesis.

Influence of pH

If the antibiotics of *C. gramineum* are active in vitro only when the medium is acid (Bruehl et al., 1964; Hopp, 1972) and if antibiotics are important in defense of the substrate, *C. gramineum* should survive longer and stronger in straws in acid than in basic soils. Survival after 11 months in artificially infested straw is greatest in moist silt loam between pH 3.9 and 5.5 and very poor or absent at pH 8.4 to 9.3. The soil pH becomes more important with the passage of time.

Hopp (1972) buried straws infested with isolates that produced abundant antibiotic (producers) and some that produced little (nonproducers) in moist silt loam adjusted to pH's from 4.5 to 8.3 and incubated them for 5 months at 15°C. At 5 months the producer isolates survived well in acid soil but had diminished at pH 8.3 to the same level as nonproducer isolates. Hopp considered this evidence that antibiotics aid *C. gramineum* surviving saprophytically. Liming an acid soil in Kansas greatly reduced the incidence of *Cephalosporium* stripe in the field (Bockus and Claasen, 1985).

Mathre and Johnston (1979) buried naturally infested straws of three wheat cultivars in three soils with pH values of 5.3, 6.4, and 7.5 and incubated them at 9°C. The wheats were infected with isolates differing in virulence and/or antibiotic production. After 12 months all straws in all soils had decomposed at essentially equal rates, and there was little difference in survival of *C. gramineum* in them. At 24 months straw decomposition was greatest in the alkaline soil, but the data indicate this difference was not due to protection of the straw from vigorous saprophytes in the acid soil by *C. gramineum*. No evidence of a role for antibiotics in survival was found.

Mathre and Johnston (1979) believed the use of autoclaved, artificially infested straw by Bruehl and his associates led to an unrealistic degree of pathogen establishment in the substrate and that this led to misleading results. Murray and Bruehl (1983) agreed that substrate artificially colonized leads to exaggerated and possibly distorted results and that experiments using such materials are imperfect.

Composition of the Substrate

Depending on how early the onset of disease occurs, normal metabolism and translocation within the host are disrupted by *C. gramineum*. Foods are not polymerized into structural elements or translocated to the inflorescence as occurs during normal maturity and grain filling. Straws from diseased plants should decompose faster than those of healthy plants, because they enter the soil with higher concentrations of easily digested foods. Straw of wheat killed or severely affected by Cephalosporium stripe is higher in nitrogen than straw of healthy wheat and is richer in soluble foods (carbohydrates, starch, hemicellulose, proteins) (Murray and Bruehl, 1983). The pH of all parts of soybean straw, with the exception of the pods, is raised by brown stem (*Phialophora gregatum*) (Lai and Dunleavy 1969). Studies comparing biological reactions of residues of diseased vs. healthy plants should consider that the substrates being compared are not the same. Organisms alter the physical and chemical nature of a substrate they invade.

Effect of Water

Cephalosporium gramineum grows most rapidly on corn meal agar adjusted with KC1 to water potentials of -2 to -8 bars (Fig. 10.1). Its growth then decreases rather rapidly to about -30 bars, after which growth declines gradually to near -98 bars. Antibiotic production, when judged by the ratios of inhibition zone/colony diameters, is greatest from between about -20 to -70 bars, with some production at all water potentials tested (Bruehl et al., 1972) (Fig. 10.1). Nonproducer and producer isolates in infested straw survived equally under aerobic conditions at 0 bars water potential at 15°C for 124 to 137 days, which was the duration of the test. In wet conditions the antibiotic confer no advantage for fungal survival. Antibiotic producers survived better than nonproducers between -10 and -67 bars (Fig. 11.2). At -196 and -258 bars (very dry soil), producers and nonproducers were equal. In the intermediate to moderately dry soil (-10 to -67 bars $-$, antagonism to this fungus was strong. The value of antibiotic production in this case is remarkable, because at pH 5.9-6.0 the antibiotic should have minimal effect. (The soil pH was measured in water, and its true pH was probably lower by an unknown amount). We do not know the pH of the straw itself but assume it to be near that of the surrounding soil. The vigor of *C. gramineum* in very wet soil indicates that bacteria are not major antagonists (Fig. 11.2).

Effects of Heat and Drought

C. gramineum survived best for 17 months in naturally infested straw on the soil surface at Pullman and Vancouver and poorest at Lind, Washington (Lai and Bruehl, 1966). Pullman and Vancouver are wetter and cooler sites than Lind, and possibly, heat at the surface of dry soil rather than biological antagonism destroyed the pathogen at the dry site. Dematiaceae (fungi with dark cell walls) thrived in the surface straws at all three locations. (Fungi with thick dark walls survive best in desert sands (Nicot, 1960). The relatively thin-walled, lighter cells of *C. gramineum* are ill-adapted to such exposure.) The straws were secured in a single layer to the soil surface with no

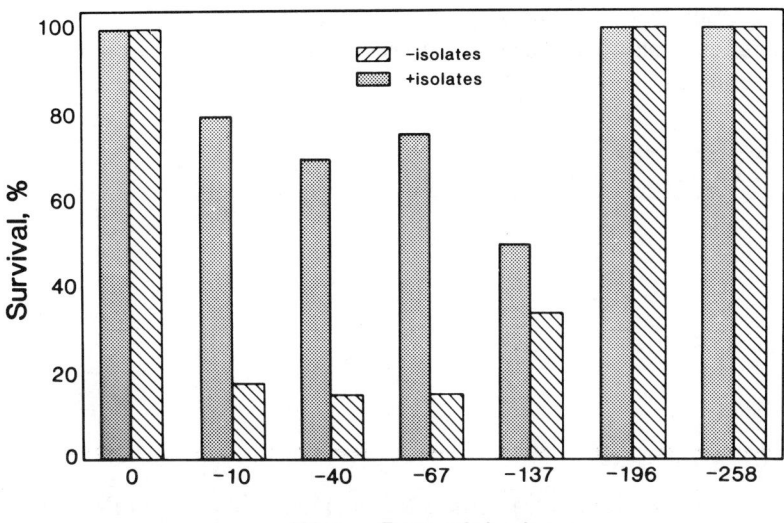

Figure 11.2. Isolates of *Cephalosporium gramineum* that produce little antibiotic (−isolates) survive as well as isolates that produce normal amounts (+ isolates) in very wet (aerobic) and in very dry conditions. In intermediate to fairly dry conditions, antibiotic-producers survive more strongly than nonproducers. After Bruehl et al., 1972.

shade so they were exposed to the extremes of temperature at the soil surface. In nature, many straws would not be appressed firmly to the soil surface and there would be some shade.

C. gramineum is most vigorous, as judged by spore production, in straw on the soil surface in Michigan (Wiese and Ravenscroft, 1975). Straw on the soil surface produces more spores and produces them longer than when buried. In northern Idaho straw on the soil surface results in the greatest amount of disease (Latin et al., 1982). Working straw into the soil, especially plowing it under, hastens its decomposition and reduces survival of *C. gramineum* (Bockus et al., 1983).

Burgess and Griffin (1968) studied the influence of temperature and moisture on the longevity of *Gibberella zeae* in wheat straw. Straw segments colonized by *G. zeae* were dried and placed on natural field soil equilibrated to 100, 87.5, 76.5, and 34% RH at 10 and 25°C. At 10°C survival of *G. zeae* was excellent at all humidities. At 25°C, *G. zeae* survived well at 100% RH and at 33% RH. *G. zeae* grew at 100% RH, but not at 87, 76, or 33% RH. At 33% RH no fungus grows. Thus the test organism maintained possession of the substrate when it could grow and at 35% RH, when it survived dormant with no competition because nothing else grew either. At 87 and 76% RH xerophytic penicillia invaded the straw (Table 11.2).

Table 11.2. Survival of *Gibberella zeae* in wheat straws incubated on natural soil at different temperatures and relative humidities

	10°C		25°C	
RH	24 weeks	40 weeks	24 weeks	40 weeks
100	100	94	100	94
87	100	100	0	0
76	100	100	6	0
33	100	97	100	85

Source: Burgess and Griffin, 1968.

EFFECT OF NITROGEN

Gaeumannomyces graminis var. *tritici* is dependent on host colonization for saprophytic survival, and is subject to control by crop rotation. Garrett (1944) reviewed the early experiments and experiences of farmers in Australia. Survival in straw is shortened in moist, well-aerated soil at moderate-to-warm temperatures, conditions favoring decomposition. Very dry or compact soil increases longevity. This simple concept is basically correct, but it does not explain the differing effects of certain soil amendments on survival. Saprophytic survival is reduced by nitrogen-poor amendments such as glucose. Ground rape (5.8% N) has no influence, and dried blood (13% N) increases saprophytic survival. Garrett expected nitrogen to hasten straw decomposition and to shorten the life of *G. graminis*. Instead, added nitrogen prolonged the life of the pathogen in the straw. Garrett concluded that *G. graminis* survives by growing actively in the substrate and that it requires nitrogen for saprophytic growth. This fungus apparently has little ability to recycle nitrogen, as do some fungi that live in wood (Table 11.3).

Table 11.3. Effect of nitrogen upon the survival of *Gaeumannomyces graminis* var. *tritici* in straw

	Straws with viable pathogen, %			
	Weeks			
Treatment	7	13	19	25
Control	92	72	20	3
+ 1.56 mg N	80	82	37	8
+ 3.125 mg N	92	76	76	44
+ 12.5 mg N	95	84	80	74

Source: Garrett, 1944.

Scott (1969a) confirmed the effect of nitrogen in prolonging the saprophytic survival of *G. graminis* var. *tritici* in straw. Scott's experiments further illustrate the complexities of saprophytic survival. When Scott added 14.1 mg of nitrogen to 100 g of straw prior to establishing *G. graminis* in the straw, the nitrogen increased the percentage of straws from which the pathogen was recovered. It also increased the rate of straw decomposition and, after 30 weeks, the total number of straws from which *G. graminis* was recovered was actually decreased. Nitrogen resulted in recovery of the pathogen from a greater percentage of straws recoverable from soil, but it increased decomposition, so that fewer straws were recoverable, resulting in a net decrease in inoculum. Scott emphasized the importance of distinguishing between death from failure to utilize the substrate (nitrogen starvation) and death from substrate exhaustion (copious growth with nitrogen enrichment). A deficiency of nitrogen reduces microbial competition for the straw, but it reduces the vigor of *G. graminis* proportionately more. Scott (1969b) stated that in England rotovating the refuse into surface soil creates good aeration and a nitrogen deficiency, conditions leading to rapid weakening of this fungus.

Garrett (1976) summarized studies at Cambridge University on the saprophytic survival of fungi in straw. If addition of nitrogen enhances survival, it is evidence that the fungus grows actively in the straw (and that it requires added nitrogen to grow). Garrett calculated S-50 values from several papers, with S-50 meaning that period of time after burial at which the test organism is recovered from 50% of the samples (50% survival values). Survival of *Fusarium culmorum* and *G. graminis* var. *tritici* is enhanced by nitrogen, survival of *Phialophora radicicola* var. *radicicola* and *Cochliobolus sativus* is decreased by added nitrogen, and survival of *Curvularia ramosa* and *Pseudocercosporella herpotrichoides* is unaffected by nitrogen added to straw.

The ability of a fungus to digest cellulose divided by growth rate in culture (Garrett's cellulolysis adequacy index) is correlated with increased longevity in straw supplemented with nitrogen. A fungus capable of rapid growth in culture and capable of digesting intermediate amounts of cellulose has greater need for nitrogen. It has a low cellulolysis adequacy index (CAI), and it responds positively to nitrogen. *G. graminis* is "exuberant" within the substrate when nitrogen is abundant, possessing the straw strongly for a period but reducing ultimate longevity by exhausting its supply of carbon.

Effect of Nitrogen on Decomposition

Several important plant pathogens persist by saprophytic survival in crop residues and die with decomposition of that residue. In general, the more immature the structure, such as green leaves, the quicker decomposition occurs. Mature residues, such as corn stalks, soybean refuse, or wheat straw, vary in chemical composition with conditions under which they were produced, and these differences affect the rate of decomposition.

The rate of decomposition of wheat straw increases as its nitrogen content increases. Douglas et al. (1980) studied straws containing 0.78, 0.49, and 0.19% nitrogen. The straws were exposed at Pendleton, Oregon, at the height of standing stubble, on the soil surface, and 15 cm below the surface. After 26 months above the soil surface, the low and high nitrogen straws lost 20 and 35% of their weight, respectively. Results on the soil surface were comparable. When buried, the high nitrogen straw lost 35% of its weight in 2 months, and the low nitrogen straw lost 10%. At 14 months the low nitrogen straw had lost 55% of the original weight, and the high nitrogen straw had lost >70%. At 26 months both were mostly (>85%) decomposed, and the original nitrogen content was no longer significant. In practical agriculture, if the straw contains normal amounts of nitrogen and normal amounts of straw are incorporated into the soil, fertilizing with nitrogen does not materially hasten its decomposition (Smith and Douglas, 1970).

SURVIVAL IN WOOD

The preceding discussion dealt with survival of fungi in wheat straw, a transient substrate of limited duration in the field, with examples of the effects of differing temperatures, pH, nitrogen and water on competitive relationships and on survival. A fungus dependent on saprophytic survival in transient substrates, if lacking a wide host range, is vulnerable to control by short rotations. The substrate itself will disappear. In addition, most infested debris is on or in the surface layers of soil, subjecting organisms within it to wide fluctuations in temperature and moisture. In contrast, a pathogen in wood is in a substantial substrate of considerable potential duration. Fox (1970), working in tropical forests where biological reactions are accelerated, reported that *Fomes lignosus* persisted 4 to 5 years, *F. noxius* 10 years, and *Ganoderma pseudoferreum* about 30 years in roots deep in the soil. Fox related longevity to duration of the substrate. Nematodes (obligate parasites) remain alive for years in grape vineyards deep in the soil after the tops have been removed because the roots retain life for long periods. Roots deep within soil are not subject to rapid and frequent fluctuations of temperature and moisture, so fungi in them are not frequently stressed. Longevity is possible without the specialized resting structures so important to organisms in surface layers of soil.

Menzies (1963), in his review of "Survival of microbial plant pathogens in soil," did not treat wood-rotting organisms in detail, because they were not exposed directly to the influences of the soil itself and because their ability to digest cellulose and lignin lessened competition. Another factor favoring longevity of some parasites of woody tissues is their ability to grow in tissues low in nitrogen. Very high C:N ratios exist in wood, and some wood-rotters have the ability to recycle nitrogen, sustaining growth when it would otherwise be impossible. Some of these fungi have special means of conserving nitrogen (Cowling, Merrill, and Levi, cited in Garrett, 1970, pp. 163–164). Another factor that favors some pathogens of trees is tolerance to tannic acids. Tannic acid, which inhibits growth of most fungi, stimulates growth of *Armillaria mellea* in culture by about 12 times (Cheo, 1982).

A factor of unknown importance in survival of parasites in substrates is the effect of commensal associations. Bacteria and yeasts associated with a wood-rotting fungus, *Coriolus (Polyporus) versicolor,* digest areas of cells parallel to the hyphae, forming troughs in the cell walls (Blanchette and Shaw, 1978). They found bacteria and yeasts only in these lysed zones, which indicates that they utilized some of the hydrolysed substances. Wood chips lose weight faster when inoculated with wood-rotting fungi plus *Enterobacter* spp., *Saccharomyces bailii,* and *Pichia pinus* (a bacterium and two yeasts, respectively) than when inoculated with the basidiomycete alone. It is possible that mutualistic relationships of this type may aid fungi in many substrates. Bacteria and yeasts may produce vitamins, growth-promoting substances, and enzymes of use to another organism in the substrate. Associations of organisms do not have to be harmful.

Placing fungi in categories based on presumed capabilities to attack specific substrates may lead to errors. Swift (1970) grew *Armillaria mellea* on sawdust of a tropical hardwood that contained 3.8% sugars and 17.8% starch. Rhizomorphs formed abundantly on unaltered sawdust and less abundantly when the sugar and part of the starch was removed. On lignin alone rhizomorph development was near zero. Sugar and starch are important to this pathogen, even though it degrades cellulose and lignin. Swift concluded that *A. mellea* was more like a "sugar" fungus than a cellulose or lignin fungus and that sugars and starch provided energy to support the attack on cellulose and lignin. Swift paired *A. mellea* with four fungi that grew well on sawdust. If *A. mellea* was given a 5-day start, it competed well when starch was present. It did less well when starch was removed. When starch was removed from the sawdust and the fungi were paired simultaneously, *A. mellea* was completely noncompetitive. These experiments of Swift on nutritional requirements of *A. mellea* explain why girdling trees in the tropics and allowing them to die prior to felling control the fungus when tea is planted immediately on cleared forests. Girdling depletes the roots of starch and sugars so that *A. mellea* does not sustain itself in old roots.

The nature of the substrate is also important. Redfern (1975) explained why *A. mellea* is more devastating to pine plantations in Britain after broad-leaved trees than after conifers. Redfern grew *A. mellea* on hard and on soft wood and used the infested wood as inoculum. Virulence and rhizomorph production were increased by the wood of broad-leafed (hard wood) trees. The wood of the broad-leaf trees was a richer energy source. Redfern cautioned, however, that even though stump-for-stump, broad-leaved trees result in greater inoculum potential, stumps and roots of conifers are adequate to sustain significant losses.

COMMENTS

Survival as a saprophyte in host tissue is precarious in insubstantial substrates, such as rootlets or herbaceous stems and leaves. Longevity is curtailed by rapid decomposition of the substrate. The pathogen, dependent on continued metabolic activity, is stressed in surface soils by variations in temperature and moisture. Most pathogens of this type, if not seed-borne to any extent and if their host range is relatively narrow, are easily controlled by rotation.

The work with *Gaeumannomyces graminis* illustrates the complexity of saprophytic survival. Abundant nitrogen supports the saprophytic activity of this fungus, making the pathogen stronger in the short-term. If, however, a host is not available within a year or less, high nitrogen causes the fungus to exhaust the substrate rapidly and leads to its death.

Because isolates of *Cephalosporium gramineum* fresh from nature produce antibiotics, it is logical to assume that nature has favored their selection. Antibiotics serve no direct purpose, since they are nonessential metabolites. Their production utilizes materials that could be used for cell synthesis. Their importance in nature is still not clear, but my belief is that the price they cost the producer must result in a net gain.

Saprophytic survival is crucial to many root pathogens of woody plants. Many fungi that attack trees have no long-lived, dormant resting structures. Their saprophytic survival in less-easily digested substrates, deep in the soil where environmental stresses are minimal, is highly effective.

Fungi are usefully classified as sugar fungi if they digest readily available foods and are noncompetitive in woody tissues (cellulose, lignin). The danger of placing too much emphasis on such categorization is evident in the importance of sugars and starches to *Armillaria mellea*, a pathogen primarily of trees (Swift, 1970). *Pythium* spp. are considered typical sugar fungi, thriving in succulent, delicate host tissues. Deacon (1979) cautioned that many *Pythium* spp. possess some ability to hydrolyse cellulose. It is doubtful, however, that these observations reduce the practical value of the original concepts of these substrate groupings of organisms.

KEY REFERENCES

Bruehl, G. W. 1975. Systems and mechanisms of residue possession by pioneer fungal colonists. Pages 77–83 in *Biology and Control of Soilborne Plant Pathogens*. G. W. Bruehl, ed. Am. Phytopathol. Soc., St. Paul, Minn.

Garrett, S. D. 1970. *Pathogenic Root-Infecting Fungi*, pp. 164–174. Cambridge Univ. Press., London.

Scott, P. R. 1969. Effects of nitrogen and glucose on saprophytic survival of *Ophiobolus graminis* in buried straw. *Ann. Appl. Biol.* **63**:27-36.

Swift, M. J. 1970. *Armillaria mellea* (Vahl ex Fries) Kummer in Central Africa: Studies on substrate colonization relating to the mechanism of biological control by ring barking. Pages 150-152 in *Root Diseases and Soil-Borne Pathogens*. T. A. Toussoun, R. V. Bega, and P. E. Nelson. Univ. of California Press, Berkeley.

REFERENCES

Barton, R. 1960. Antagonism amongst some sugar fungi. Pages 160-167 in *The Ecology of Soil Fungi*. D. Parkinson and J. S. Waid, eds. Liverpool Univ. Press, Liverpool, UK.

Blanchette, R. A. and C. G. Shaw. 1978. Associations among bacteria, yeasts and basidiomycetes during wood decay. *Phytopathology* **68**:631-637.

Bockus, W. W. and M. M. Claasen. 1985. Effect of lime and sulfur application to low pH soil on incidence of *Cephalosporium* stripe under continuous winter wheat production. *Plant Dis.* **69**:576-578.

Bockus, W. W., J. P. O'Connor, and P. J. Raymond. 1983. Effect of residue management method on incidence of cephalosporium stripe under continuous winter wheat production. *Plant Dis.* **67**:1323-1324.

Bruehl, G. W. 1963. *Hymenula cerealis*, the sporodochial stage of *Cephalosporium gramineum*. *Phytopathology* **53**:205-208.

Bruehl, G. W., B. Cunfer, and M. Toivianen. 1972. Influence of water potential on growth, antibiotic production, and survival of *Cephalosporium gramineum*. *Can. J. Plant Sci.* **52**:417-423.

Bruehl, G. W. and P. Lai. 1966. Prior colonization as a factor in the saprophytic survival of several fungi in wheat straw. *Phytopathology* **56**:766-768.

Bruehl, G. W., P. Lai, and O. Huisman. 1964. Isolation of *Cephalosporium gramineum* from buried, naturally infested host debris. *Phytopathology* **54**:1035-1036.

Bruehl, G. W., R. L. Millar, and B. Cunfer. 1969. Significance of antibiotic production by *Cephalosporium gramineum* to its saprophytic survival. *Can. J. Plant Sci.* **49**:235-246.

Burgess, L. W. and D. M. Griffin. 1968. The recovery of *Gibberella zeae* from wheat straws. *Aust. J. Exp. Agric. Anim. Husb.* **8**:364-370.

Cheo, P. C. 1982. Effects of tannic acid on rhizomorph production by *Armillaria mellea*. *Phytopathology* **72**:676-679.

Deacon, J. W. 1979. Cellulose decomposition by *Pythium* and its relevance to substrate-groups of fungi. *Trans. Brit. Mycol. Soc.* **72**:469-477.

Douglas, C. L., Jr., R. R. Allmaras, P. E. Rasmussen, R. E. Ramig and N. C. Roager, Jr. 1980. Wheat straw composition and placement effects on decomposition in dryland agriculture of the Pacific Northwest. *Soil Sci. Soc. Am. J.* **44**:833-837.

Fox, R. A. 1970. A comparison of methods of dispersal, survival, and parasitism in some fungi causing root diseases of tropical plantation crops. Pages 179-187 in *Root Diseases and Soil-Borne Pathogens*. T. A. Toussoun, R. V. Bega, and P. E. Nelson, eds. Univ. of Calif. Press, Berkeley.

Garrett, S. D. 1944. *Root Disease Fungi*. Chronica Botanica Co., Waltham, Mass.

Garrett, S. D. 1956. Biology of Root-Infecting Fungi. Cambridge Univ. Press, London.
Garrett, S. D. 1976. Influence of nitrogen on cellulolysis rate and saprophytic survival in soil of some cereal foot-rot fungi. *Soil Biol. Biochem.* **8**:229–234.
Hopp, A. D. 1972. The influence of antibiotic production and soil pH on survival of *Cephalosporium gramineum* in infested wheat straw. Ph.D. thesis, Wash. State Univ., Pullman.
Kobayashi, K. and T. Ui. 1979. Phytotoxicity and antimicrobial activity of Graminin A, produced by *Cephalosporium gramineum*, the causal agent of *Cephalosporium* stripe disease of wheat. *Physiol. Plant Pathol.* **14**:129–133.
Lai, P. and G. W. Bruehl. 1966. Survival of *Cephalosporium gramineum* in naturally infested wheat straws in soil in the field and in the laboratory. *Phytopathology* **56**:213–218.
Lai, P. Y. and J. M. Dunleavy. 1969. Sporulation and spore germination of *Cephalosporium gregatum* as influenced by host substrate and soil moisture. *Phytopathology* **59**:1646–1649.
Latin, R. X., R. W. Harder and M. V. Wiese. 1982. Incidence of Cephalosporium stripe as influenced by winter wheat management practices. *Plant Dis.* **66**:229–230.
Macer, R. C. F. 1961. The survival of *Cercosporella herpotrichoides* Fron in wheat straw. *Ann. Appl. Biol.* **49**:165–172.
Mathre, D. E. and R. H. Johnston. 1979. Decomposition of wheat straw infested by *Cephalosporium gramineum*. *Soil Biol. Biochem.* **11**:577–580.
Menzies, J. D. 1963. Survival of microbial plant pathogens in soil. *Bot. Rev.* **29**:79–122.
Murray, T. D. and G. W. Bruehl. 1983. Composition of wheat straw infested with *Cephalosporium gramineum* and implications for its decomposition in soil. *Phytopathology* **73**:1046–1048.
Nicot, J. 1960. Some characteristics of the microflora in desert sands. Pages 94–97 in *The Ecology of Soil Fungi*. D. Parkinson and J. S. Ward, eds. Liverpool Univ. Press, Liverpool, UK.
Nisikado, Y., H. Matsumoto and K. Yamanti. 1934. Studies on a new *Cephalosporium*, which causes the stripe disease of wheat. *Ber. Ohara Inst. Landwirtsch. Forsch., Kurashiki, Jap.* **6**:275–306.
Redfern, D. B. 1975. The influence of food base on rhizomorph growth and pathogenicity of *Armillaria mellea* isolates. Pages 69–73 in *Biology and Control of Soil-Borne Plant Pathogens*. G. W. Bruehl, ed. *Am. Phytopath. Soc.*, St. Paul, Minn.
Scott, P. R. 1969. Control of survival of *Ophiobolus graminis* between consecutive crops of winter wheat. *Ann. Appl. Biol.* **63**:37–43.
Smith, J. H. and C. L. Douglas. 1970. Influence of silica and nitrogen contents and straw application rate on decomposition of Gaines wheat straw in soil. *Soil Sci.* **109**:341–344.
Walker, A. 1941. The colonization of buried wheat straw by soil fungi, with special reference to *Fusarium culmorum*. *Ann. Appl. Biol.* **28**:333–350.
Wiese, M. V. and A. V. Ravenscroft. 1975. *Cephalosporium gramineum* populations in soil under winter wheat cultivation. *Phytopathology* **65**:1129–1133.

Chapter 12

Survival as Spores, Hyphae, and Microsclerotia

Dormant survival is accomplished by fungi primarily in the form of spores, sclerotia, or dormant hyphae. Sclerotia and resting spores are physiologically and morphologically differentiated from hyphae, and most of them have greater inoculum potential and greater longevity than dormant hyphae. *Verticillium dahliae*, which forms microsclerotia, persists longer in soil in a wider geographic range and is more difficult to control by rotation than *V. albo-atrum*, which forms dark, dormant hyphae (dauermycelium). Hyphae of phycomycetes are ephemeral and lyse rapidly, and these fungi depend on resistant sporangia, chlamydospores, or oospores for survival.

Nematodes differ widely as to which stage of the life cycle is best adapted for dormant survival. In *Meloidogyne* spp., the egg (in egg masses) is the most resistant stage. In cyst-nematodes, eggs in the dead female body (the cyst) are the most resistant stage. In *Ditylenchus dipsaci*, the third or fourth stage larvae are most resistant, and in *Anguina tritici* the second stage larvae within dry wheat galls are most resistant. Fourth-stage larvae of *Ditylenchus dipsaci* and second-stage larvae of *Anguina tritici* can live for years in an anhydrobiotic condition.

GERMINABILITY

Oospores of *Pythium* spp. are held dormant in soil by fungistasis, but dormancy of oospores is more complicated than that. Oospores of *P. graminicola, P. debaryanum, P. ultimum,* and *P. aphanidermatum* are endogenously dormant when first produced in host tissues within soil (Stanghellini, 1974). The nature of this endogenous dormancy in oospores of most species of *Pythium* is not known, but Burr and Stanghellini (1973) believed that enzymes alter the oospore wall of *P. aphanidermatum,* changing its permeability and rendering the oospores germinable.

Oospores of *P. ultimum* have walls 2.1 μm thick when first formed, but after exposure to nonsterile soil extract the oospore walls are 0.5 μm thick (Lumsden and Ayers, 1975). The thick-walled oospores will not germinate, but thin-walled oospores are germinable; thin-walled oospores stain deeply with lactofuchsin, like sporangia, but thick-walled oospores do not stain. Conversion of thick-walled to thin-walled oospores in nonsterile soil extract is 3% complete in 2 weeks, 25% in 4 weeks, and 65% in 6 weeks. Germination of thin-walled oospores is rapid, with 97 to 100% germination within 2 hr on a selective medium, and germination is nearly 100% in distilled water after 3 or more hr (they are not energy deficient).

When beans are planted in soil treated with aerated steam and inoculated with thick-walled oospores, no damping-off occurred. When thin-walled oospores are used, all bean seedlings damp-off within 2 weeks. When the soil inoculated with thick-walled oospores is fallowed for 2 more weeks and seeded to beans, 63% of the seedlings damped-off.

Oospores produced in culture become thin-walled in sterile or nonsterile water, as well as in nonsterile soil extract. Neither sporangia nor thin-walled oospores germinate unless nutrients are supplied or the propagules are in sterile water. Lumsden and Ayers believe germination in water and conversion of oospores from thick-walled to thin-walled in water is evidence that a germination inhibitor is removed by soaking in water and that removal of the inhibitor permits the oospore to hydrolyse part of its own wall, changing the spore from endogenously dormant to an exogenously germinable state. Their experiments indicate that conversion is accomplished by the oospores themselves.

Variation in conditions favoring germination occurs among species of a genus such as *Pythium*. The preceding discussion was of *P. ultimum*. Ruben, et al. (1980) found that oospores of *P. aphanidermatum* produced in culture are rendered germinable by drying. After drying, soaking in dilute potassium permanganate or soaking in water at 39°C further increased germination. After the above treatments, placing them in a 0.01% solution of lecithin resulted in germination of over 90% of the oospores. Few spores germinated in the absence of lecithin. If substantial amounts of food were supplied, germination was by germ tube (direct); otherwise, germination was primarily by means of zoospores (indirect). The difference in type of germination, direct or indirect, depending on environmental conditions, is an example of phenotypic plasticity (see Chapter 24).

In nature, oospores would be liberated gradually over time from infested host debris as the latter disintegrates, and they would accumulate in various degrees of readiness for germination. The germination of oospores produced in culture is greatly increased by passage through snails and earthworms, and Stanghellini and Russell (1973) suggested that this process may be significant in nature.

Requirements for Germination

Absorption of calcium and an energy source into the oospore of *Pythium aphanidermatum* initiates germination, and calcium absorption initiates digestion of the endospore wall, after which these spores are germinable (Stanghellini and Russell, 1973). Germination does not procede if an exogenous energy source is absent in the germination medium. Several carbon sources support germ tube elongation. These observations were all made in the absence of soil microflora. These oospores, in the absence of microflora, would not germinate without an exogenous energy source. Calcium would be present in the soil solution, and all that would be lacking would be the carbon source. Stanghellini and Russell considered this support for Lockwood's energy-deficiency hypothesis on the nature of fungistasis.

The requirement of a carbon source by *P. aphanidermatum* to complete germination is further complicated by the observation of Stanghellini and Burr (1973) that both oospores (30%) and sporangia (90%) in soil flooded with water germinate indirectly, by producing zoospores. Where is the need for food for completion of germination? Fungistatic factors in the soil may have been diluted to ineffective levels by flooding, and in their absence exogenous food is not essential. Also, Lumsden and Ayers (1975) germinated oospores of *P. ultimum* in water.

Inhibitors of Germination

An unidentified species of *Bipolaris* obtained from soil from Laysan Island produces substances that inhibit germination of its own conidia (Brooks, in Palm and Goos, 1980). The inhibitor is not destroyed by freezing or autoclaving and is stronger from vegetative hyphae than from conidia. The inhibitor produced by the *Bipolaris* sp. is similar to one produced by *Fusarium oxysporum,* which has been identified as nonanoic acid. There is evidence that nonanoic acid is produced by several fungi (Garrett and Robinson, 1969). Autoinhibitors have survival value, in that they prevent germination of conidia still attached to parent hyphae (Gottlieb, 1978).

TOUGHNESS AND LONGEVITY OF STRUCTURES

Pythium spp.

Rapid drying kills both thin-walled oospores and sporangia of *P. ultimum* but not thick-walled oospores (Lumsden and Ayers, 1975). The sporangia of *P. ultimum* resist drying (Stanghellini and Hancock, 1971). Lumsden and Ayers, however, subjected sporangia and oospores in mycelial mats to rapidly moving air at room temperature for 30 minutes. Only thick-walled oospores survived. Such rapid drying is highly unlikely in nature.

Bipolaris sorokiniana

Wheat preceded by 0, 1, 2, 3, and 5 years of crops not susceptible to *Bipolaris sorokiniana* has 68, 64, 37, 34, and 14%, respectively, of infected plants (Ledingham, 1961). Only a long rotation would be of much value, and in dryland agriculture economically feasible rotation crops are few. Chinn (1976) collected soil in northeastern Saskatchewan that had been in wheat or barley, fallow, or rape the previous years. Samples of these soils were seeded to wheat, and the wheat was rated for lesions caused by *B. sorokiniana* and for frequency of isolation of the pathogen. Disease following the 1973 crop was greater than following the 1971 crop, reflecting the greater number of viable conidia present, but rotation had little effect (Table 12.1). Chinn concluded that the number of conidia per gram of soil exceeded the threshold for maximum disease development, that the limited rotations economically possible in Saskatchewan will not control the disease. The longevity of conidia in the soil is too great. Conidia survived in a fine sandy loam from southwestern Nebraska for at least 490 days outdoors (over 60% germinable), but survival was greatly reduced in less than 6 months in a heavy soil of southeastern Nebraska (Boosalis, 1962). Meronuck and Pepper (1968) observed chlamydospores within the conidia of *B. sorokiniana* buried in soil.

Simmonds (1961) retrieved specimens of diseased wheat that had been

Table 12.1. Severity of common root rot of wheat in soil of different histories, and the number of viable conidia of *Bipolaris sorokiniana* per gram of soil

	Crop	Cereal fields	Fallow fields	Rape fields
Mean disease ratings	1971	18	23	18
	1973	33	32	32
Viable conidia	1971	75	45	27
	1973	235	137	46

Source: Chin, 1976.

stored for 18 months in a herbarium. The specimens had soil around the crowns and roots. He removed bits of crown and individual roots, dislodging as little soil as possible, moistened them, and after 7 days, recorded sporulation by *B. sorokiniana*. Sporulation occurred on 47 roots out of 152. Severely diseased crown pieces supported heavy sporulation. The dormant mycelium remained alive when highly desiccated within host tissue, and antagonistic effects of microorganisms present on the unwashed tissues did not prevent sporulation.

Verticillium spp.

Repeated germination reduces the ability of microsclerotia of *V. dahliae* to withstand adversity. Microsclerotia of *V. dahliae* from dead stems of diseased potatoes were added to moist soil incubated at 26 to 28 °C (Menzies and Griebel, 1967). The number of germinable propagules increased 3 to 5 times during the first 10 days in soil. The increase is attributed to production of conidia by some of the microsclerotia. The population of germinable propagules in the soil declined thereafter. Sporulation by microsclerotia is supported mainly by reserves within the microsclerotia, because added nutrients result in relatively little increase in propagule counts. Gentle cycles of wetting and drying produce repeated crops of conidia, but the spore crops decline as the microsclerotia become exhausted. If the moist soils in which germination experiments have been conducted for 3 weeks or more were air-dried, the counts of viable propagules decline sharply, evidence that many of the microsclerotia have germinated during the moist period and that germination destroyed their ability to withstand desiccation.

Conidia of *V. dahliae* survive for weeks or days in soil, not long enough to serve as survival structures. Green (1969) inoculated soil with *V. dahliae* at the rates of 1000, 5000, 10,000, 50,000, and 100,000 conidia or 10, 25, 50, 100, 500, and 1000 microsclerotia/g of soil. Tomato plants were transplanted into the soils at 0, 1, 3, 5, and 7 weeks later. Microsclerotia at 100/g or more resulted in 100% diseased tomatoes at all dates of transplanting with no evidence of loss of infectivity during the 7 weeks pretransplant time. Conidia at 100,000/g resulted in 100% diseased plants at 0 to 3 weeks, about 60% at 5 weeks, and 20% at 7 weeks. After 5 weeks the ability to produce disease was not detected at the 50,000 conidia/g of soil rate or by lower rates. Conidia were effective for short periods only.

Wilhelm (1954) observed clumps of conidia transformed into microsclerotia. When gradual drying of watery droplets of conidia in spore heads occurs extramatrically, the conidia may swell and germinate by fine germ tubes and the germ tubes anastomose. The connected mass of conidia becomes a small, brown extramatrical microsclerotium. Wilhelm observed these structures on straw and in culture. He suggested that they could be wind-disseminated if they form above the soil surface on refuse. McKeen and Thorpe (1981) observed heavy sporulation of *V. dahliae* on dead, blackened potato stems after heavy rains and profuse formation of microsclerotia on the surface of dead leaves and stems. Whether these microsclerotia originated from fused conidia, as observed by Wilhelm, was not stated.

V. albo-atrum, which lacks microsclerotia, rarely survives the winter in southern Ontario, Canada (McKeen and Thorpe, 1981). *V. dahliae*, in contrast, persists for years as microsclerotia. *V. albo-atrum* reduced yields of potatoes every year in six yearly inoculation trials, whereas *V. dahliae* did so only once, but the dormant mycelium of *V. albo-atrum* is a poor survival structure in Canada.

LIBERATION OF PROPAGULES FROM INFESTED DEBRIS

A single 2.5-cm piece of a potato stem in central Washington can contain 20,000 to 50,000 viable sclerotia of *Verticillium dahliae*, but the microsclerotia bound within the host tissues will not reach their full inoculum potential until the host tissues are decomposed, freeing them from the tissues (Menzies, 1970). When freed, subsequent tillage operations will disperse them within tillage layers of the soil, increasing the chances for their encounters with host roots. Soil counts could increase a year or two after death of the diseased crop, not because of saprophytic increase, but because of inoculum becoming individual units rather than clusters of units.

Ashworth et al. (1974) assayed infested soils for microsclerotia after a cotton crop without tillage, after rototilling, and after irrigation. After the cotton was harvested, the plants were shredded and disked into the soil. Irrigation was applied to part of the soil during the winter. Stem decomposition in moist soil was more effective in liberating propagules than tillage. Ashworth et al. supported Menzie's idea that when microsclerotia bound in host tissues are freed, propagule counts increase even though there is no production of new microsclerotia.

RHIZOCTONIA SOLANI IN HOST DEBRIS

Rhizoctonia solani survives in sugar beet debris as dark-brown, irregularly branched hyphae *within* infested host fragments (Boosalis and Scharen, 1959). These hyphae contained living protoplasm 7 months after destruction of the diseased sugar beets. The debris was retrieved by wet sieving, after which Boosalis and Scharen broke the fragments to finer fragments by gentle pressure against screens preparatory to microscopic examination. Heavy, dark *surface* (extramatrical) hyphae did not grow when plated on water agar containing streptomycin sulfate. About 60% of the beet fragments contained dark brown hyphae within the tissues, and cultures developed from about 35% of such fragments. Sclerotia, varying in size from 10 to 500 μm in diameter were found. Many of these sclerotia germinated in tapwater.

SEEDS

Witchweed (*Striga asiatica*) is a parasitic plant important on maize, sorghum, and some other grasses. The age, size, and weight of the seed affects their potency as inoculum (Bebawi et al., 1984). Weight of maize shoots is severely reduced by seed 1 to 4 years old, with damage decreasing from seed 6 years old to no loss from seeds 8 years old. Damage from large seeds is greater than from small seeds. Seed germinability is greatest with large heavy seeds and least with small light seeds. These same relationships would probably exist with many types of propagules.

COMMENTS

The complexities of dormancy are apparent in that oospores of several *Pythium* spp. as formed within host tissues could not be germinated, but after appropriate incubation, they became germinable. In most *Pythium* spp. secondary spread is accomplished by zoospores. There may be few times when germination of an oospore within a diseased root might be advantageous to the fungus. Endogenous dormancy of freshly formed oospores within host tissues provides inoculum for a different generation of hosts, not for reinvasion of the same host generation. The development of being born with endogenous dormancy, and of a system of release from endogenous dormancy is surely not fortuitous.

Long-lived propagules often make practical rotations for disease control economically unfeasible, as with *Bipolaris sorokiniana* in the Canadian prairies. The advantages of a specialized dormant structure (microsclerotium) in survival over modified, dark hyphae (dauermycelium) is apparent in differential survival of *V. dahliae* and *V. albo-atrum* in southern Canada.

Some aspects of spore physiology are treated in this chapter. Most interesting to me is the work with oospores of *Pythium* species. The lysis of part of the spore wall by the fungus itself, rendering its oospores germinable, is basic biology. Hatching factors in the case of some cyst-nematode eggs (Chap. 7) and cell wall changes in oospores of some *Pythium* spp. support the importance of permeability, as does scarification of seeds of some higher plants.

Some propagules (chlamydospores of *Fusarium* spp., sporangia and oospores) of some species can germinate in response to exudates, fail to encounter a host, and gather the protoplasm into a new (replacement) resting structure. Sporangia of *Pythium ultimum* can form replacement sporangia (Stanghellini and Hancock, 1971) but those of *P. aphanidermatum* do not (Stanghellini and Burr, 1973). The ability to germinate, form a germ tube, fail to establish a parasitic relationship, and save the protoplasm is a remarkable development. This phenomenon is probably most significant within *Fusarium spp.* in which chlamydospores respond to exudates from many sources, but can form replacement chlamydospores and assume apparent immortality if the stimulus is from a nonhost.

Spores are treated throughout this book, especially under phenotypic plasticity in the chapter on evolution. A good, brief treatment with thorough referencing is provided by Griffin (1981, pp. 219-239).

Careful reading of this chapter reveals some contradictions I cannot explain.

KEY REFERENCES

Green, R. J., Jr. 1969. Survival and inoculum potential of conidia and microsclerotia of *Verticillium albo-atrum* in soil. *Phytopathology* **59**:874-876.

Lumsden, R. D. and W. A. Ayers. 1975. Influence of soil environment on the germinability of constitutively dormant oospores of *Pythium ultimum*. *Phytopathology* **65**:1101-1107.

Stanghellini, M. E. 1974. Spore germination, growth and survival of *Pythium* in soil. *Proc. Am. Phytopathol. Soc.* **1**:211-214.

REFERENCES

Ashworth, L. J., Jr., O. C. Huisman, D. M. Harper, and L. K. Stromberg. 1974. Free and bound microsclerotia of *Verticillium albo-atrum* in soils. *Phytopathology* **64**:563-564.

Bebawi, F. F., R. E. Eplee and R. S. Norris. 1984. Effect of age, size, and weight of witchweed seeds on host, parasite relations. *Phytopathology* **74**:1074-1078.

Boosalis, M. G. 1962. Precocious sporulation and longevity of conidia of *Helminthosporium sativum* in soil. *Phytopathology* **52**:1172-1177.

Boosalis, M. G. and A. L. Scharen. 1959. Methods for microscopic detection of *Aphanomyces euteiches* and *Rhizoctonia solani* and for isolation of *Rhizoctonia solani* associated with plant debris. *Phytopathology* **49**:192-198.

Burr, T. J. and M. E. Stanghellini. 1973. Propagule nature and density of *Pythium aphanidermatum* in field soil *Phytopathology* **63**:1499-1501.

Chinn, S. H. F. 1976. Influence of rape in a rotation on prevalence of *Cochliobolus sativus* conidia and common root rot. *Can. J. Plant Sci.* **56**:199-201.

Garrett, M. K. and P. M. Robinson. 1969. A stable inhibitor of spore germination produced by fungi. *Arch. Mikrobiol.* **67**:370-377.

Gottlieb, D. 1978. *The Germination of Fungus Spores*. Meadowfield Press, Durham, U. K.

Griffin, D. H. 1981. *Fungal Physiology*. Wiley, New York.

Ledingham, R. J. 1961. Crop rotations and common root rot of wheat. *Can. J. Plant Sci.* **41**:479-486.

McKeen, C. D. and H. J. Thorpe. 1981. Verticillium wilt of potato in southwestern Ontario and survival of *Verticillium albo-atrum* and *V. dahliae* in field soil. *Can. J. Plant Pathol.* **3**:40-46.

Menzies, J. D. 1970. Factors affecting plant pathogen populations in soil. Pages 16-21 in: Root Diseases and Soilborne Pathogens. T. A. Toussoun, R. V. Bega, and P. E. Nelson, eds. Univ. of Calif. Press, Berkeley.

Menzies, J. D. and G. E. Griebel. 1967. Survival and saprophytic growth of *Verticillium dahliae* in uncropped soil. *Phytopathology* **57**:703-709.

Meronuck, R. A. and E. H. Pepper. 1968. Chlamydospore formation in conidia of *Helminthosporium sativum*. *Phytopathology* **58**:866–867.

Palm, L. and R. D. Goos. 1980. Autoinhibition of conidium germination in an isolate of *Bipolaris*. *Mycologia* **72**:937–949.

Ruben, D. M., Z. R. Frank, and J. Chet. 1980. Factors affecting behavior and devlopmental synchrony of germinating oospores of *Pythium aphanidermatum*. *Phytopathology* **70**:54–59.

Simmonds, P. M. 1961. Effects of inoculating wheat seedlings with *Helminthosporium sativum* and spread of the fungus to external substrate. *Can. J. Plant Sci.* **41**:791–798.

Stanghellini, M. E. and T. J. Burr. 1973. Germination in vivo of *Pythium aphanidermatum* oospores and sproangia. *Phytopathology* **63**:1493–1496.

Stanghellini, M. E. and J. G. Hancock. 1971. Radial extent of the bean spermosphere and its relation to the behavior of *Pythium ultimum*. *Phytopathology* **61**:165–168.

Stanghellini, M. E. and J. D. Russell. 1973. Germination in vitro of *Pythium aphanidermatum* oospores. *Phytopathology* **63**:133–137.

Wilhelm, S. 1954. Aerial microsclerotia of *Verticillium* resulting from conidial anastomosis. *Phytopathology* **44**:609–610.

Chapter 13

Sclerotia

Sclerotia are vegetative, multicellular compact aggregates of fungal hyphae. At maturity they are hard and rich in stored foods. Sclerotia vary in structure from those with a well-developed, differentiated rind, such as in *Typhula* spp. and *Sclerotium rolfsii,* to those without a differentiated rind as in *Rhizoctonia solani*. In some fungi, such as *Typhula* spp., the rind may develop before maturity and stretch during final development. In *Botrytis* spp. the rind is thick, does not permit expansion, and is formed in final stages of sclerotial development. Internal to the rind is the cortex and/or the medulla. Sclerotia of *Sclerotinia* spp. are internally differentiated into a cortex of thin-walled compact cells surrounding loosely packed hyphae that form the medulla (Townsend and Willitts, 1954). In well-developed, differentiated sclerotia, rind cells do not germinate.

Coley-Smith (1979) grouped sclerotial fungi according to the method of germination (myceliogenic = by hyphae, sporogenic = by conidiophores and conidia, carpogenic = by production of ascocarps or basidiocarps) (Fig. 13.1), presence or absence of a well-developed rind, and whether a species had a wide or narrow host range. He attempted to relate these relationships with longevity. Sclerotia of the *Botrytis* spp. produce conidia on germination

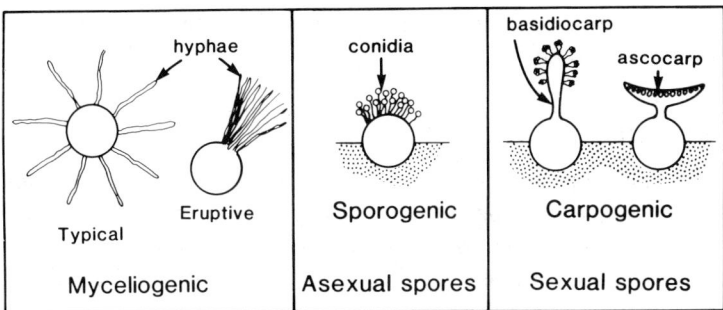

Figure 13.1. Myceliogenic, sporogenic, and carpogenic germination of sclerotia: germination by vegetative hyphae, conidiophores, or production of sporocarps, respectively.

(sporogenic germination), and the sclerotia rot rapidly after spore production. Sclerotia that germinate carpogenically usually disintegrate within a month or two after germination. Sclerotia of *Typhula* spp. can germinate either carpogenically (produce a sporophore), myceliogenically, or both, depending on conditions.

Sclerotia that respond to host-specific host exudates or volatiles are the most long-lived in soil (Coley-Smith, 1979). Most sclerotia of *Sclerotinia cepivorum* and *Stromatinia gladioli*, which respond to host specific stimulants, are germinable after 10 years in field soil. These sclerotia are in a cryptobiotic condition with a very low level of metabolism.

The role of a well-developed sclerotial rind in longevity is obscure. *Sclerotinia cepivorum* and *Stromatinia gladioli*, both very long-lived, have well-developed rinds. Some sclerotia with well-developed rinds are less long-lived. Coley-Smith lists *Sclerotium delphinii, S. rolfsii*, and *Phymatotrichum omnivorum* as examples of the latter. These fungi have wider host ranges, so extreme longevity may not have been selected for in nature. Sclerotia of most species survive for years in a desiccated state, but sclerotia of *Phymatotrichum omnivorum* die within 90 minutes when dried on a laboratory bench (King et al., 1931).

Hashiba and Mogi (1975) studied developmental changes in sclerotia of the rice sheath blight fungus, *Rhizoctonia solani* AG 1. The mature sclerotium of *R. solani* is loosely interwoven, with no well-defined internal zones. Sclerotia of the rice sheath blight fungus in early development sink in water; later they float. At initiation sclerotia are white, but within 30 hr they begin to turn brown and are brown within 40 hr at 25°C. By 30 hr sclerotia have reached full size. By 15 days the peripheral cells (about a 20-cell deep layer) are essentially devoid of contents, while in the central mass the cells are dense and filled with living protoplasm. Buoyancy is detected at 10 days and is complete in 15 days. Buoyant sclerotia have outer layers of empty cells. Internal differentiation into an outer ring of empty cells is known only in rice-forms of *Rhizoctonia solani*, and buoyancy is an adaptation to the aquatic habitat. Infection occurs when floating sclerotia encounter rice plants at the water's surface.

SIZE AND FUNCTION

Garrett (1944) observed that large sclerotia, such as those of *Claviceps purpurea* and *Sclerotinia borealis,* produce ascocarps (carpogenic germination) and that infection results from spores. In contrast, small sclerotia frequently germinate vegetatively by hyphae (myceliogenic germination) and act directly as infectious units. Although this concept is generally true, there are exceptions. Microsclerotia of *Verticillium dahliae* can produce conidia. Sclerotia of *Typhula* spp. (0.3 to 1 mm diam) frequently germinate carpogenically but also germinate by hyphae alone (myceliogenically). Sclerotia of *Phymatotrichum omnivorum,* which are relatively large (usually 1 to 2 mm, but up to 1 cm), germinate only myceliogenically.

Henis and Ben-Yephet (1970) produced *Rhizoctonia solani* sclerotia in culture. When mixed with soil in a concentration of four per gram of soil, propagules smaller than 250 μm caused no disease in bean seedlings within 21 days. Larger sclerotia produced infection of all seedlings within 5 days and symptoms within 7. Because the small propagules produced no disease when randomly distributed in natural soil, they placed them in direct contact with the seed. No disease resulted from propagules smaller than 150 μm, even when as many as 36 were next to the seed. When 13 propagules 150 to 250 μm in diameter were next to the seed, 22% became infected. When eight propagules 500 to 1000 μm in diameter were used, 89% of the seedlings were infected. The power to infect increases with propagule size. They estimated the energy level of the propagules by germinating them on glass slides in the absence of external nutrients. Sclerotia 500 to 1000 μm in diameter produced colonies 1.28 cm in diameter. Sclerotia 250 to 500 μm in diameter produced colonies 0.68 μm in diameter with fewer hyphal extensions per propagule. Smaller propagules produce smaller colonies with few hyphal extensions.

Naiki and Ui (1977) found about 2 sclerotia per gram of dry rhizosphere soil and 0.9 per gram of nonrhizosphere soil about severely diseased sugar beets, and about 0.08 and 0.03 sclerotia per gram of dry soil about healthy sugar beets in rhizosphere and nonrhizosphere soil, respectively. Viability of the sclerotia increases as the severity of the rot increases, and size of the sclerotia increases with severity of rot. Sclerotia obtained furthest from the root have the lowest viability and are smaller and weaker than those formed close to the root.

FORMATION

The ability to form sclerotia may be lost in culture by improper transfers (Humpherson-Jones and Cooke, 1977). Sclerotia form normally when young hyphae near the margin of a colony of *Sclerotinia sclerotiorum* are transferred, but transfers of older hyphae from near the center of the dish may fail to form sclerotia. If the fungus is abused habitually, it may lose its sclerotium-forming ability altogether. They believe that cytoplasmic factors are important, that older hyphae lose some competency, that a balance of extranuclear elements is required for normal sclerotium formation.

Trevethick and Cooke (1973) grew *Sclerotium rolfsii* in culture at 25°C under continuous light or in continuous darkness. In the light 884 and in the dark 205 sclerotia were formed per dish. The average dry weight of each sclerotium produced in the light and dark was 0.81 and 0.95 mg, respectively. The total weight in the light was 0.716 g, and in the dark, 0.195 g. *Sclerotium delphinii* responded to light even more strongly, producing 570 sclerotia per dish in light and 28 in darkness. When sclerotia were removed from cultures of *Sclerotinia sclerotiorum* grown in light at room temperature for 6 or 7 days, the time at which sclerotium numbers reach maximum in undisturbed cultures, a new crop of sclerotia was produced. The second crop was over half as large as the first. Trevethick and Cooke believe that the second crop came from sclerotial initials already present on the culture that had been suppressed by the first crop. The medium or mycelium had not been exhausted by producing the first crop of sclerotia. They grew *Sclerotinia sclerotiorum* in dishes of different sizes, from 6.7 to 18.5 cm in diameter. The larger the dish, the faster the fungus grew and the farther from the inoculum piece the first ring of sclerotial initials appeared. Once initials formed, however, each successive ring of sclerotia formed 0.5 cm from the last regardless of the size of the dish.

Wheeler and Walker (1965) grew *Sclerotium rolfsii* on medium in Petri dishes of differing diameters. Except in the largest dishes (13 cm in diameter), sclerotial initials did not form until the mycelium reached the edges of the dishes. Wheeler and Walker concluded that hyphal tips of rapidly growing colonies inhibit sclerotium formation by some mechanism and that competition for essential metabolites is involved. All efforts to demonstrate stimulation of sclerotium production by staling products or accumulation of some stimulator have failed, supporting the concept of exhaustion of nutrients as the initiator of sclerotium formation, as shown in subsequent work by Christias and Lockwood (1973).

Translocation

How efficient is the process of food storage in resting structures? Christias and Lockwood (1973) grew *Sclerotinia sclerotiorum, Sclerotium rolfsii, Sclerotium cepivorum,* and *Rhizoctonia solani* in potato dextrose broth. The washed mycelia were incubated on natural soil or on glass beads. Some mycelia on glass beads were exposed to constant leaching. At first the mycelia grew, but growth ceased within 24 hr except for *R. solani* on soil. Within 18 to 24 hr sclerotial initials formed in all fungi, and sclerotia matured within 4 days. Formation of sclerotia was accompanied by lysis of vegetative hyphae. On leached glass beads *S. rolfsii* converted 42% of carbohydrate into sclerotia, 25% was leached away, 26% was lost as CO_2, and 7% remained in the spent mycelium. For nitrogen, 44% was incorporated into sclerotia, 43% lost in exudates, and 13% remained in the mycelium. Conservation of nutrients by all four fungi on soil without leaching was somewhat higher.

If the mycelia were "leached" with potato dextrose broth, sclerotia did not form until growth was so heavy the broth did not penetrate the mycelium. If left in potato dextrose broth, sclerotia did not form until the mycelium covered the broth and reached the edges of the flasks. The evidence is strong that nutrient deprivation stimulates sclerotium formation and that materials are translocated from existing hyphae to form the sclerotia. The conservation of substance is so efficient that even hyphal cell walls are partially used.

Christias and Lockwood provided quantitative proof of the ability of *Rhizoctonia solani* to derive nutrients from soil. The mycelium from liquid culture started with a dry weight of 58 mg and ended with 114 mg after incubation on soil. The other fungi did not gain weight on soil (Table 13.1).

E. H. Ellis in 1929 (in Garrett, 1956) found rhizomorphs of *Armillaria mellea* 80 ft below the soil surface in a mine. They had traversed a pile of soil 1.8 m thick and grown down tunnels for the remaining distance. Butler in 1943 reported rhizomorphs of *Merulius lacrymans* that had grown along a steel beam for 3 m. Significant translocation of nutrients occurs from a substantial food base under favorable conditions (in Garrett, 1956). The ability of hyphae of most root pathogens to grow through soil depends on translocation from a food base. Bruehl et al. (1966) found sclerotia of *Typhula idahoensis* on stones and wooden stakes among wheat plants affected by snow mold. They caution that finding sclerotia on a substrate may not prove a host range.

Table 13.1. Conversion of dry matter in mycelia to sclerotia

Fungus	Original mycelium, mg	Dry matter in sclerotia	
		on soil, mg	in leaching system, mg
Sclerotinia sclerotiorum	120.6	66.6	48.0
Sclerotium cepivorum	226.0	154.2	118.3
Sclerotium rolfsii	380.3	193.4	142.4
Rhizoctonia solani	58.0	114.0	22.5

Source: Christias and Lockwood, 1973.

Exudate during Formation

Sclerotial growth involves excretion of water (increase in % dry weight) and loss of some carbohydrate in the exudate (Cooke, 1969). Appearance of water droplets occurs early in the formation of sclerotia of *Sclerotinia sclerotiorum* and *S. trifoliorum*. Relatively large amounts of carbohydrate occur in the droplets. It is not a passive process, because of qualitative and quantitative differences between carbohydrates in the exudate and within the sclerotia.

Much of the dry material in the exudate on sclerotia of *Sclerotinia sclerotiorum* reenters the sclerotium and is metabolized or stored there in some fashion. As the liquid declines in volume, if the substances in solution are not withdrawn, the solute content increases. The dry matter in the exudate decreases 55% between the fifth and tenth day. Protein during the same period decreases 76%. Potassium declines 79% from the sixth to tenth day. Magnesium retains a relatively constant level, and sodium increases 206%. In resorption, potassium is favored, and sodium is not (Colotelo, 1973).

Christias (1980) analyzed the exudate that accumulates in droplets as sclerotia of *Sclerotium rolfsii* mature. The droplets contained amino acids, but 99% of the nitrogen was in the form of proteins. Protein was more abundant than carbohydrate in the exudate. Christias tested the exudate against a wide array of soil microorganisms and found no evidence of antibiotic activity. He concluded that any materials on or leaking from these sclerotia should be readily available to soil microorganisms.

Density

Differences in density of propagules have been observed and used in plant pathology for some time. Sclerotia of rice pathogens (*Sclerotium oryzae*, *Rhizoctonia solani* AG-1) and teliospores of *Neovossia (Tilletia) horrida* (kernel bunt of rice) float and germinate on water, adaptations to an aquatic habitat. Sclerotia of *Typhula* spp. are dense. Floating would have little value to them and could be disadvantageous.

Sir John Sinclair, near the beginning of the nineteenth century, found that farmers in the Netherlands used copper acetate dissolved in human urine in a concentration adequate to float bunt balls (*Tilletia* spp.) from wheat. The heavy seed sank, and the bunt balls floated. By 1856, this method was widely used in Britain. Galls containing *Anguina tritici* and ergot sclerotia can be removed from wheat by vigorously stirring the seed in a salt solution (18.1 kg sodium chloride in 94.6 l of water). Nematode galls, ergot sclerotia, light kernels, and trash rise to the surface, and sound kernels sink.

Rogers (1936) mechanized the screening of sclerotia of *Phymato-*

trichum omnivorum to the point that 1815 to 3630 kg of soil could be processed in a single day. To facilitate removal of sclerotia from the screenings, a sugar solution of specific gravity 1.15 to 1.25 was used to float the sclerotia which were slightly heavier than water. The sugar solution plus debris is stirred and then let stand for a few seconds to allow the sclerotia to float to the top.

Chinn et al. (1960) separated spores of *Bipolaris sorokiniana* from infested soil by flotation on mineral oil. Watanabe et al. (1970) used an ammonium sulfate solution to remove *Macrophomina phaseoli* sclerotia from soil. Sclerotia of *Ramulispora sorghi* (Odvody and Dunkle, 1973) were removed from the soil by wet-screening, ammonium sulfate flotation, and centrifugation on a 79% (w/v) sucrose shelf. Ben-Yephet and Pinkas (1976) used a cesium chloride flotation technique for the isolation of *Verticillium dahliae* microsclerotia from soil. No differences in density were observed between sclerotia produced in culture or from nature. Rodriguez-Kabana et al. (1974) determined sclerotial populations of *Sclerotium rolfsii* in soil by a rapid flotation-sieving technique. No difference in density was observed between sclerotia produced in culture and natural sclerotia.

Rice sheath blight, caused by *Rhizoctonia solani* AG 1, (= *Thanatephorus cucumeris*) is important in the southern areas of the US (Lee and Rush, 1983). Sclerotia are nonbuoyant when first produced but become buoyant in about 30 days. Sclerotia from previous crops float on water, and most initial infection occurs at the water surface. Sclerotia can germinate by hyphae several times. Japanese workers use the numbers of floating sclerotia in paddy water as a means of forecasting disease.

Leakage after Maturity

The idea that exudates from mature sclerotia encourage microflora about the sclerotium, enforcing fungistasis on the propagule, has been suggested. Long-term dormancy would require long-term exudation, which would exhaust the propagule, depending on the rate of exudation and the quantity of food reserves. This type of dormancy in most cases would preclude a long-term survival function.

Leachates from sclerotia of *S. rolfsii* stimulate the growth of some bacteria, but not of actinomycetes or of fungi (Gilbert and Linderman, 1971). The increase is detected within 24 hr of contact between the sclerotia and moist soil. The bacteria favored by materials leaching from the sclerotia are more tolerant to streptomycin and to oxgall bile salts than the general soil bacterial microflora. Gilbert and Linderman referred to the special environment about sclerotia as the *mycosphere*.

Coley-Smith and Dickinson (1971) seeded a nutrient-deficient agar medium with bacteria. Without a source of food the bacteria did not develop.

Washed sclerotia from culture that had been air-dried for 6 weeks were then placed in groups on the medium for 72 hr at 4°C to permit diffusion, and then the medium was inoculated with bacteria. Bacterial growth was stimulated in a 1-cm radius around the sclerotia. Laboratory sclerotia of the same source, but surface disinfested in calcium hypochlorite solution, stimulated bacterial growth 2.15 cm beyond the sclerotia. Sclerotia removed from soil after 3 and 24 months and disinfested, stimulated growth of the bacteria 0.6 and 0.4 cm in each direction, respectively. No zone of stimulation was detected about sclerotia from soil that were not disinfested.

Coley-Smith and Dickinson did not say how many sclerotia were in the "group" on the assay dish, but the quantitative loss of nutrients after surface disinfection must be unsustainable. They considered that the absence of a zone of stimulation about unsterilized sclerotia recovered from soil is evidence that they leak very little under "normal conditions," that surface disinfection stimulates leakage. They assumed that the 72-hr diffusion period at 4°C prior to incubation at 25°C minimized antagonistic or competitive effects by bacteria on the sclerotia, but many bacteria grow at 4°C.

Leakage from some sclerotia is stimulated by sequential exposure to wetting and drying. Coley-Smith (1979) cited such data for *Stromatinia gladioli, Sclerotium cepivorum, S. rolfsii, S. delphinii,* and *Rhizoctonia tuliparum*. All but *R. tuliparum* have differentiated rinds, and among these examples, only *R. tuliparum* suffers excessive weight loss with alternating wetting and drying (43% loss, cycle 1, 17% loss in cycle 2, and 11% loss in cycle 3). The lowest weight loss given is by *Stromatinia gladioli,* with 2.4, 0.9, and 0.3% loss in each of three successive wet-dry cycles. In spite of the substantial weight loss effected by alternate wetting and drying of sclerotia of *R. tuliparum* in these experiments, sclerotia of this species survived 7 years on the soil surface in the field. Sclerotia on the soil surface must be subject to wetting and drying, and how sclerotia so prone to leak can survive so long is a mystery. In general, however, Coley-Smith associates leakage with reduced longevity.

Sclerotia of *R. tuliparum* leak a wide spectrum antibiotic, which accounts for about 10% of the dry weight of the sclerotia (Gladders, 1976, in Coley-Smith, 1979). Whether this antibiotic plays a role in survival of this species is not known, because sclerotia of all isolates possess the antibiotic. Gladders moistened sclerotia of *R. tuliparum,* killed them by heating to 100°C, which also destroyed the heat-labile antibiotic, and buried them in soil. The sclerotia remained physically intact for at least 2 years, which indicates that the antibiotic may not be important in longevity.

Coley-Smith (1979), using *Bacillus subtilis* as the assay organism, found that sclerotia of *Rhizoctonia carotae* and *R. cerealis* exude antibiotic but that those of *R. solani* do not. He also reported the existence of antibiotic in sclerotia of *Typhula incarnata*.

Water Regulation

Rind cells are melanized, and Trevethick and Cooke (1973) attempted to determine whether the rinds of *Sclerotium rolfsii*, *S. delphinii*, or *Sclerotinia sclerotiorum* restrict water loss. They removed mature sclerotia from the surface of cultures, weighed them, then placed them in a desiccator. *S. delphinii* lost most of its water in 20 hr. Most of the water was gone from *S. rolfsii* in 30 hr, and from *Sclerotinia sclerotiorum* in 70 hr. After drying to equilibrium, the sclerotia were rehydrated in distilled water. Water uptake was extremely rapid (1 to 3 hr) in *Sclerotium delphinii* and *Sclerotinia sclerotiorum*. The small size of sclerotia of *Sclerotium rolfsii* prevented Trevethick and Cook from obtaining hydration data for this species. They sliced sclerotia in half and rehydrated the halves. Rehydration of whole sclerotia was as rapid as of sliced sclerotia, indicating that the rind in these species did not restrict water movement. *S. rolfsii* retained more water in the dehydrated state (about 20% of its original water) than the others. Drastic dehydration (4 weeks at 0% relative humidity) did not reduce germination. The authors concluded that water content of these sclerotia in soil is determined by the wetness of the soil and that they possess no means of regulating their water content.

Sclerotia of *S. rolfsii* survive rapid drying. Punja and Grogan (1981a) dried sclerotia in a laminar flow chamber. About 30% of the moisture was lost in the first hour, and by 2 hr about 60% was gone. The ability of the sclerotia to germinate eruptively went from 28% at zero time to 100% in 1 hr [Fig. 13.2, (1)]. Drying for 7 hr reduced the water content by 92%, and the sclerotia were still viable. It is unlikely that drying in nature could be this fast. Drying is not a threat to survival of sclerotia of this species, in contrast to those of *Phymatotrichum omnivorum*.

EFFECT OF ENVIRONMENT ON LONGEVITY AND GERMINATION

Sclerotia of *Sclerotium rolfsii* survive in and on natural field soil at -15 bars for 6 months at 15, 25 and 35°C with no loss in viability (Beute and Rodriguez-Kabana, 1981). Their state of dormancy must be great to resist exhaustion by respiration at 35°C for 6 months. Viability on and in wet soil($-1/3$ bar) is maintained for 6 months at 15°C at the 100 to 97% level, but it declines gradually with higher temperatures, until at 35°C it is about 60%. Viability is reduced a small but consistent amount by burial 5 to 6 cm deep when compared with sclerotia on the soil surface. Beute and Rodriguez-Kabana concluded that sclerotia of *S. rolfsii* are well adapted to survive under normal field conditions.

Rice stem rot in California is correlated with the population of viable sclerotia in the soil at planting time. The number of sclerotia of *Sclerotium oryzae* remaining buoyant and germinable is inversely related to the wetness between field capacity and air dryness, 79% remain buoyant, and 62% germinate. When sclerotia are incubated in soil at field capacity at 24°C for 2 to 10 weeks, 10% remain buoyant and 17% germinate. When water is in excess (60% by weight), only 2% remain buoyant, and 1% germinate.

Sclerotia of *S. oryzae* form in rice stems of diseased plants and are returned to the soil with the straw. In California, some heavy, poorly drained soils are essentially unprofitable for crops other than rice. The longevity of the sclerotia [up to 6 years in soil in Arkansas, according to Tullis and Cralley (1941)], the lack of profitable rotation crops in part of California and the lack of adequately resistant varieties make cultural controls important. Burning the straw after harvest greatly reduces inoculum levels, especially if done repeatedly.

Weakened Sclerotia

Davey and Leach (1941) drenched soil infested with *Sclerotium rolfsii* with formalin (37% formaldehyde) solutions of varying strengths (0 to 1:1000, formalin:water) and found that as the concentration of formalin increases, the germinability of the sclerotia decreases, and as germinability decreases, invasion of sclerotia by *Trichoderma* increases (Table 13.2).

Eruptive Germination

Eruptive germination (germination by a mass of hyphae from a portion of a sclerotium, a process that exhausts the sclerotium) of sclerotia of *Sclerotium rolfsii* is suppressed by the presence of nutrients (Punja and Grogan, 1981a). When eruptive germination occurs and no source of nutrients is available, secondary (replacement) sclerotia develop. Eruptive germination on quartz sand gradually increases with the age of undried sclerotia from 7% for sclerotia from PDA cultures 2 weeks old to 42% for sclerotia from 10-week-old cultures. If the sclerotia are dried, eruptive germination of sclerotia of all ages (>2 weeks) is essentially 100%. Washing to remove nutrients from the

Table 13.2. Effect of weakening sclerotia of *Sclerotium rolfsii* on invasion by *Trichoderma* spp. Davey and Leach, 1941

Formalin:water	Germinability, %	Trichoderma, %
Control	98	0
1:1000	87	3
1:400	65	16
1:200	31	21
1:100	9	57

Source: Davey and Leach, 1941.

sclerotia increases eruptive germination. Punja and Grogan suggested that membrane changes and enzyme activities are affected by drying.

When fresh, undried sclerotia are placed adjacent to bean seedlings or young sugar beet plants, no infection occurs. When a dead leaf and a fresh sclerotium are placed adjacent to the plant, 57% of the beans and 40% of the sugar beets are infected. When dried sclerotia are used without a dead leaf, 95% of the beans and 82% of the sugar beets are attacked. Adding a dead leaf does not increase infection when dried sclerotia (conditioned for eruptive germination) are placed next to the hosts. Punja and Grogan (1981b) concluded that many infections at the soil surface may result because of drying of sclerotia in the surface soil.

Carpogenic versus Myceliogenic Germination in Epidemiology

Snow mold, caused by *Typhula idahoensis* and *T. ishikariensis*, occurs erratically in Washington. It may develop every 2 years or so, or it may fail to develop for 3 or 4 years, depending on snow conditions. In dry autumns no basidiocarps form, and infection is accomplished only by myceliogenic germination of sclerotia. In wet autumns, basidiocarps form on some sclerotia, so both carpogenic and myceliogenic germination occurs.

We wondered why the sexual state persisted, why it was preserved by selection. In the 1982–1983 season, after 3 to 4 years with little snow mold, and consequently with little or no reproduction, snow mold developed uniformly on large areas of winter wheat. Every leaf of every plant in local areas was diseased. The residual sclerotial population in the soil could not possibly have accomplished this uniform infection. The autumn was prolonged and wet, favorable for sporulation. A significant spore shower from basidiospores must have produced the uniform infection. The sexual state appeared to have "rejuvenated" the population numerically (D. Jacobs, personal communication).

The aggressiveness of mycelium from sclerotia exceeds that of infections originating from basidiospores. When the sclerotium population is low, basidiospore-infections would experience little competition from sclerotia germinating myceliogenically, and they would function in a significant way (Cunfer and Bruehl, 1973). The sexual state is significant in epidemiology in ways independent of genetic interchange. Sclerotial numbers in soil are not always correlated with disease incidence.

COMMENTS

Preparing this chapter impressed me with the need for more physiologic studies of propagules. Size, shape, and density influence the inoculum potential (mass) and physical properties, and they are essential aids to identification of pathogens, but physiology is the key to understanding. Do they respond to fungistasis? How long do they survive under different conditions? Do they germinate all at the same time, some now, some later, or only after a resting period? Characteristics of this type demand persistent and imaginative experimentation, but they influence epidemiology and are worth much study.

Elucidation of the effect of drying sclerotia of *Sclerotium rolfsii* on the mode of germination (by a few hyphae or eruptively) is a real step forward. Sclerotia conditioned to germinate eruptively can infect without a food base. Few unconditioned sclerotia can. The mass (energy content) of the two are equal, yet their effects differ greatly.

Evidence is clear that large propagules are more effective than small ones of the same type in causing infection, yet another effect of size was observed by Wijetunga and Baker (1979). In a study of suppression of *Rhizoctonia solani*, they found that suppression develops quickly when only small (<250 μm) propagules are used but not when the propagules are large (>589 μm). They inoculated natural soil with these propagules and then planted and removed radishes at weekly intervals thereafter. During the first 3 weeks disease was the same. At 4 weeks disease began to decline in the small-propagule pots, and by 6 weeks it was almost completely suppressed. In contrast, disease suppression was undetectable at 10 weeks in the large-propagule series, at which time the experiment was terminated. Davey and Leach (Table 13.2) had shown that weakened sclerotia of *Sclerotium rolfsii* are attacked by *Trichoderma* spp. It may be that the relatively weak hyphae of *R. solani* emerging from small propagules are highly susceptible to the suppressive agent, and that stronger hyphae are not (See *R. solani*, Chapter 18).

The sclerotia of all the important sclerotial diseases of paddy rice float on water, as do teliospores of the rice bunt fungus. This is evidence of the adaptability of pathogens to particular environments. It is especially impressive in that sclerotia of *R. solani* of the rice sheath-blight fungus develop buoyancy, whereas close relatives, even members of the same anastomosis group, do not.

The evidence that sclerotia have little control of their water content and are in equilibrium with their environment shows that they must have effective means of maintaining a low level of metabolism during prolonged periods of survival in moist soil. Otherwise, use of food reserves in respiration would quickly exhaust them.

What is the significance of the droplets of liquid that form on sclerotia of several species? Is the exudate the means of maintaining translocation into the young sclerotia at a rate faster than materials can be polymerized? The resorption of materials in the exudate on sclerotia of *Sclerotinia sclerotiorum* with changes in its composition during resorption (Colotelo, 1973) in-

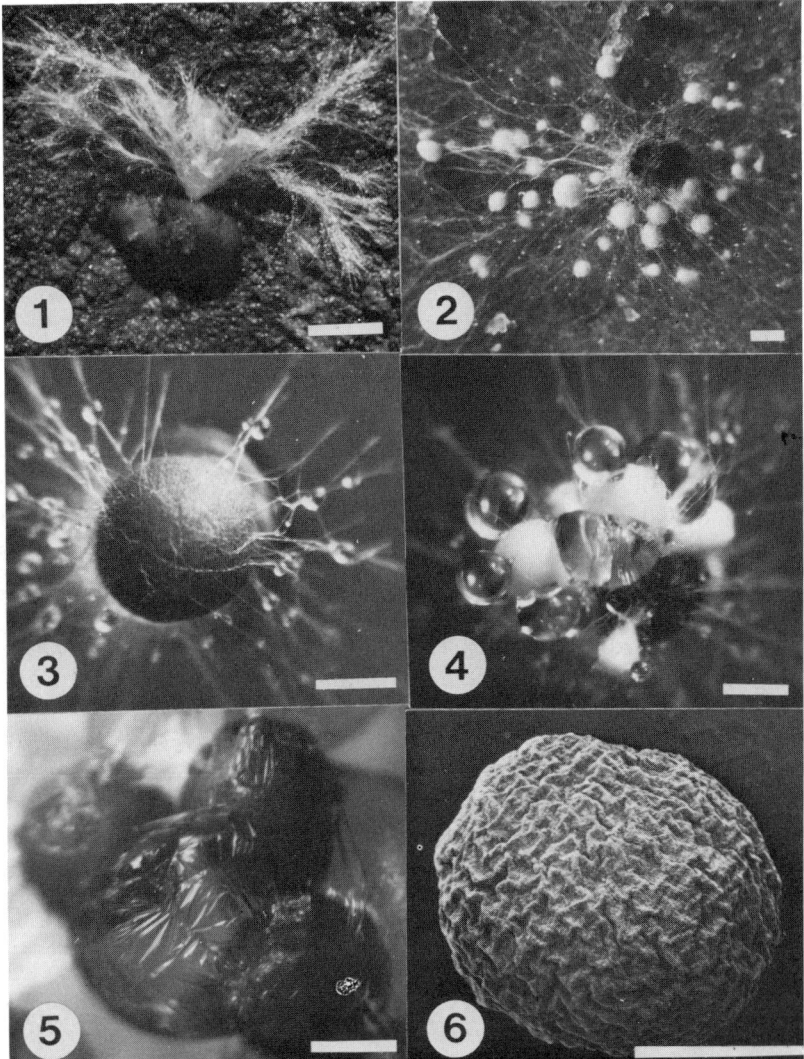

Figure 13.2. Eruptive germination of a sclerotium of *Sclerotium rolfsii* (1), formation of secondary sclerotia (2), myceliogenic or hyphal germination (3), droplets of exudate (4), evaporation of the exudate to form a membranous deposition (5), and the surface (rind) of a mature, dried sclerotium (6). Punja and Grogan 1981a.

dicates that the process is more fundamental than the discharge of excess fluids as the dry matter content of sclerotia increases. Resorption could be significant to sclerotia formed on plant surfaces, but what would be the fate of exudate from sclerotia within soil? Punja and Grogan (1981a) estimated a

3.9% loss in total weight when undried sclerotia of *Sclerotium rolfsii* are washed. They attributed this loss to materials on the surface of the sclerotia [Fig. 13.2, (4), (5)]. Water soluble residues on the surface would support an active microflora and initiate the mycosphere effect of Gilbert and Linderman (1971).

The physiology of translocation into resting structures seems to be neglected. Oospores are certainly more dense than hyphae of most phycomycetes, and cytoplasmic streaming is remarkable. Is liquid lost to the medium during sporangium, oogonium, and antheridium formation? Considerable concentration gradients are overcome during these events.

KEY REFERENCES

Christias, C. and J. L. Lockwood. 1973. Conservation of mycelial constituents in four sclerotium-forming fungi in nutrient-deprived conditions. *Phytopathology* **63**:602–605.

Coley-Smith, J. R. 1979. Survival of plant pathogenic fungi in soil in the absence of host plants. Pages 39–57 in *Soilborne Plant Pathogens*. B. Schippers and W. Gams, eds. Academic Press, New York.

Hashiba, T. and S. Mogi. 1975. Developmental changes in sclerotia of the rice sheath blight fungus. *Phytopathology* **65**:159–162.

Punja, Z. K. and R. G. Grogan. 1981b. Mycelial growth and infection without a food base by eruptively germinating sclerotia of *Sclerotium rolfsii*. *Phytopathology* **71**:1099–1103.

REFERENCES

Ben-Yephet, Y. and Y. Pinkas. 1976. A cesium chloride flotation technique for the isolation of *Verticillium dahliae* microsclerotia from soil. *Phytopathology* **66**:1252–1254.

Beute, M. K. and R. Rodriguez-Kabana. 1981. Effect of soil moisture, temperature, and field environment on survival of *Sclerotium rolfsii* in Alabama and North Carolina. *Phytopathology* **71**:1293–1296.

Bruehl, G. W., R. Sprague, W. R. Fischer, M. Nagamitsu, W. L. Nelson, and O. A. Vogel. 1966. Snow molds of winter wheat in Washington. *Agric. Exp. Sta. Bull.* 677.

Chien, C. C., S. C. Jong, and C. L. Chu. 1963. Studies on the number of sclerotia of the rice sheath blight fungus dropped on the paddy field and difference of germinability between natural and cultured ones. *Int. Taiwan Agric. Res.* **12**:7–13 (Engl. summary).

Chinn, S. H. F., R. J. Ledingham and B. T. Sallons. 1960. Population and viability studies of *Helminthosporium sativum* on field soils. *Can. J. Bot.* **38**:533–539.

Christias, C. 1980. Nature of the sclerotial exudate of *Sclerotium rolfsii* Sacc. *Soil Biol. Biochem.* **12**:199–201.

Coley-Smith, J. R. and D. J. Dickinson. 1971. Effects of sclerotia of *Sclerotium cepivorum* Berk. on soil bacteria. *Soil Biol. Biochem.* **3**:27–32.

Colotelo, N. 1973. Physiologic and biochemical properties of the exudate associated with developing sclerotia of *Sclerotinia sclerotiorum* (Lib.) DeBary. *Can. J. Microbiol.* **19**:73–79.

Cooke, R. C. 1969. Changes in soluble carbohydrates during sclerotium formation by *Sclerotinia sclerotiorum* and *S. trifoliorum*. *Trans. Brit. Mycol. Soc.* **53**:77–86.

Cunfer, B. M. and G. W. Bruehl. 1973. Role of basidiospores as propagules and observations on sporophores of *Typhula idahoensis*. *Phytopathology* **63**:115–120.

Davey, A. E. and L. D. Leach. 1941. Experiments with fungicides for use against *Sclerotium rolfsii* in soils. *Hilgardia* **13**:523–547.

Garrett, S. D. 1944. *Root Disease Fungi*. Chronica Botanica Co., Waltham, Mass.

Garrett, S. D. 1956. *Biology of Root-Infecting Fungi*. Cambridge Univ. Press. London.

Gilbert, R. G. and R. G. Linderman. 1971. Increased activity of soil microorganisms near sclerotia of *Sclerotium rolfsii* in soil. *Can. J. Microbiol.* **17**:557–562.

Henis, Y. and Y. Ben-Yephet. 1970. Effect of propagule size of *Rhizoctonia solani* on saprophytic growth, infectivity, and virulence on bean seedlings. *Phytopathology* **60**:1351–1356.

Humpherson-Jones, F. M. and R. C. Cooke. 1977. Changes in sclerotium formation of *Sclerotinia sclerotiorum* in culture. *Trans. Brit. Mycol. Soc.* **68**:459–461.

Keim, R. and R. K. Webster. 1974. Effect of soil moisture and temperature on viability of sclerotia of *Sclerotium oryzae*. *Phytopathology* **64**:1499–1502.

King, C. J., H. F. Loomis and C. Hope. 1931. Studies on sclerotia and mycelial strands of the cotton root rot fungus. *J. Agric. Res.* **42**:827–840.

Lee, F. N. and M. C. Rush. 1983. Rice sheath blight: A major rice disease. *Plant Dis.* **67**:829–832.

Naiki, T. and T. Ui. 1978. Ecological and morphological characteristics of the sclerotia of *Rhizoctonia solani* Kuhn produced in soil. *Soil Biol. Biochem.* **10**:471–478.

Odvody, G. N. and L. D. Dunkle. 1973. Overwintering capacity of *Ramulispora sorghi*. *Phytopathology* **63**:1530–1532.

Punja, Z. K. and R. G. Grogan. 1981a. Eruptive germination of sclerotia of *Sclerotium rolfsii*. *Phytopathology* **71**:1092–1099.

Rodriguez-Kabana, R., P. A. Backman and E. A. Wiggins. 1974. Determination of sclerotial populations of *Sclerotium rolfsii* in soil by a rapid flotation-sieving technique. *Phytopathology* **64**:610–615.

Rogers, C. H. 1936. Apparatus and procedure for separating cotton root rot sclerotia from soil samples. *J. Agric. Res.* **52**:73–79.

Townsend, D. B. and H. J. Willetts. 1954. The development of sclerotia of certain fungi. *Trans. Brit. Mycol. Soc.* **37**:213–221.

Trevethick, J. and R. C. Cooke. 1973. Non-nutritional factors influencing sclerotium formation in some *Sclerotinia* and *Sclerotium* species. *Trans. Brit. Mycol. Soc.* **60**:559–566.

Tullis, E. C. and E. M. Cralley. 1941. Longevity of sclerotia of the stem rot fungus, *Leptosphaeria salvinii*. *Phytopathology* **31**:279–281.

Watanabe, T., R. S. Smith, Jr. and W. C. Snyder. 1970. Populations of *Macrophomina phaseoli* in soil as affected by fumigation and cropping. *Phytopathology* **60**:1717–1719.

Wheeler, B. E. J. and J. M. Walker. 1965. The production of sclerotia by *Sclerotium rolfsii*. II. The relationship between mycelial growth and initiation of sclerotia. *Trans. Brit. Mycol. Soc.* **48**:303–314.

Wijetunga, C. and R. Baker. 1979. Modeling of phenomena associated with soil suppressive to *Rhizoctonia solani*. *Phytopathology* **69**:1287–1293.

Chapter 14

Survival of Bacteria in Soil

Plant pathogenic bacteria are poorly equipped to survive in soil in the absence of a host or outside of host tissues invaded while the host was alive. This seems strange, because soil is the home of most of the bacteria of the world. Spores are absent among the main genera of Eubacteriales pathogenic to plants (*Pseudomonas, Xanthomonas, Agrobacterium, Corynebacterium, Erwinia*). Spores form in *Streptomyces* and *Bacillus*, and the latter genera persist indefinitely in soil. *Streptomyces ipomea* attacks sweet potatoes and *S. scabies* attacks tubers of *S. tuberosum. Bacillus cereus* produces toxins in soil that cause frenching of tobacco and deformation of some other plants. Bacteria that form spores are not discussed here.

The competitive ability of bacteria is governed, probably more than for any other group of microorganisms, by subtle physiologic differences. An indication of the significance of physiologic differences in competitive relationships is the ease with which bacterial plant pathogens lose pathogenicity in culture. The genome of bacteria is limited, and relatively small changes in it have profound effects. No true, nonspore-forming bacterium is fully competitive both as a saprophyte in soil and as a parasite of living plants.

The number of soil bacteria within free water films around moist soil particles, on the basis of free liquid volume, is about the same as in sour milk or fresh feces (Clark, 1967). In general, the numbers of soil bacteria are proportional to available nutrients within the soil. Nutrition is, therefore, the factor that generally limits reproduction. Bacteria are abundant on root surfaces. They may coat fungal hyphae and clay particles. They occur mostly in clusters, or mini-colonies, rather than as individuals. Bacteria can move significant distances (up to 2.5 cm/day) in soil water by use of flagella under favorable conditions, and they are moved by fungal hyphae or growing roots, or by anything else that moves in soil. The bacterial flora of the rhizosphere typically consists of *Pseudomonas* (13.6%), *Agrobacterium* (1.4%), and *Xanthomonas* spp. (1.2%) (Clark, 1967). The rhizosphere is higher in simple sugars and amino acids than bulk soil, enabling some bacteria that cannot compete in bulk soil to thrive in a rhizosphere. The vast majority of bacteria pathogenic to plants, with the exception of *Bacillus cereus,* diminish in soil in the absence of a host or a favorable rhizosphere. Clean propagating materials and crop rotation with few weeds are useful in the control of most bacterial diseases, evidence that these bacteria are not soil inhabitants and that they are poorly adapted to survive in soil.

Serious effort should be made to determine why some species survive for years in sterile water but not free in moist soil. *Pseudomonas solanacearum* remains alive for several years at room temperature in sterile water (Lucas, 1975). *P. tabaci* likewise survives for $3\frac{1}{2}$ years (Valleau et al., 1943) and *Agrobacterium tumefaciens* for 4 years (J. Ogawa, Univ. Calif., Davis, personal communication). These cells, therefore, resist starvation for many years. Do compounds in soil solutions stimulate loss of bacterial cell contents over time? Within bacterial masses, as in infested host materials, there may be chemical or physical factors that maintain a hypobiotic state or protect the bacteria from substances in the soil solution that reduce longevity.

There is little (or no?) evidence that plant pathogenic bacteria multiply in natural soil, except in the rhizospheres of favorable plants or in other special situations. *Agrobacterium tumefaciens* is usually listed as a soil inhabitant, but is it? This bacterium has a very wide host range. What is known about its prevalence in the rhizospheres of plants? A true soil inhabitant should be able to survive indefinitely in bulk soil and should not be dependent on rhizosphere effects. Dickey (1961) found that *A. tumefaciens* survives longer in fine- than in coarse-textured soil, and survival is greater in cool (2.5°C) than in hot 34°C) soil. This bacterium survives in soil outdoors in the absence of host plants for at least 250 days, by which time an initial count of 10^8 is reduced to about 10^3 viable cells per gram of soil. Dickey concluded that *A. tumefaciens* is a soil invader but not a soil inhabitant. If *A. radiobacter* is a true soil inhabitant, then the Ti plasmid that renders *A. tumefaciens* pathogenic must also reduce the saprophytic competitive ability of receptor cells in nature. Buddenhagen (1965) discussed the relation of bacteria to the soil and placed them in categories of relative adaptiveness. His paper is highly recommended.

EXAMPLES

Pseudomonas solanacearum

For many years *P. solanacearum,* the species responsible for southern bacterial wilt of many solanaceous plants (race 1, broad host range; race 3 primarily on potatoes) and moko disease of bananas (race 2), was thought to persist indefinitely in soil. Recent evidence indicates that *P. solanacearum* is not a soil-inhabiting bacterium. In Georgia, Dukes et al. (1965) reported severe bacterial wilt of tomato on land newly cleared from forest. The bacterium is considered indigenous. Weed hosts could account for the bacterium's presence in virgin soils. McCarter et al. (1969) sampled Georgia soils to 90 cm depth and found a highly variable distribution with depth. In five locations *P. solanacearum* is abundant in the top 30 cm, below which the species is rare or absent. In one site the bacterium is not detected in the top 15 cm, but significant numbers exist from that depth to the 60 to 75 cm zone. The topsoil of this site was dry at the time of collection. They believed the bacterium is reduced in surface soil to undetectable levels by drying. Kelman (1953) recorded low tolerance of this bacterium to desiccation.

Rotations are of little value in reducing the population of race 1, the broad host-range race, in Costa Rica, where temperatures are moderate and soil moisture is maintained by frequent rains (Jackson and Gonzalez, 1981). No control is obtained with maize, cowpeas, wilt-resistant tomatoes, or sweet potatoes. The only treatment that reduces the inoculum level is resistant potatoes, in which the soil is kept free of weeds. Races 2 and 3, which have restricted host ranges, are more effectively reduced by rotation.

When bean and maize, both presumed nonhosts, are root-dip inoculated with race 1, high populations are obtained from surface-disinfected roots 12 weeks after inoculation by comminuting the root systems in water (Granada and Sequeira, 1983). Because root-dip inoculation could favor infection more than is possible in nature, Granada and Sequeira started maize, bean, sorghum, peas, soybean, and rice (all "nonhosts") from seed in soil infested with four isolates. Bacteria were recovered from the roots of some plants of all species after surface disinfection, proving that *P. solanacearum* invades roots of symptomless carriers. Race 1 is better adapted to survive in "nonhosts" than race 2 or 3.

Nesmith and Jenkins (1983) studied survival of *P. solanacearum* in four North Carolina soils, two in which bacterial wilt is a consistent problem (persistent soils) and in two in which bacterial wilt is not a consistent problem (nonpersistent soils). They introduced large quantities of race 1, biotype 1 in September and monitored its decline. The four soils are similar in pH. Because weeds are significant for survival, part of each area was fumigated with methyl bromide to kill weed seeds prior to inoculation. In addition, all microplots were cultivated 2 times per week. In spite of these precautions, bacterial wilt developed the following spring on tomato transplants in the two favorable (persistent) soils. The bacteria declined more rapidly in the unfavorable soils than in the favorable soils. After 60 days bacteria were undetectable on a selective medium, but when susceptible tomatoes were transplanted, wilt developed in all fumigated soils and in the nonfumigated favorable soils but not in the nonfumigated unfavorable soils. Some intrinsic soil factors, probably of biologic origin but not attributable to weeds, affect survival of the bacteria. Differences in bacterial wilt within short distances under the same cultural practices and the same cultivars are common in North Carolina, and they are often associated with soil type.

The two soils in which *P. solanacearum* persists are classified as well to moderately well drained, and the two soils in which the bacterium is nonpersistent are either poorly or excessively drained (Nesmith and Jenkins, 1985). Survival is good in all four soils at -0.5 to -1 bar. The soil type affects moisture relationships which in turn affects the populations of antagonistic organisms. Antagonists, though present, are not important at 0.5 to -1 bar. In drier soils actinomycetes are important, in wetter soils, bacteria. Survival is poorest in saturated soil and at -15 bars in all four soils. Their data implicate antagonism as a significant factor. This explains, to me at least, why these bacteria survive for long periods in water but not in saturated soil. Survival is augmented by reproduction. In the wettest treatments, from flooded to -1 bar, *P. solanacearum* increases for a while in all soils, but under very wet conditions avirulent isolates increase relative to virulent isolates. Kelman and Hruska, in Nesmith and Jenkins (1985), reported that the greater motility of avirulent strains without heavy polysaccharide are more motile and more capable of aerotactic orientation.

Race 2 does not persist long in soil in the absence of a suitable host (Stover, 1972), and differences in survival exist within this race. Being fallow for 12 months eliminates the B-strain (regular banana wilt strain) and for 6 months eliminates the SFR (small, fluidal, round) banana substrain (see Chapter 24 for a discussion of these strains). An isolate of the potato race (race 3) of *Pseudomonas solanacearum* survives less than 23 weeks in a half-sand, half soil-mix at 56% field capacity (-2 bars). Race 2 (the banana race) survives less than 20 weeks, and some bacteria of race 1 are alive after 20 weeks but not at 24 weeks at 28°C (Granada and Sequeira, 1983).

Pseudomonas syringae

Survival of *P. syringae* pv. *tomato* in infested debris was tested at Tifton on a loamy sand in the coastal plain of Georgia, and at Athens on a sandy loam in the Piedmont area (McCarter et al., 1983). On an annual basis, it is about

4.6°F warmer in the coastal plain than in the Piedmont. The bacterium disappeared from infested host material buried 20 cm deep in both soils within 24 days in the summer. Within 60 days the bacterium was not recovered from debris on the soil surface. The bacterium introduced directly into soil and incubated outdoors was not recovered after 15 days in the coastal plain or after 24 days in the Piedmont. Longevity of *P. syringae* pv. *tomato* is very short in the humid, warm Georgia summer. In soil temperature tanks, bacteria free in the soil die within 7 days at 38°C and by 27 days at all temperatures between 18 and 38°C. Within host tissue some bacteria were detected at 18°C after 81 days.

P. syringae pv. *tomato* was isolated from leaves of *Stellaria media*, *Chenopodium album*, and *Plantago lanceolata* in March in the coastal plain, and it survived the summer on roots of *Xanthium pennsylvanicum*, *Lepidium, virginicum, Digitaria sanguinalis*, and *Erigeron canadensis*, even though it did not survive on tomato. In Georgia, where production of tomato transplants stops in June and the beds are worked at that time, high temperature leads to rapid death of *P. syringae* pv. *tomato* in soil and debris. In this area, seed and weed hosts are the main source of *P. syringae* pv. *tomato* primary inoculum in transplant beds.

Valleau et al. (1944) concluded that *P. syringae* pv. *tabaci* is a rhizosphere inhabitant of many plants in Kentucky and that its appearance on tobacco as the incitant of wild fire or black fire is only incidental to its survival. Schneider and Grogan (1977) concluded that pv. *tomato* exists similarly in parts of California and that its spurts of activity in tomato fields has its origin in unseen phases on other plants. They did not find bacteria in the seeds, but they found *P. syringae* pv. *tomato* in rhizosphere soil and on roots and leaves of several plants growing in soil that had never grown tomatoes or that had been in orchard for at least 40 years. In California, especially in the cooler regions, *P. syringae* pv. *tomato* survives as a resident on leaves and roots of nonsuscepts. Most of the tomatoes for processing are produced by direct-seeding: tomato seeds are drilled in the field so there are no transplants and no spread of disease in a transplant bed. Nevertheless, bacterial speck is important in some seasons.

Erwinia carotovora

E. carotovora ssp. *carotovora*, the main incitant of soft rot of vegetables, is widespread and has a wide host range, including many fleshy vegetables and potato. *E. carotovora* ssp. *atroseptica* is primarily a pathogen of potato, associated with both black leg and tuber rot. In their review of the "Ecology of the soft rot erwinias," Perombelon and Kelman (1980) stressed the importance of temperature for the occurrence of these diseases. In potato tuber rot, ssp. *carotovora* and *atroseptica* are important below 15°C. Above that temperature *Closteridium* and *Bacillus* spp. become more important. In causing black leg of potato, ssp. *atroseptica* is more prevalent in cool soil, ssp. *carotovora* at higher temperature, and a mix of both is likely at intermediate temperatures. In potatoes, tubers are a source of inoculum, but true seed would seldom be.

Root maggots are larvae of flies belonging to the family Anthomiidae, genus *Delia (Hylemyia)*. Leach (1940) studied the relationships of *H. platura*, the seed-corn maggot, and black leg and soft rot of potatoes in Minnesota during the 1920s and 1930s. Kloepper et al. (1979) found sspp. *atroseptica* and *carotovora* associated with nine families of diptera near potato cull piles, lettuce and potato fields in Colorado. Phillips and Kelman (1982) identified *E. carotovora* sspp. *carotovora* and *atroseptica* in or on insects. Bacteria within an insect, even though the insect is in soil, should not be classified as soil invaders, because the bacteria are not in direct contact with soil.

Kikumoto (1968) studied the occurrence of *Erwinia* sp. in the rhizosphere of Chinese cabbage near Sendai, Japan. Soft rot bacteria were detected in bulk soil only once (on 20 October), and then in small numbers. Soft-rot bacteria were detected in rhizosphere soil 45 to 48 days after sowing, with the number increasing with the age of the plant. Rhizospheres of Chinese cabbage increased the *Erwinia* sp. from undetectable initial levels to easily detectable levels on the rhizoplane and in the rhizosphere of mature cabbage. Kikumoto and Sakamoto (1969) sampled the rhizospheres of several plants 67 to 82 days after sowing. They detected soft rot bacteria in the rhizospheres of weeds (*Sonchus oleraceus, Chenopodium album,* and *Commelina communis*) and about the roots of *Brassica pekinensis, B. Chinensis, Euchlaena mexicana, Allium tuberosum,* and *Lycopersicum esculentum*. The rhizospheres of most species assayed negative, but the bacterium is favored to some extent within the soil by a variety of roots. Togashi (1981) ranked 14 cultivars of Chinese cabbage (*Brassica pekinensis*) for reproduction of *E. carotovora* in their rhizospheres. Pathogen numbers in the rhizospheres were not correlated with resistance to soft rot, and the increase in numbers was not detected until the plants were past the seedling stages.

Using a selective medium, Meneley and Stanghellini (1976) could not detect *E. carotovora* in soils containing fewer than 1000 bacteria/g of soil. They enriched soil with sodium pectate and mineral nutrients and incubated it anaerobically for 48 hr. They detected what were originally 10 cells/g of soil 100% of the time and 2 to 7 cells/g of soil 78% of the time. They suggested that soft rot bacteria may exist in bulk soils at levels undetectable by most means. Burr and Schroth (1977) enriched soil with a polygalacturonic acid, Tergitol, and mineral broth, and incubated it 48 hr at 28°C under anaerobic or aerobic conditions. Soil enrichment with anaerobic incubation provides the greatest sensitivity in detecting the pathogen in soil. Even with this method, however, Burr and Schroth were unable to isolate *E.*

carotovora from fallow potato fields free of weeds and undecomposed debris. They found soft rot bacteria within the rhizospheres of a diverse group of plants and concluded that these bacteria cannot persist over long periods free in the soil.

Most studies (Perombelon and Kelman, 1980) indicate that *E. carotovora* is not a soil inhabitant, and that it persists longer in infested refuse than free in the soil. Rotation, other than to allow decomposition of host remains, is usually not listed as a control measure unless in relation to an insect vector, such as the seed-corn maggot.

Xanthomonas spp.

The late J. G. Dickson, in lectures on bacterial diseases, stated that many foliage-blighting bacteria are more difficult to control in regions of intermediate rainfall than in dry or wet regions. Rain facilitates spread and infection, but in moist climates infested debris rots quicker and the bacteria die more quickly when outside the living host. Dickson apparently based this statement, made in the 1940s, mainly on *translucens* on barley. But the observation is well founded. If pathogen-free seed is used and residues of diseased crops are the main source of inoculum, climate becomes important in devising rotations.

Brinkerhoff and Fink (1964), in reviewing the literature on survival of *X. malvacearum,* stated that infested debris is important in dry climates such as in Arizona or in the Sudan. In the Sudan, which is dry when irrigation is not underway, bacteria live from one season to the next in dried cotton carpels but not in delicate tissues such as leaves. In the Lake Province of Tanganyika, a relatively wet part of Africa, the bacteria are reduced to very low levels between cotton crops. Brinkerhoff and Fink detected *X. malvacearum* in carpels of 7 of 50 infested bolls on the soil surface that overwintered in Oklahoma. They survived longer than necessary to infect a new planting. They concluded that survival declines as the infested debris decomposes. Undecomposed debris, whether in or on the soil, contains live pathogenic bacteria. Smith (1962) obtained *X. malvacearum* from the rhizospheres of grasses, alfalfa, grain sorghum, and an *Amaranthus* sp. in or near diseased cotton during summer and autumn months. All plants collected in the winter or in fields not containing diseased cotton were negative. He concluded that survival in the rhizosphere of miscellaneous plants is of no importance in New Mexico.

X. campestris, the cabbage black rot organism, dies when free in soil within 14 days in the summer in Georgia and in 42 days in the winter (Schaad and White, 1974). In contrast, they survive at least 244 days in large numbers in infected cabbage stems, and Schaad and White estimated that some bacteria would remain viable for at least 615 days. How does entrapment within host tissue influence survival to this extent? Does bacterial exudate still surround them? Does reduced respiration in masses increase their longevity? Why do they die so quickly when free within soil?

EXUDATE

As a generality, bacterial blights produce masses of bacteria within host tissues. These masses escape to the outside in bacterial exudate, which is a mixture of extracellular polysaccharides and other substances plus live bacterial cells. Hedges (1926) reported that *Corynebacterium flaccumfaciens* survives in dry exudate 5 years. Several workers reported that *Erwinia amylovora* survives for some years in dry exudate from the natural host but that its life is shorter in exudate produced in culture. The greatest longevity of bacterial cells that I am aware of was reported by Murray (1982); *Corynebacterium agropyri* survived in dried host exudate 37 years in the laboratory. Individual bacterial cells of nonspore-forming plant pathogens die quickly when dried under ordinary conditions, but bacterial exudate increases the period of survival. Leach et al. (1957) concluded that bacterial exudate protects bacteria from rapid desiccation and exposure to light. Bacterial colonies in a dormant condition surrounded by mucilage have been observed among clay particles in soil by means of the transmission electron microscope (Russell, 1977).

Many pathogenic fungi whose spores are dispersed by water form spores within a slimy matrix. *Fusarium* does so in sporodochia and *Colletotrichum* in acervuli, and *Hymenula cerealis* (sporodochial stage of *Cephalosporium gramineum*) accumulates spores within a matrix. Nicholson and Moraes (1980) found that spores of *Colletotrichum graminicola* produced in a polysaccharide-protein slime become dried and powdery within 48 hr at RH below 70%. Spores within the matrix survive 4 weeks, but when free of the matrix most are dead in 24 hr, and all die within 48 hr. The matrix protected the conidia during rapid desiccation. Most conidia produced in slimes are hyaline and unprotected by melanin, a condition which gives greater significance to the function of slime. Eggs of root-knot nematodes retain life longer within the egg mass than when free in the soil, so the product of the female rectal glands also has protective value. Slimes (bacterial, fungal, nematode) tend to increase the longevity of propagules.

Rapid Stress

Few organisms in an active or nondormant condition can survive rapid changes in temperature or humidity. *Pseudomonas putida,* a species of potential use in biologic control of diseases, is no exception. Dupler and Baker (1984) added 10^7 cells/g to soil adjusted to $-0.3, -2, -15,$ and -100 bars matric potential, including the water containing the bacterial suspension. *P. putida* died almost immediately when added directly to soil at -100 bars and

27°C. In 5 weeks, when the experiment terminated, the soil originally at -15 had dried to -100 bars and 5×10^5 cells/g of soil were alive. The cells added to moderately dry soil were able to accommodate to some extent to gradual drying, even when not within exudate. When *P. putida* was added to soil at -0.3 bars, survival for 5 weeks was essentially 100%.

Leben (1974) also stressed that active bacterial cells are susceptible to desiccation. There seems little doubt that unspecialized bacterial cells, when outside host tissue and not in masses in exudate, are severely stressed by desiccation. Likewise, tobacco mosaic virus persists in dried host materials for many years, yet the virus particles free of host tissue are disrupted in dry soil (Allen, 1984). But drying does not account for the death of the SFR strain of *Pseudomonas solanacearum* in banana soils in Central America. The surface soil may dry somewhat, but surely no severe moisture stress develops deeper in the soil profile.

COMMENTS

Evidence is strong that bacteria pathogenic to plants survive in soil in infested host debris, in symptomless infections in nonsuscepts, and in conjunction with rhizospheres of favorable plants. Some are harbored as residents on leaves of several plants without visible symptoms. But the evidence is strong that they do not survive for long periods in nonrhizosphere soil when free within the soil. Why? It is difficult to believe that bacteriophages are sufficiently abundant to destroy the bacteria so generally. Most bacterial plant pathogens possess some fatal inability to cope with life free in soil.

Valleau et al. (1943, 1944) discovered the importance of the rhizospheres of many plants, including nonhosts, to survival of plant pathogenic bacteria. Modern studies have reinforced the importance of the rhizosphere, and this phase of survival accounts for earlier beliefs that some bacteria persist indefinitely in soil.

Another phase of survival, as residents on the surface of leaves, explains the occurrence of bacterial plant pathogens not otherwise explainable on the basis of being seedborne or in refuse in or on soil. Leben (1974) defined a resident bacterium as one that multiplies on aerial parts of healthy plants. *Pseudomonas syringae* spread from leaves of hairy vetch to common bean, doing no damage to vetch but severely damaging the beans (Ercolani et al., 1964). Leben (1974) concluded that, "Plant pathogenic bacteria are poorly adapted for survival away from plant tissues," and that ". . . the saprophytic phase seems to contribute few cells which survive."

Plant pathogenic bacteria, though most of them are unable to survive for long free in soil, survived in nature before agriculture. In the absence of crop rotation and because many are seedborne, the presence of favorable rhizospheres, phyllospheres, and infested host debris sufficed to perpetuate the species. Ability to survive free in bulk soil was unnecessary.

This chapter completes the section on survival (saprophytic, dormant spores, sclerotia, bacteria). The next section deals with complex interorganism interactions (competition, microbialization, hyperparasites, mycorrhizae, soilborne viruses).

KEY REFERENCES

Buddenhagen, I. W. 1965. The relation of plant-pathogenic bacteria to the soil. Pages 269–282 in *Ecology of Soilborne Plant Pathogens*. K. F. Baker and W. C. Snyder, eds. Univ. of California Press, Berkeley.
Granada, G. A. and L. Sequeira. 1983. Survival of *Pseudomonas solanacearum* in soil, rhizosphere, and plant roots. *Can. J. Microbiol.* **29**:433–440.
Leach, J. G. 1940. *Insect Transmission of Plant Diseases*, pp. 168–181. McGraw-Hill, New York.
Schaad, N. W. and W. C. White. 1974. Survival of *Xanthomonas campestris* in soil. *Phytopathology* **64**:1518–1520.

REFERENCES

Allen, W. R. 1984. Mode of inactivation of TMV in soil under dehydrating conditions. *Can. J. Plant Pathol.* **6**:9–16.
Brinkerhoff, L. A. and G. B. Fink. 1964. Survival and infectivity of *Xanthomonas malvacearum* in cotton debris and soil. *Phytopathology* **54**:1198–1201.
Burr, T. J. and M. N. Schroth. 1977. Occurrence of soft-rot *Erwinia* spp. in soil and plant material. *Phytopathology* **67**:1382–1387.
Clark, F. E. 1967. Bacteria in soil. Pages 15–49 in *Soil Biology*. A. Burges and F. Raw, eds. Academic Press, London.
Dickey, R. S. 1961. Relation of some edaphic factors to *Agrobacterium tumefaciens*. *Phytopathology* **51**:607–614.
Dukes, P. D., S. F. Jenkins, Jr., C. A. Jaworski and D. J. Morton. 1965. The identification and persistence of an indigenous race of *Pseudomonas solanacearum* in a soil in Georgia. *Plant Dis. Reptr.* **49**:586–590.
Dupler, M. and R. Baker. 1984. Survival of *Pseudomonas putida*, a biological control agent, in soil. *Phytopathology* **74**:195–200.
Ercolani, G. L., D. J. Hagedorn, A. Kelman and R. E. Rand. 1974. Epiphytic survival of *Pseudomonas syringae* on hairy vetch in relation to epidemiology of bacterial brown spot of bean in Wisconsin. *Phytopathology* **64**:1330–1339.
Hedges, F. 1926. Bacterial wilt of beans (*Bacterium flaccumfaciens* Hedges), including comparisons with *Bacterium phaseoli*. *Phytopathology* **15**:1–22.
Jackson, M. T. and L. C. Gonzales. 1981. Persistence of *Pseudomonas solanacearum* (race 1) in a naturally infested soil in Costa Rica. *Phytopathology* **71**:690–693.
Kelman, A. 1953. The bacterial wilt caused by *Pseudomonas solanacearum:* a literature review and bibliography. *N. C. Agric. Exp. Sta. Tech. Bull.* 99.
Kikumoto, T. 1968. Ecological studies on the soft-rot bacteria of vegetable. V. The variation of the rhizosphere microflora in relation to the development of Chinese cabbage. Pages 355–365 in *Jubilee Publication in Commemoration of*

the 60th Birthday of Prof. M. Sakamoto. Inst. Agric. Res., Tohoku Univ., Sendai, Japan.
Kikumoto, T. and M. Sakamoto. 1969. Ecological studies of the soft-rot bacteria of vegetables. VII. The preferential stimulation of the soft-rot bacteria in the rhizospheres of crop plants and weeds. *Ann. Phytopathol. Soc. Jap.* **35**:36-40.
Kloepper, J. W., M. D. Harrison and J. W. Brewer. 1979. The associate of *Erwinia carotovora* var. *atroseptica* and *Erwinia carotovora* with insects in Colorado. *Am. Potato J.* **56**:351-361.
Leach, J. G., V. G. Lilly, H. A. Wilson and M. R. Purvis, Jr. 1957. Bacterial polysaccharides: the nature and function of the exudate produced by *Xanthomonas phaseoli*. *Phytopathology* **47**:113-120.
Leben, C. 1974. Survival of plant pathogenic bacteria. *Ohio Agric. Exp. Sta. Special Cir.* 100.
Lucas, G. B. 1975. *Diseases of Tobacco,* 3rd ed. Biological Consulting Associates, Raleigh, N. C.
McCarter, S. M., P. D. Dukes and C. A. Jaworski. 1969. Vertical distribution of *Pseudomonas solanacearum* in several soils. *Phytopathology* **59**:1675-1677.
McCarter, S. M., J. B. Jones, R. D. Gitaitis, and D. R. Smitley. 1983. Survival of *Pseudomonas syringae* pv. *tomato* in association with tomato seed, soil, host tissue, and epiphytic weed hosts in Georgia. *Phytopathology* **73**:1393-1398.
Meneley, J. C. and M. E. Stanghellini. 1976. Isolation of soft-rot *Erwinia* spp. from agricultural soils using an enrichment technique. *Phytopathology* **66**:367-370.
Murray, T. D. 1982. Recovery of *Corynebacterium agropyri* from ooze on naturally infected *Agropyron smithii* after 37 years. *Phytopathology* **72**:992. (Abstract).
Nesmith, W. C. and S. F. Jenkins, Jr. 1983. Survival of *Pseudomonas solanacearum* in selected North Carolina Soils. *Phytopathology* **73**:1300-1304.
Nesmith, W. C. and S. F. Jenkins, Jr. 1985. The influence of antagonists and controlled matric potential on survival of *Pseudomonas solanacearum* in four North Carolina soils. *Phylopathology* **75**:1182-1187.
Nicholson, R. L. and W. B. C. Moraes. 1980. Survival of *Colletotrichum graminicola:* importance of the spore matrix. *Phytopathology* **70**:255-261.
Perombelon, M. C. M. and A. Kelman. 1980. Ecology of the soft rot Erwinias. *Ann. Rev. Phytopathol.* **18**:361-387.
Phillips, J. A. and A. Kelman. 1982. Direct fluorescent antibody stain procedure applied to insect transmission of *Erwinia carotovora*. *Phytopathology* **72**:898-901.
Russell, R. S. 1977. *Plant Root Systems: Their Function and Interaction with the Soil.* McGraw-Hill, London.
Schneider, R. W. and R. G. Grogan. 1977. Bacterial speck of tomato: sources of inoculum and establishment of a resident population. *Phytopathology* **67**:388-394.
Smith, T. E. 1962. A variant culture of *Xanthomonas malvacearum* obtained from weed roots. *Phytopathology* **52**:1313-1314.
Stover, R. H. 1972. *Banana, Plantain, and Abaca Diseases.* Commonwealth Mycol. Inst., Kew, Surrey, England.
Togashi, J. 1981. Studies on relationship between varietal difference of Chinese cabbage in resistance to the soft rot disease and growth of the causal organism

(Erwinia carotovora) in rhizosphere soils. *Bull. Yamagata Univ. Agric. Sci.,* **8**:657–663.

Valleau, W. D., E. M. Johnson and S. Diachun. 1943. Angular leafspot and wildfire of tobacco. *Kentucky Agric. Expt. Sta. Bull.* 454.

Valleau, W. D., E. M. Johnson and S. Diachun. 1944. Root infection of crop plants and weeds by tobacco leaf-spot bacteria. *Phytopathology* **34**:163–174.

Chapter 15

Competition

Competition denotes a striving for the same object or prize. Antagonism is the state of being in active opposition. A predator plunders, robs, or kills another organism. The most useful way to discuss and to think of interorganism relationships that are not beneficial to the participants is to treat these three states separately. As in most biologic relationships, they are not always clear cut or distinct, and gradations among them exist.

Clark (1965) discussed competition from many viewpoints. He listed five factors of microbial activity given by Waksman in 1952: "competition, as for nutrients; production of unfavorable environmental conditions, as alcohols or acidity; the production of specific antibiotics; parasitism; and predation." Various workers use various terms to refer to these relationships. Most workers excluded antibiosis or antagonisms from competition, thereby reserving competition for more benign relationships. Competition should be limited to organisms striving for the same substrate. The same substrate means substances digestible by the organisms involved. *Pythium* spp. do not compete for lignin because they do not digest lignin.

Competition for food can exist at the level of host-parasite interaction. Plant-parasite fungi enjoy their greatest competitive advantage over the general microflora when within a host, where they develop in almost pure culture with no or little competition from other microbes. They enjoy their greatest advantage in the saprophytic state in host tissues colonized through parasitism, or by colonizaion of the substrate before the host tissue dies. This attribute was called noncompetitive colonization by Trujillo (1969), and prior colonization earlier in this book.

Competition among plant pathogens is lessened by many factors. *Rhizoctonia solani* clones differ in tolerance to CO_2, so some are favored within the soil and others are favored at or above the soil surface. These adaptations presumably reduce competition among clones of this fungus. Pathogens also differ in response to temperature; some are active during cool seasons, others when higher temperatures prevail. Some pathogens tolerate or escape the intense bacterial activity in a spermosphere, and others cannot tolerate it. Some pathogens are capable of attacking delicate rootlets of a tree, and others invade large roots. Some are favored by succulent tissues and some by tissues undergoing senescence. Differences in adaptation reduce direct competition among pathogens.

The niche concept states that two organisms can not long exist if they both occupy the same niche or do the same thing to the same substrate at the same time. If we assume that Gause's principle (1934) is correct, most plant pathogens differ significantly in some way from each other, whether we know how they differ or not.

"Competition is occurring if the growth rate (increase in numbers) of an organism is greater by itself than in combination with another organism" (Norton, 1978). This is the way we interpret competition between a crop plant and weeds, or of inidividuals of the same species. But even this straightforward, simple concept is not so easy to interpret. If root-knot nematodes invade a young plant and stunt it, total numbers of ectoparasitic nematodes would be reduced by the smaller total root system. A wheat plant severely stunted by take-all will have fewer total rust pustules than a plant without take-all. In these cases reduced reproduction of one species by the presence of another is not a result of direct competition.

GENERAL SOIL MICROFLORA

Wilhelm (1965) emphasized the importance of biological balance in natural soil (meaning nonsterilized soil), whether cultivated or not. Additions of fresh substrate upset balance; balance returns, and stability develops as the fresh substrate is exhausted. The multitude of species in soil with their diverse physiologic capabilities is a major cause of balance. The effect of a complex microflora in maintaining balance was illustrated by Wilhelm by the limitation of growth of *Verticillium dahliae* through soil after various treatments. When conidia are added to natural soil, no growth is detected, and the conidia die in a few weeks. In soil fumigated with methyl bromide, slight growth from conidia occurs. After chloropicrin, the fungus is relatively active, and after steam sterilization it colonizes the soil to 3 cm from the inoculum, the maximum distance tested. Its growth is inversely proportional to the population and/or diversity of microflora present after each treatment. A complete microflora enforces balance in natural soil.

It is also true that more nutrients are available in steamed soil than in the other treatments, as well as reduced competition for those nutrients. Steaming, air-drying, or any other severe treatment increases available nutrients by killing all or part of the microbiota. The dead bodies return nutrients to the soil. A common procedure among plant pathologists is to sterilize soil, inoculate it, and to inoculate nonsterile soil with the same amount of inoculum. Comparison of disease severity are than made. In most cases disease is more severe in inoculated, sterilized soil, and the difference is usually attributed to microbial competition, or antagonism. *Rhizoctonia solani, Fusarium* spp., *Pythium* spp., and others can be increased in sterile fertile soils, but few people measure this growth. Ho et al. (1941) grew *Pythium graminicola* in autoclaved soil in culture tubes. At 25°C it grew in soil more than half as fast as on agar media.

Rangaswami and Rangarajan (1966) enumerated *Xanthomonas musicola,* a pathogen of banana leaves, in sterilized and natural soil. In unsterile soil the counts went 21, 6, 4, 3, and 2×10^5 at 0, 3, 7, 14, and 21 days, respectively. Note the large drop (21 to 6) during the first 3 days. In sterile soil the counts were 21, 40, 32, 27, and 19×10^5 at the same time intervals. In the first 3 days the populations of *X. musicola* doubled in sterile soil, and it was not until 14 days after inoculation that the population decreased to that at the time of inoculation. If this were a soilborne pathogen that attacks plants soon after sowing, the disease should have been more severe in sterilized, inoculated soil, not only because of any antagonistic effects that may exist in natural soil, but also because of the increased inoculum in sterile soil. *X. musicola,* a soil-invader, declines rapidly in natural soil; it is not a true soil bacterium. The addition of the pathogen (the invader) had little or no effect on the resident soil population (inhabitants), and the resident, background microflora remained constant and quiescent during the trial.

The citrus nematode, *Tylenchulus semipenetrans,* is important in southern California. Mankau and Minteer (1962) tested several materials for their effects on the survival of *T. semipenetrans* larvae in fallow soil, including castor pomace from which the ricinine was extracted. The survival of larvae of *T. semipenetrans* in fallow soil was reduced by the additions of organic matter, and the number of microphagous (mainly bacterial feeders) nematodes was increased. The test period was 84 days (Table 15.1). Castor pomace was highly effective in eliminating the pathogen, but microphagous nematodes thrived. If nematode parasites or predators were important, they were not effective against the microphagous nematodes, because the latter increased. Nematological literature is full of examples of reduced survival of pathogenic nematodes subsequent to intense microbial activity. Even though the mechanisms of action are essentially unknown, the importance of background microbiota is demonstrated.

Table 15.1. Effect of organic amendments upon survival of *Tylenchulus semipenetrans* and upon the increase of microphagous nematodes in fallow soil

Treatment	T. semipenetrans	Microphagous nematodes	Final soil pH
Nontreated	9,795	380	7.7
Steer manure	7,337	2,510	7.9
Chicken manure	3,915	5,205	7.4
Cotton waste	3,075	9,915	7.8
Sugar beet pulp	2,235	13,055	7.1
Alfalfa pellets	1,780	5,910	6.9
Alfalfa hay	924	4,705	6.9
Castor pomace	0	43,275	5.7

Source: Mankau and Minteer, 1962.

The importance of background microflora in pathogen suppression is illustrated by the frequent devastation that occurs when seedlings in sterilized soil are attacked by a pathogen that contaminates such soil. In the absence of the usual soil microbiota, the introduced pathogen, particularly rapid growers such as *Pythium* spp. and *Rhizoctonia solani,* are often much more destructive than after no soil treatment at all. The work of Baker (1962) and others in developing use of aerated steam for treating soil for greenhouse or plant bed use is particularly significant. Aerated steam, at 60° C for 30 minutes, kills fungal pathogens but not the heat-resistant soil inhabitants. The population of surviving microbes increases following the gentle heat treatment and prevents the uninhibited increase of pathogens which may be introduced into such soil.

Phymatotrichum omnivorum

Phymatotrichum root rot is included in many courses as an example in which biologic control of a plant pathogen is demonstrable. A wide host range and long-lived sclerotia deep within the soil profile make *Phymatotrichum omnivorum* hard to control. Among the controls are 4-year rotations with 3 years of grasses (maize, sorghum, small grains) between cotton or alfalfa crops. Deep plowing, particularly when the soil is dry and warm, assists in controlling the disease. Many have observed benefits from barnyard manure and other organic amendments, but, in general, results from these materials have not been dependable.

Sweet clovers, when grown in the winter following cotton, reduces the disease (Lyle et al., 1948). The cotton crop is harvested and plowed down early and the sweet clover planted in early October. The clover is plowed under about March 15. Cotton is seeded in mid-April. Peas grown as a winter legume in Arizona produce similar results. The winter legumes add nitrogen to the soil, but the main effect is microbiological. Most of the sclerotia are deep in the soil, and it is strange that so simple a treatment directly involving only the plow layer can be effective. Perhaps a favorable microflora becomes established on the cotton seedling roots that follows the roots downward. The biologically active surface soil may limit development of fungal strands on roots in the plow layer, reducing or delaying girdling of the host. Cotton produces more branch roots in surface soils rich in organic matter, with less dependence on the tap root, and this may also contribute to the benefit.

Cotton was once left standing in the field after harvest. It is now known that plowing to a depth of 7 to 9 in. in autumn reduces root rot on subsequent cotton. *P. omnivorum* continues to develop on cotton roots as long as nutrients remain in them. Cutting the root or girdling the plant reduces survival of the pathogen by decreasing the availability of soluble nutrients in the roots (Ezekiel, 1940). Mitchell et al. (1941) isolated bacteria and fungi from the root surfaces of uninjured cotton, from roots cut below the soil surface, and from girdled roots. The surface microflora increased greatly on injured roots as compared to intact roots. An intense surface microflora could reduce ectotrophic activity of *P. omnivorum* on the roots of cotton, resulting in a reduction in total sclerotium production.

Mitchell et al. (1941) buried heat-killed and viable young and old sclerotia in soil heavily amended with superphosphate, chopped green alfalfa, straw, or manure. After 17 days the sclerotia were recovered. Superphosphate did not reduce the number of dead or live sclerotia. The organic amendments reduced the number of recoverable live sclerotia by an average of 37%, and of dead sclerotia by 19%. Under sterile conditions, organic amendments did not reduce sclerotial survival. Viable sclerotia were buried by Clark (1942) in soil amended with ground alfalfa and incubated for 30 days. Sclerotial decomposition was not affected by temperature in unamended soil, but sclerotial numbers declined precipitously in warm soil rich in organic matter. Clark incubated sclerotia in soil amended with several organic amendments, and destruction of sclerotia was proportional to the amount of amendments. Phymatotrichum root rot is reduced by the increase in general biologic activity in the soil because there is no single group of organisms to which sclerotial decomposition can be attributed.

RADIATE EVOLUTION

Fusarium oxysporum is one of the most widespread soil fungi in the world, and it is logical to assume that the highly pathogenic formae speciales originated from it in regions where the specific hosts were common. Little morphologic or general physiologic differentiation occurred, but subtle physiologic differences enabled these fungi to become systemic pathogens in different hosts. There is little or no evidence of significant interformae speciales competition among them. Very similar fungi can coexist for long periods if they do not occupy the same niche. They escape direct competition by developing within different plants.

A significant study of interformae relationships was made by Kraft and Burke (1974) with *Fusarium solani* f. sp. *phaseoli* and f. sp. *pisi*. They selected two fields that had produced 6 consecutive crops of peas or 12 crops of beans. Kraft and Burke split each field and grew peas and beans on each half for 3 more years. For 2 years peas yielded well on former bean land, and beans yielded well on former pea land, but in the third year yields of both were depressed. Beans were attacked by f. sp. *phaseoli* in soil infested with f. sp. *pisi,* and peas were attacked by f. sp. *pisi* in soil infested with f. sp. *phaseoli*. No effective competition existed between the special forms. The two pathogens were noncompetitors, separated by host preference.

INTRASPECIES SELECTION

Many organisms mutate frequently in culture, and they presumably mutate in nature, yet many are remarkably uniform when isolated from diseased tissues. Miller (1945) observed that pionnotal types of *Fusarium oxysporum* f. sp. *melonis* were common in culture and pathogenic when added to sterile soil, yet he did not isolate them from wilted muskmelons. Isolates of *F. solani* f. sp. *phaseoli* are apparently under such severe selection pressure that some clones can be recognized, whether isolated in Australia or in the United States (Cook et al., 1968). I have seen hundreds of fresh cultures of *F. culmorum* (wild-type) from central Washington, and they were essentially alike. It is obvious that intraspecies selection quickly eliminates unfit individuals.

When a pathogen diverges from a saprophytic state, physiologic changes are involved. Chlamydospores of saprophytic *F. oxysprum* germinate in soil with lower levels of exogenous nutrients than those of pathogenic relatives (*F. oxysporum* f. spp. *batatas, cubense* and *lycopersicj*) (Smith and Snyder, 1972). The pathogens, having lost much of their competitive saprophytic ability, preserve inoculum by only germinating in a richer environment, such as in the rhizosphere of a potential host, or in a site in which they can compete.

COMPETITIVE SAPROPHYTIC ABILITY

Garrett (1970) listed attributes contributing to the success of fungi as competitive saprophytes. In abridged form they are, (1) rapid germination of propagules and rapid hyphal growth—to reach the substrate soon after it enters the soil; (2) producing appropriate enzymes to utilize at least part of the substrate; (3) excreting substances either fungistatic or bacteriostatic to reduce the growth of competitors; and (4) tolerance to deleterious substances produced by other organisms within the substrate.

The Cambridge System

A widely used method of assessing saprophytic competitive-colonizing ability, called the Cambridge system, gives relative rankings of competitive saprophytic ability to different fungi. The system employs graded dilutions of soil and inoculum. Butler (1953) grew *Fusarium culmorum* in 100 parts sand, 3 parts maize meal, and 13 parts water by weight for 1 month at 25°C. This inoculum was added to natural soil at 0, 2, 10, 50, 90, 98 and 100%, the 0 rate being pure soil and the 100% rate being pure inoculum. Sterilized wheat straws were buried in these soils immediately after the soil-inoculum mixtures were made. The soil was maintained at a moisture content and temperature favorable for *F. culmorum* for 4 weeks. The straws were then assayed for colonization by *F. culmorum*. In pure inoculum 100% of the straws were colonized and, at 98 to 50% inoculum levels, 92% of the straws were colonized. The colonization at 10, 2 and 0% inoculum levels was 44, 40 and 4%, respectively. Butler concluded that *F. culmorum* is a vigorous competitor of the soil-inhabiting type.

There was no increase in *F. culmorum* following several years of continuous wheat in the northern Great Plains of the United States and southern Canada (Gordon and Sprague, 1941). Cook and Bruehl (1968) found little evidence for the saprophytic increase of *F. culmorum* in central Washington, where wheat is grown without rotation, a system that provides a perpetual supply of wheat straw. Nyvall and Kommedahl (1973) reported no increase in *F. culmorum* following 7 consecutive years of wheat near St. Paul, Minnesota. Many laboratory techniques give results at odds with field observations, and until the results of laboratory techniques are verified by field observations, caution is required in extrapolating laboratory results to the field.

Condition of Substrate

Competitive saprophytic ability is not measured by the increase of a pathogen on living substrates freshly plowed into moist soil. Papaya stems, pineapple leaves, sugar cane stalks, etc., retain life for a time, and some soilborne plant pathogens with quick response invade these structures, not as saprophytes but as parasites. Trujillo and Hine (1965) found that green tissues of pineapple were rapidly colonized by *Thielaviopsis paradoxa* and the immature, white tissues by *Phytophthora cinnamomi*. Papaya stems incorporated in soil were colonized aggressively by *Pythium aphanidermatum*, and the inoculum density was increased by more than 100 times. These are not examples of competitive saprophytic ability, because these tissues, living when incorporated, were colonized within 48 hr. This is comparable to using carrot discs to bait *Thielaviopsis basicola* or apples or other living substrate as baits for other pathogens.

Gindrat and Pilloud (1976) incorporated 20 g of fresh soybean stems and leaves into 1 l of soil and 1 day later seeded the soil with cucumbers. The fresh soybean tissues were colonized by *Pythium* spp. within 24 hr. The leaves increased the natural inoculum of *Pythium* spp. Without soybean tissue, 64% of the cucumbers were alive 1 month after seeding. With the soybean amendment, 10% of the cucumbers were alive. In another experiment, 64% survived without amendment and 2% with amendment. When the soil was amended with soybeans plus diazoben, 32 of 45 survived, but without the fungicide, only 4 survived. Gindrat and Pilloud suggested assaying natural soils for their potential for Pythium seedling blight by using only the natural inoculum of the soil plus fresh soybean tissues incorporated 1 to 2 days prior to seeding with test plants.

Young trees in forest nurseries are subject to attack by pre- and postemergence blights. Wall (1984) reviewed the benefits and dangers of adding organic matter to nursery soils. He indicated that the time interval between incorporation and planting and the nature of the organic matter are important. Decomposed sawdust is superior to fresh sawdust in most cases. Green manures generally increase the damage by *Pythium* spp., and disease from *Rhizoctonia solani* is increased by rye. Wall concluded that the effects of organic amendments are too complex for generalizations, that considerable time should elapse between additions of organic substrates and seeding, and that peat and sawdust are less dangerous than green manure crops. No biological analysis of the soil was made, but this paper illustrates the complexities that exist within soil undergoing marked microbial activity.

In nature a continuum of substrate conditions exists beneath growing plants. Some roots are alive and healthy, some are dead, and many (most?) are in some intermediate condition between dead and healthy. Stack and Millar (1985) question the wide use of "competitive saprophytic ability" and suggest the use of "competitive colonization of organic matter" to express the results of colonization of substrates of varied condition.

Weathering of Substrate According to the Cambridge system, *Fusarium culmorum* has high competitive saprophytic ability. Something reduces this capability where wheat is grown in monoculture in central Washington. Barley is the only important alternate crop in this region. This soil was subjected to continuous additions of wheat or barley straw for at least 50 years, yet Cook and Bruehl (1968) were unable to detect *F. culmorum* in soil dilutions of 58 of 80 random fields. Saprophytism had done little to increase *F. culmorum* in these fields. Cook (1970) buried bright straw collected right after harvest, slightly weathered straw, weathered straw, and severely weathered straw in a naturally infested silt loam containing 800 propagules/g of *F. culmorum*. The longer straw stood in the field before incorporation into soil, the less it was colonized by the pathogen. The low populations of *F. culmorum* in most fields, despite the abundance of straw, indicates the importance of colonization of the straw by miscellaneous molds on the weathered straw before *F. culmorum had* a chance to colonize it.

Carbon:Nitrogen Ratios

Papavizas and Davey (1961) proved that *Rhizoctonia* spp. could colonize buckwheat and oat stem pieces in competition with the general soil microflora. The C:N ratio of the buckwheat straw was 95:1, and that of oat straw was 83:1. Enriching the residue with sodium nitrate favors colonization and increased longevity of *R. solani* in the baits. Increasing the C:N ratio by adding glucose to the stem pieces (both buckwheat and oat) tends to reduce colonization by *R. solani* and to reduce its longevity in the bait pieces. *R. solani* is recovered most frequently from baits 2 to 4 days after burial, after which the recovery rate declines rapidly. *R. solani* in effect attempts sustained colonization and fails more times than it succeeds. Competition rather than exhaustion of substrate is acting because of the short period of occupancy. *R. solani* is recovered from about 90% of the bait pieces at 2 to 4 days. By 10 days it is recovered from fewer than 50% of the baits, but *R. solani* is still alive in about 30% of the baits at 120 days. Competitive saprophytic colonization in the latter cases was successful, and either sclerotia or tough hyphae had formed within the tissues.

Why does nitrogen favor saprophytic competitive colonization and increase longevity of the fungus in the substrate? Straw soaked in sodium nitrate surely contains excess nitrogen, because all the carbon of the stems could not have been immediately available. Why does excess carbon (the glucose-amended stem pieces) reduce saprophytic colonization and longevity? Papavizas and Davey (1961) suggested several explanations, but they tended to attribute the response of *R. solani* to added nitrogen to induced carbohydrate deficiency that reduced bacterial competition.

Davey and Papavizas (1963) adjusted cellulose powder to a wide range of C:N ratios, incorporated it in soil, and sampled the microflora after 3 weeks. Fungal populations were increased and bacterial populations were reduced by low C:N ratios (5:1). Bacteria were most abundant at C:N ratios from about 20 to 60:1, and actinomycetes were most abundant at about a 60:1 ratio. Fungi favored by nitrogen may be relatively poor competitive saprophytes. Green plants have lower C:N ratios than most crop residues, and the basic physiology of most pathogenic fungi is more adapted to live than to dead substrates.

Nutrients from the Soil

Evidence for the ability of pathogenic fungi to obtain nutrients from unamended soil rather than from bits of organic debris within soil is limited.

The experiments of Blair (1943) and Christou and Lockwood (cited in Chapter 13, Table 13.1) prove that *Rhizoctonia solani* assimilates nutrients from soil. Blair put agar media blocks of *R. solani* culture at one end of soil tubes on unamended, moist soil and measured the growth of the fungus from the inoculum blocks. In some tubes the inoculum blocks were left in place; in others, after giving *R. solani* a 2-day start, the mycelium was severed and the blocks were removed. On sand, with the inoculum blocks in place, the fungus grew about 5 cm. On soil, growth was over 20 cm during the same incubation period. A similar amount of growth on soil occurred when the inoculum block was removed after a 2-day start, confirming the ability of *R. solani* to compete saprophytically in soil. Nutrients obtained from the inoculum block sustained 5 cm of growth, and the additional 15 cm of growth on soil was supported by nutrients obtained from the soil.

Jacobs (1984) used this technique with *Typhula* species. He germinated sclerotia on moist soil, allowed the hyphae to grow from the sclerotia over the soil surface, and then severed the hyphae from the sclerotia. Little growth occurred after hyphae were severed from the sclerotia. Pathogenic *Typhula* spp. were poor competitors for nutrients in soil, in contrast to *R. solani*.

COMMENTS

Competition in the parasitic stage is limited by differing host ranges, types of tissues attacked, response to environment, and intrinsic differences in parasitic capabilities of organisms. Clark (1965) reviewed competition primarily as it relates to life within the soil. After a thorough review of the literature, he concluded that competition within soil is for food, that availability of nutrients is the major limiting factor within soil, that even in saprophytism, competition for nutrients is moderated by differences among organisms in their abilities to digest different fractions of the substrate, and that cooperation as well as competition exists. Soils rich in organic matter sustain a large and varied microbiota that stabilizes life within soil. A rich microbiota reduces available food in soil to low levels, the reason most propagules are dormant in soil.

Most plant pathogens are poor saprophytic competitors. Many experiments have exaggerated their saprophytic capabilities. True saprophytes are better at saprophytism. Timing, being established in the substrate first, through parasitism, is the greatest advantage most pathogens capable of saprophytic growth have in competition with other organisms (Fig. 11.1).

Among pathogens of similar parasitic capabilities, such as among seed-rotters and seedling-blighters of wide host ranges, physical factors (temperature and water) influence relative success or failure at any given time. *Pythium ultimum* is a major seed pathogen in cold, wet soil. At 7°C, it grows 7 mm in 24 hr (Middleton, 1943). *P. aphanidermatum*, a warm-hot soil pathogen, makes no growth at 7°C. At 40°C, *P. ultimum* does not grow. In contrast, *P. aphanidermatum* grows 44 mm in 24 hr at that temperature (40°C). If a seed is equally susceptible to both species, temperature and relative growth rates of the pathogens determine which fungus kills the seed.

Seed-rotters compete in what is comparable to a 100-m dash. Attacks on hypocotyls or epicotyls are a different event, a 200-m dash. In peas or beans, *Fusarium solani* or *Rhizoctonia solani* may be outstanding competitors in the latter event. Competition should be studied within events (seed-rotting, post-emergence damping-off, hypocotyl rots, rots of fine rootlets, rots of coarse roots, rots of stems, etc.). A competitor may be competitive in more than one event (*F. solani* may attack the hypocotyl, the tap root, and fine rootlets).

Competition within a species is apparent (Chapter 5, "Nematode Reproduction") when maize seedlings are inoculated with increasing numbers of root-knot nematode larvae. When inoculum is sparse, most larvae establish within the host and reproduce greatly. When inoculum is excessively high, most larvae in the inoculum die, because there are too few root tips to sustain them.

What about competition within the host? I present a hypothetical case. *Verticillium dahliae* and *Fusarium oxysporum* f. sp. *vasinfectum* both cause vascular wilt in cotton. In acid soils Fusarium wilt dominates in the southeastern United States (race 1 of f. sp. *vasinfectum*). In alkaline soils Verticillium wilt dominates (southwestern United States). Now, at some intermediate soil pH, each should be equally competitive in a cotton equally susceptible to both. *V. dahliae* develops at lower temperature than *F. oxysporum* on cotton, so theoretically, at the right soil condition, *V. dahliae* should occupy the vascular elements and win the competition, just as *F. oxysporum f* f. sp. *pisi* race 1 attacks peas earlier in the spring than race 2 (near-wilt). One organism has an advantage over the other, by getting started earlier. This influences the outcome, but is it really competition? If we follow this line of reasoning, then when *Pythium ultimum* attacks cotton seed and kills it, it has out-competed all other pathogens, because no plant remained to be attacked by anything else. I again recommend the analysis of competition as presented by Clark (1965).

It is may opinion that there are no "super" organisms (with the possible exception of *Rhizoctonia solani*), that a well-adapted parasite is unlikely to be a well-adapted saprophyte, and vice versa. The physiologic attributes that lead to high competitive saprophytic ability, whatever they are, contribute little to parasitism. There must be physiologic adaptations that are not combinable in one individual.

Many adaptations equip the fungus *Hirsutella rhossiliensis* to parasitize certain nematodes. When a host, *Criconemella xenoplax,* was exposed in both natural and sterile soil 67 and 77%, respectively, were parasitized. When the nematode was killed by heat and then introduced into natural and sterile soil, 18 and 100%, respectively, were colonized. The fungus was recovered from autoclaved wheat seeds buried in sterile soil but not from seeds in natural soil. Jaffee and Zehr (1985) conclude that this fungus is adapted to parasitism, not to saprophytism (see Chapter 17). Fortunately, this relationship is the rule and not the exception.

KEY REFERENCES

Blair, J. D. 1943. Behavior of the fungus *Rhizoctonia solani* Kuhn in the soil. *Ann. Appl. Biol.* **30**:118-127.
Clark, F. E. 1965. The concept of competition in microbial ecology. Pages 339-345 in *Ecology of Soilborne Plant Pathogens*. K. F. Baker and W. C. Snyder, eds. Univ. of Calif. Press, Berkeley.
Nyvall, R. F. and T. Kommedahl. 1973. Competitive saprophytic ability of *Fusarium roseum* f. sp. *cerealis* 'Culmorum' in soil. *Phytopathology* **63**:590-597.
Papavizas, G. C. 1970. Colonization and growth of *Rhizoctonia solani* in soil. Pages 108-122 in *Rhizoctonia solani: Biology and Pathology*. J. R. Parmeter, Jr., ed. Univ. of Calif. Press, Berkeley.

REFERENCES

Baker, K. F. 1962. Principles of heat treatment of soil and planting material. *J. Aust. Inst. Agric. Sci.* **28**:118-126.
Butler, F. C. 1953. Saprophytic behavior of some cereal root rot fungi. I. Saprophytic colonization of wheat straw. *Ann. Appl. Biol.* **40**:284-297.
Clark, F. E. 1942. Experiments toward the control of the take-all disease of wheat and the *Phymatotrichum* root rot of cotton. *U.S. Dept. Agric. Tech. Bull.* **835**:1-27.
Cook, R. J. 1970. Factors affecting saprophytic colonization of wheat straw by *Fusarium roseum* f. sp. *cerealis* 'Culmorum'. *Phytopathology* **60**:1672-1676.
Cook, R. J. and G. W. Bruehl. 1968. Relative significance of parasitism versus saprophytism in colonization of wheat straw by *Fusarium roseum* 'Culmorum' in the field. *Phytopathology* **58**:306-308.
Cook, R. J., E. J. Ford and W. C. Snyder. 1968. Mating type, sex, dissemination and possible sources of clones of *Hypomyces (Fusarium) solani* f. *pisi* in South Australia. *Aust. J. Agric. Res.* **19**:253-259.
Davey, C. B. and G. C. Papavizas. 1963. Saprophytic activity of *Rhizoctonia* as affected by the carbon-nitrogen balance of certain organic soil amendments. *Soil Sci. Soc. Am. Proc.* **27**:164-167.
Ezekiel, W. N. 1940. Relation of age of cotton plants to susceptibility to field inoculations with Phymatotrichum root rot. *Phytopathology* **30**:704 (Abstract).
Garrett, S. D. 1970. *Pathogenic Root-Infecting Fungi*. Cambridge Univ. Press. London.
Gause, G. F. 1934. *The Struggle for Existence*. Williams and Wilkins, Baltimore.
Gindrat, D. and R. Pilloud. 1976. Une methode simple de stimulation des Pythium dans le sol par incorporation de residus verts de soja. *Rech. Agron. Suisse* **15**:129-136.

Gordon, W. L. and R. Sprague. 1941. Species of *Fusarium* associated with root rots of Gramineae in the Northern Great Plains. *Plant Dis. Reptr.* **25**:168-180.

Ho, W. C., C. H. Meredith and I. E. Melhus. 1941. *Pythium graminicola* Subr. on barley. *Iowa Agric. Exp. Sta. Res. Bull.* 287.

Jacobs, D. L. 1984. Identification, distribution and saprophytic ability of *Typhula* species in Washington and Idaho. Ph.D. thesis. Washington State University, Pullman.

Jaffee, B. A. and E. J. Behr. 1985. Parasitic and saprophytic abilities of the nematode-attacking fungus *Hirsutella rhossiliensis*. *J. Nemat.* **17**:341-345.

Kraft, J. M. and D. W. Burke. 1974. Behavior of *Fusarium solani* f. sp. *pisi* and *Fusarium solani* f. sp. *phaseoli* individually and in combination on peas and beans. *Plant Dis. Reptr.* **58**:500-04.

Lyle, E. W., A. A. Dunlap, H. O. Hill and B. D. Hargrove. 1948. Control of cotton root rot by sweet clover in rotation. *Texas Agric. Exp. Sta. Bull.* **699**:5-21.

Mankau, R. and R. J. Minteer. 1962. Reduction of soil populations of the citrus nematode by addition of organic materials. *Plant Dis. Reptr.* **46**:375-378.

Middleton, J. T. 1943. The taxonomy, host range and geographic distribution of the genus *Pythium*. *Mem. Torrey Bot. Club* **20**:1-171.

Miller, J. J. 1945. Studies on the *Fusarium* of muskmelon wilt. I. Pathogenic and cultural studies with particular reference to the cause and nature of variation in the organism. *Can. J. Res.* **23**:16-43.

Mitchell, R. B., D. R. Hooton, and F. E. Clark. 1941. Soil bacteriological studies on the control of Phymatotrichum root rot of cotton. *J. Agric. Res.* **63**:535-547.

Norton, D. C. 1978. *Ecology of Plant Parasitic Nematodes*. Wiley, New York.

Papavizas, G. C. and C. B. Davey. 1961. Saprophytic behavior of *Rhizoctonia* in soil. *Phytopathology* **51**:693-699.

Rangaswami, G. and M. Ragarajan. 1966. Studies on the survival of plant pathogens added to the soil. IV. *Xanthomonas musicola*. *Ind. Phytopath.* **19**:294-297.

Smith, S. N. and W. C. Snyder. 1972. Germination of *Fusarium oxysporum* chlamydospores in soils favorable and unfavorable to wilt establishment. *Phytopathology* **62**:273-277.

Trujillo, E. E. 1969. Relationship of crop residues to increased persistence and inoculum density of soilborne pathogens. Pages 23-25 in: Nature of the Influence of Crop Residues on Fungus-Induced Root Diseases. R. J. Cook and R. D. Watson, eds. *Wash. Agric. Exp. Sta. Bull.* 716.

Trujillo, E. E. and R. B. Hine. 1965. The role of papaya residues in papaya root rot-caused by *Pythium aphanidermatum* and *Phytophthora parasitica*. *Phytopathology* **55**:1293-1298.

Wall, R. E. 1984. Effects of recently incorporated organic amendments on damping-off of conifer seedlings. *Plant Dis.* **68**:59-60.

Wilhelm, S. 1965. Analysis of biological balance in natural soil. Pages 509-517 in *Ecology of Soil-Borne Plant Pathogens*. K. F. Baker and W. C. Snyder, eds. Univ. of Calif. Press, Berkeley.

Chapter 16

Microbialization

Simmonds (1947) reviewed early work on bacteria on plant surfaces and seeds and stated that Duggeli in 1904 found *Bacterium herbicola* and *B. fluorescens (Pseudomonas)* on wheat seeds and seedlings. The early workers and Simmonds believed that surface microflora are not chance contaminants but are adapted to life on plant surfaces. Novogrudski in 1936 (in Simmonds) developed methods for isolating and testing bacteria lysogenic to *Fusarium graminearum* and to *F. oxysporum* f. sp. *lini*. Beresova and Navumova in 1939 (in Simmonds) applied bacteria, mainly *Pseudomonas* and *Achromobacter* spp., to wheat seed during efforts to reduce disease severity, and they named this procedure *bacterization*.

The potential benefit of applying bacteria to seeds was thought to be great, and by 1958 in Russia 10^7 ha were seeded with microorganisms added to the seed (Brown, 1974). Soon "bacterization" was applied more broadly than for disease control. Bacterization, according to Brown, is "the treatment of seeds or seedling roots with cultures of bacteria that will improve plant growth; such preparations are frequently called bacterial fertilizers."

Kenneth Baker, according to T. Kommedahl, suggested the use of *microbialization*, because this term could logically include application of fungal spores or other propagules, as well as bacteria, to seed or other plant structures. I therefore used this term because of the extensive amount of investigations with fungi as well as with bacteria.

More than 200 investigators are working in microbialization as it relates to plant pathology (Kommedahl and Windels, 1981). About 25% of the organisms being applied to seed belong to three genera of fungi: *Trichoderma, Penicillium,* and *Gliocladium.* Among the bacteria, *Agrobacterium, Bacillius,* and *Streptomyces* are used most frequently. Four commercial applications have been found: (1) crown gall, (2) annosus rot of pine, (3) carnation stem rot, and (4) Verticillium wilt of mushrooms. Kommedahl and Windels accepted the definition of biologic control used by Sewell, "the induced or natural, direct or indirect, limitation of a harmful organism, or its effects, by another organism or group of organisms." The subject of this book is not biologic control, but microbialization experiments are experiments in microbial ecology.

Brown (1974) cited 107 papers dealing with seed and root bacterization. The yield responses in general are smaller on cereals than in horticultural crops, and best results are obtained in fertile soil to which mineral fertilizers has been applied (Mishustin and Naumova, 1962). *Azotobacter chroococcum* is thought to increase growth by fixing nitrogen and *Bacillus megaterium* by mineralizing organic phosphorus compounds. Subsequent studies prove that growth responses could not be attributed to action as bacterial "fertilizers." Mishustin and Naumova listed 26 bacteria and fungi that in various experiments stimulate crop growth. Indian workers, using bacteria supplied from the Soviet Union and local isolates of *Pseudomonas* and *Beijerinckia,* reported increased yields in some field experiments. Brown among others in England and several workers in other countries have obtained similar results. In general, these later workers, as those in the Soviet Union, reported best results on fertile, fertilized soils high in water content.

Under natural conditions neither *Azotobacter chroococcum* nor *Bacillus megaterium* are true rhizosphere bacteria *A. chroococcum* applied to wheat seed in the field colonize the root system, including adventitious roots arising above the seed. The population increases initially, holds static, and then disappears after harvest (Brown et al., 1962). The growth response to *A. chroococcum* is due, at least in part, to its production of small amounts of gibberellins. Many of the positive growth responses involve production of growth regulators, antibiotics, or both, according to Brown.

Schroth and Hancock (1982) prefer the term rhizobacteria for those which thrive as epiphytes on plant roots, to distinguish them from bacteria that do not aggressively colonize root surfaces. Rhizobacteria can be beneficial, harmful, or neutral. Rhizobacteria implies a stronger relationship with roots than rhizoplane or rhizosphere bacteria—the latter might be transient opportunists. Most of the beneficial rhizobacteria are *Pseudomonas fluorescens* or *P. putida,* and only a small percentage of the isolates aggressively colonize roots.

The term plant *growth-promoting rhizobacteria* (PGPR) was used by Kloepper and Schroth (1978) to specify rhizobacteria having beneficial growth-promoting effects on plants in natural soil in the field. When plants are grown under gnotobiotic conditions with PGPR, no increased growth results, indicating that hormones are not responsible for the effect. When mutants of PGPR with no antibiotic-(or siderophore?) producing ability are used in the field, growth is not affected. The wild-type PGPR reduces total fungi in the rhizosphere by 23 to 64%, and gram-positive bacteria by 25 to 93%. A few growth-promoting bacteria do not produce antibiotics, so this subject requires more study.

There are several ways in which a high population of nonpathogens on a plant surface can reduce infections. If organisms are present in great numbers and if they respond quickly to exudates, they can use the exudates before these substances can overcome the fungistasis of propagules in the soil. If applied organisms produce antibiotics rapidly, they can inhibit some pathogens. If applied organisms are slightly "pathogenic," they can elicit an incompatible host response and render the plant temporarily resistant to virulent pathogens. They can compete for essential growth factors other than exudates, and they can occupy infection sites.

NATURAL MICROBIALIZATION

Surface disinfection of wheat seed increases lesions caused by *Bipolaris sorokiniana* (Simmonds, 1947). Ledingham et al. (1949) subjected seed to hot water and to hot water and formalin. Removal of natural bacteria on the seed increased disease when seeds were inoculated with conidia of *B. sorokiniana,* and inoculating wheat seeds with bacteria proved that the bacteria reduces disease development. Spring wheat was grown in a dry atmosphere in the greenhouse with no water applied to the foliage. A week or two after the grain was ripe, portions of the wheat were subjected to 1 day of mist, to 2 days of intermittent moisture and high humidity, or to 3 days of such treatment. Seeds from these plants were then inoculated with *B. sorokiniana*. Three days of weathering increased the natural microflora on the wheat seeds, resulting in some protection against *B. sorokiniana* seedling blight (Table 16.1). Seed disinfection destroyed the surface microflora and led to increased disease severity and incidence. Ledingham et al. concluded that natural seed bacterization could influence results of experiments, and they recommended surface disinfection of seed prior to some types of varietal tests. Giha (1976) and Lang and Kommedahl (1976) made similar observations.

Bipolaris victoriae devastates oats of Victoria crosses in the United States, yet oats do not suffer greatly from Victoria blight in Brazil, even though *B. victoriae* has been isolated from oat seed sent from Rio Grande do Sul, Brazil. The isolates from Brazilian oats are highly virulent on suscepti-

Table 16.1. Disease ratings (incidence × severity) on seedlings of Thatcher wheat

Days of weathering	Seed not disinfected with formalin	Seed disinfected with formalin
0	69	63
1	59	71
2	67	77
3	25	73

Source: Ledingham et al., 1949.

ble oats. Much of the oat seed from Brazil is infected with three species of *Chaetomium*. Tveit and Moore (1954) suspected that the naturally occurring *Chaetomium* on oat seed in Brazil protects the oats from *B. victoriae*. They treated seed naturally infected with *B. victoriae* with *Chaetomium* spp. and found that *Chaetomium* provides control as effective as that from fungicides. The "resistance" of the Brazilian oats to *B. victoriae* disappears when the seed is treated with hot water and subsequently inoculated with the pathogen (Tveit and Wood, 1955). The Brazilian oats are not resistant to the pathogen; they are protected from it by *Chaetomium* on the seed.

SOME EARLY EXPERIMENTS

Morrow et al. (1938) attempted to establish three fungi and two bacteria antagonistic to *Phymatotrichum omnivorum* in soil. The fungi *(Trichoderma lignorum, Aspergillus luchuensis,* and *Penicillium luteum)* provided the most conclusive evidence of establishment, because they are morphologically distinct and identifiable. The bacteria used (*Pseudomonas fluorescens* and *Achromobacter radiobacter)* are less satisfactory, because of difficulty in determining whether recovered isolates are of the introduced strains or of natural occurrence. Modern workers select bacterial strains resistant to certain antibiotics to which the wild population is susceptible, and tolerance to the antibiotics serves to identify the test organisms.

Morrow et al. applied organisms to cotton seed by soaking them in liquid cultures of the antagonists. The seed was then dried and planted. The organisms were also added to soil near the seedlings by pouring liquid suspensions into the soil adjacent to cotton seedlings. The three fungi were obtained more frequently from the roots and soil about diseased plants than from about healthy plants. Inoculation increased the populations of the antagonists on and about both healthy and diseased plants. Pouring inoculum about seedling roots resulted in greater establishment of the introduced organism than did seed inoculation. The disease was not controlled, and colonization of the root or rhizosphere by the introduced organisms was intermittent rather than continuous.

After World War II many experiments were made with antibiotics as a means of controlling seed rot and seedling blights. Pine in 1948 (in Gregory et al., 1952) tried to increase the efficacy of this procedure by pelleting seed in peat moss treated with antibiotics alone or with spores of *Penicillium patulum* and *Trichoderma lignorum*. The pelleting was promising, and Gregory et al. (1952) tried various pelleting procedures, using blends of peat moss, methyl cellulose, and glucose. In a silt loam inoculated with *Pythium debaryanum*, 4% of alfalfa seedlings survived with no pelleting, and 76% survived when seeds were pelleted with *T. lignorum*. Wright (1956) dusted white mustard seeds with spores of *Trichoderma viride, Penicillium nigricans, P. frequentans,* and *P. godleswkii*. Only *P. nigricans* failed to give some protection. A gliotoxin-producing strain of *Trichoderma viride* was more effective than a viridin-producing strain, and each was more effective than a strain that produced neither antibiotic. Wright concluded, however, that protection is not entirely due to antibiotics.

Pythium ultimum and *P. debaryanum* cause severe emergence and postemergence damping-off of table beets in western Oregon (Liu and Vaughan, 1965). Older seedlings are damaged by *Rhizoctonia solani*. Coating the beet seed with spores of *Trichoderma viride* or *Penicillium frequentans* gives some protection against *Pythium* The fungi, both antagonistic to the *Pythiuum* spp., give stands of 71 to 75%. Without fungal spores on the seeds the stands are 21% after 20 days. The seed coatings do not protect against attack by *Rhizoctonia solani* past the early seedling stage.

MECHANISMS

Hypersensitivity

Fusarium avenaceum attacks the cut ends of carnation when this plant is increased by vegetative cuttings in plant beds. Aldrich and Baker (1970) and Michael and Nelson (1972) found that dipping the cuttings in a suspension of nonpathogenic bacteria immediately after cutting protects them from serious damage. This appears at first to be a case of "possession of substrate," the first organism dominating the substrate and preventing the establishment of a second (*F. avenaceum*), but Baker et al. (1978) concluded that it is primarily due to host response when challenged by an incompatible (nonpathogenic) agent. They dipped cut ends of carnation stems in spore suspensions of many fungi, and most offered some degree of protection, due mainly to a hypersensitive host response to the protectants. The culture filtrate of *Fusarium roseum* 'Gibbosum' used as a protectant was effective whether it had or had not been autoclaved. This proved that the protectant need not be alive, ruling out competitive possession of substrate aspects. Windels (1981) found that *Penicillium oxalicum* applied as a seed-protectant on peas produces minute necrotic lesions on the cotyledons. She did not suggest protection via a hypersensitive host response, but this may have been an important contributing factor in the resultant disease control.

Competition for Iron

Pseudomonas fluorescens produces an array of secondary metabolites, including siderophores that sequester iron (Schroth and Hancock, 1982). Some isolates produce siderophores with a high affinity for iron, rendering iron less available for other microorganisms. Fungi produce hydroxymate siderophores with a lower affinity for iron than the bacterial siderophore. The pseudobactin-type siderophore produced by *Pseudomonas* B-10 could deprive fungi of iron.

Kloepper et al. (1980) obtained a rhizobacterium *Pseudomonas* strain B10, from potatoes grown in a take-all–suppressive soil that, when added to take-all–or Fusarium wilt-conducive soils, suppresses the development of both diseases. Pseudobactin isolated from cultures of B10 added every other day to soil suppresses take-all in a take-all–conducive soil and it suppresses flax wilt in a wilt-conducive soil. Adding the siderophore is as effective as inoculating the soil or seed with living B10 prior to planting. Both treatments controlled take-all during a 2-week test period and flax wilt during a 4-week test period. Adding chelated iron to the soils negates disease control by both the siderophore or live bacteria. They concluded that iron is made unavailable to the fungal pathogens by *Pseudomonas* strain B10 and that iron deficiency resulting from microbial action may be a basis for suppression of some diseases caused by fungi.

The water-soluble, yellow-green fluorescent pigments characteristic of many bacteria in the genus *Pseudomonas* are produced in greatest quantities on media low in iron, and these pigments from 156 isolates of *Pseudomonas* spp. inhibit *Geotrichum candidum* in culture (Misaghi et al., 1982). A pigment from an isolate of *P. fluorescens* inhibits *Pythium aphanidermatum* in a direct, quantitative manner. The inhibition is quantitatively counteracted by adding soluble iron in excess of the chelating power of the pigment. The fluorescent pigment has the properties of siderophores (iron-chelators). Misaghi et al. concluded that fluorescent pseudomonads, essentially ubiquitous in agricultural soils, could be important in suppressing fungal pathogens through induced iron deficiency, particularly in alkaline soils.

Smith and Snyder (1971) described soils in the Salinas Valley, California, in which Fusarium wilts are unimportant in spite of repeated growth of susceptible hosts. Scher and Baker (1982) obtained *Pseudomonas putida* isolate A12 from these soils that, when added to conducive soil, suppresses wilts caused by *F. oxysporum* f. sp. *lini,* f. sp. *cucumerinum,* and f. sp. *conglutinans.* Adding iron chelators to the soil also reduces Fusarium wilts. Iron deficiency does not prevent germination of microconidia of *Fusarium oxysporum* f. sp. *lini* in culture but germ tube elongation is greatly reduced. The *P. putida* isolate produces iron chelator, and Sher and Baker concluded that iron deficiency is a cause of wilt suppression.

The suppressiveness of the California soil can be reduced by lowering its pH (Scher and Baker, 1980). At pH 8, about 20% of the flax seedlings wilt; at pH 7, about 60%, and at pH 6, about 82%. Lowering the pH increases availability of iron. Fusarium wilt of cotton was historically more important in acid soils in the southeastern Cotton Belt, but less so from Central Texas westward in neutral or alkaline soils; and soil pH was considered an important ecological factor (Fig. 3.3).

Hydroxymate siderophores of the ferrichrome type can obtain iron in neutral and alkaline soils in which other compounds are ineffective as iron chelators. Powell et al. (1983) used a defective strain of *Escherichia coli* to assay for ferrichrome-type hydroxymate siderophores in soil and found them present in a vertisol soil, pH 7.9, with 7% organic matter. This study proved that siderophores exist in some soils in biologically detectable quantities.

BACTERIA VERSUS BACTERIA

Kloepper (1983) dusted potato seed pieces with dried inoculum of growth-promoting rhizobacteria. The potatoes were planted in a sandy loam, pH 7.2, near Shafter and in peat, pH 7.0, near Tulelake, California. Two weeks before harvest *Erwinia carotovora* populations were determined from root washings. These populations were reduced 95 to 100% by the plant growth-promoting rhizobacteria. Kloepper attributed this reduction to active colonization of the roots by the inoculant and by siderophore plus antibiotic activity of the PGPR in the root zone. In another trial with the same inocula, the number of daughter tubers from which *E. carotovora* was isolated was reduced. This is not an example of benign competition. The inoculant bacteria practically eliminated *E. carotovora* from the root surfaces.

SUPPRESSION OF TAKE-ALL

Dark runner hyphae of *Gaeumannomyces graminis* var. *tritici* grow ectotropically on the surface of wheat roots in advance of penetration, making this fungus subject to inhibitory effects of rhizoplane and rhizosphere microbial competition or antagonism. Antibiotic-producing fluorescent pseudomonad bacteria are more numerous on wheat roots in soil following wheat monoculture than after rotation with other crops, (Smiley, 1978) suggesting that these bacteria may be important in take-all decline (see Chapter 18). Weller and Cook (1983) selected *Pseudomonas fluorescens* antagonistic to *G. graminis* in culture and inoculated wheat seeds with these bacteria. Two isolates reduced take-all of both winter and spring wheat in the field, and there was evidence that a mixture of the two was slightly more effective than either alone. They calculated that roots of seedlings grown from inoculated seed had 10^6 colony-forming units per 0.1 g of root tissue 3 weeks after planting in natural soil in the field. These strains of the bacterium are competitive in the root environment.

Weller (1983), using strains of *P. fluorescens* resistant to both rifampin and nalidixic acid, found that after applying 10^8 colony-forming units per seed, up to 10^6 of these bacteria per 0.1 g of root were present 1 month after seeding winter wheat in October. The bacteria colonized a 7-cm segment of the root in 1 month after seeding in cold soil. By March the population had declined to 2800 cfu per 0.1 gram of root tissue. At that time the wheat was still dormant. On plants infected with *G. graminis* the bacteria multiplied 10 times after March and retained that level on the roots until the wheat was mature. The inoculant accounted for nearly 100% of the fluorescent pseudomonads on seminal roots, but in spring they accounted for only 10% of the bacteria on coronal roots. The added bacteria would be more effective in protecting seminal roots that arise from the seed than coronal roots that originate later from the crown. Even though take-all control was not complete, this procedure has promise, and Weller and Cook patented this use for these bacteria.

BROAD SPECTRUM MICROBIALIZATION

Kommedahl and Windels (1978) chose to attack a "root rot" complex rather than a single disease. They selected organisms antagonistic to *Aphanomyces euteiches, Fusarium oxysporum* f. sp. *pisi* races 1 and 2, *F. solani* f. sp. *pisi,* and *Rhizoctonia solani.* Microorganisms were compared with captan for their ability to protect peas. They started by testing 58 bacterial isolates for antagonism against *F. solani* f. sp. *pisi* and *R. solani.* Of 58, there were 22 antagonistic to one or both of these pathogens. The 21 isolates that inhibited a Minnesota isolate of *F. solani* failed to inhibit a Wisconsin isolate of that fungus. Likewise, isolates differed in their ability to inhibit races 1 or 2 of *F. oxysporum* f. sp. *pisi.* Of the 22 bacterial antagonists tested against nine isolates of four root pathogens of peas, none was antagonistic to all. In contrast, of 42 fungal isolates, 37 survived the initial screening, and 17 of them were antagonistic to the nine isolates of the four pathogens. It was easier to obtain wide-spectrum antagonists of fungal pathogens among fungi than among bacteria.

Instead of seeking more antagonists, Windels and Kommedahl (1978) chose to study ways to use their best isolate of *Penicillium oxalicum* most efficiently. Best results required 3 to 6 million spores per seed; i.e., all the spores the seed will hold. Results equal to captan were obtained in greenhouse trials in soil from a pea disease garden. The main benefit occurred within the first 2 to 3 weeks after planting. Preemergence problems with peas are effectively controlled by microbialization. Chickpeas with cream-colored seed are decimated by *Pythium ultimum* when seeded in cool, wet soil. Kaiser and Hannan (1984) obtained *P. oxalicum* from Windels and found that seeds heavily coated with spores are protected nearly as well as by captan. Spores of *Trichoderma hamatum* and *T. harzianum* are ineffective. *P. oxalicum* must have special attributes as a seed protectant.

Scanning electron micrographs reveal an almost solid covering of the pea seeds by spores of *P. oxalicum* (Windels, 1981). The spores germinate on the seed. Within 2 days a network of hyphae entwine the seed, and within 3 days sporulation is in progress. In field soil, conidia on roots had not germinated within 1.5 to 2 weeks after planting. The exudates from the seed are in quantities sufficient to stimulate and support growth of *P. oxalicum* but spores away from the seed, as on roots, did not germinate. Harman et al. (1980) believed that use of seed exudates by spores added to the seed is important for the success of this disease control strategy, and Windels has agreed with that assessment.

The work of Kommedahl and Windels in protecting pea seeds against a complex of soilborne fungi by coating the seeds with living organisms illustrates some interesting aspects of ecology (Fig. 16.1). If we consider that *Pythium* spp. are present in the pea disease nursery soil at a theoretical 300 oospores g of soil and it requires 3 to 5 million spores of *Penicillium oxalicum* to protect a seed, a ratio of 1:30,000 (target: protectant organism) is required for controlling the disease. If we figure 3000 chlamydospores of pathogenic fusaria, the ratio is 1 of the pathogen to 1000 of the protectant. These figures illustrate the remarkable power of the pathogens, not the

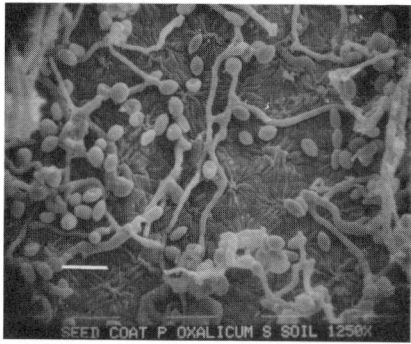 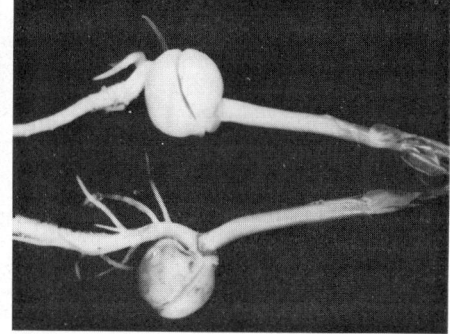

A B

Figure 16.1. (A) Germinating spores of *Penicillium oxalicum* on the surface of a pea seed. (B) The discolored surface of pea cotyledons after germination of pea seeds treated with *P. oxalicum* spores. I interpret the discoloration of the cotyledons as evidence of hypersensitivity. After Windels, 1981.

power of the protectants! Another attribute of protectants applied to seed must be little self-inhibition. The application of propagules to the seed in such concentration could interfere with germination if self-inhibitors are strongly present.

COMMENTS

The study of siderophores, or iron bearers, is an exciting field of research. Lewin (1984) explained how iron, though one of the earth's most abundant elements, became a limiting factor for the growth of microorganisms by being oxidized. Many microorganisms produce siderophores to obtain iron. No comparable group of compounds was developed to make other metals available, evidence that obtaining iron is a serious problem. Siderophores differ in power, and usually the producer has sites to accept specific siderophores that assist in transporting the rather large molecules through its own membranes. The evidence already obtained by plant pathologists strongly suggests that iron is important in the ecology of several fungi, particularly of some Fusarium wilt fungi.

For several years Australians have applied bacteria to seeds of many plants (Broadbent et al., 1977). Results with different isolates and different plants have varied from growth suppression to large increases in yield. In some cases, excellent response has resulted from spraying gently pasteurized soil with suspensions of growth-promoting bacteria in the production of bedding plants. The causes of the increased growth remain unclear. Broadbent et al. speculated that there could be a protection against nonparasitic root pathogens or the production of hormones and conversion of minerals and organic compounds to forms more available to higher plants or possibly even some nitrogen-fixation. It is obvious that this field of research is still in its formative stages.

Living seed protectants must be tested under several conditions. Chang and Kommedahl (1968) found that *Bacillus subtilis* improves stands of maize at temperatures below 20°C and reduces stands at 30°C. R. Smiley (Cornell University, personal communication) stated that *Phialophora graminicola* protects wheat from *Gaeumannomyces graminis* at near or slightly below 24°C, but that at 29°C it becomes highly pathogenic.

Stimulation of resistance by incompatibility reactions, production of antibiotics, siderophores, and bacteriocins, and competition will be exploited to control plant pathogens through microbialization as our knowledge increases.

KEY REFERENCES

Kloepper, J. W., M. N. Schroth, and T. D. Miller. 1980. Effects of rhizosphere colonization by plant growth-promoting rhizobacteria on potato plant development and yield. *Phytopathology* **70**:1078–1082.

Kommedahl, T. and C. E. Windels. 1978. Evaluation of biological seed treatment for controlling root disease of pea. *Phytopathology* **68**:1087–1095.

Lewin, R. 1984. How microorganisms transport iron. *Science* **225**:401–402.

Misaghi, I. J., L. S. Stowell, R. G. Grogan and L. C. Spearman. 1982. Fungistatic activity of water-soluble fluorescent pigments of fluorescent pseudomonads. *Phytopathology* **72**:33–36.

Scher, F. M. and R. Baker. 1982. Effect of *Pseudomonas putida* and a synthetic iron chelator on induction of soil suppressiveness to Fusarium wilt pathogens. *Phytopathology* **72**:1567–1573.

Windels, C. E. 1981. Growth of *Penicillium oxalicum* as a biological seed treatment on pea seed in soil. *Phytopathology* **71**:929–933.

REFERENCES

Aldrich, J. and R. Baker. 1970. Biological control of *Fusarium roseum* f. sp. *dianthi* by *Bacillus subtilis*. *Plant Dis. Reptr.* **54**:446–448.

Baker, R., P. Hanchey, and S. D. Dottarar. 1978. Protection of carnation against Fusarium stem rot by fungi. *Phytopathology* **68**:1495–1501.

Broadbent, P., K. F. Baker, N. Franks, and J. Holland. 1977. Effect of *Bacillus* spp. on increased growth of seedlings in steamed and in nontreated soil. *Phytopathology* **67**:1027–1034.

Brown, M. E. 1974. Seed and root bacterization. *Ann. Rev. Phytopathol.* **12**:181–197.

Brown, M. E., S. K. Burlingham, and R. M. Jackson. 1962. Studies on *Azotobacter* species in soil. II. Populations of *Azotobacter* in the rhizosphere and effects of artificial inoculation. *Plant Soil* **17**:320–323.

Chang, I. and T. Kommedahl. 1968. Biological control of seedling blight of corn by coating kernels with antagonistic organisms *Phytopathology* **58**:1395–1401.

Giha, O. H. 1976. Natural wheat seed protection by saprophytic bacteria against infection by *Helminthosporium rostratum*. *Plant Dis. Reptr.* **60**:985–987.

Gregory, K. F., O. N. Allen, A. J. Riker, and W. H. Peterson. 1952. Antibiotics and antagonistic microorganisms as control agents against damping-off of alfalfa. *Phytopathology* **42**:613–622.

Harmon, G. E., I. Chet, and R. Baker. 1980. *Trichoderma hamatum* effects on seed and seedling disease induced in radish and pea by *Pythium* spp. or *Rhizoctonia solani*. *Phytopathology* **70**:1167–1172.

Kaiser, W. J. and R. M. Hannan. 1984. Biological control of seed rot and preemergence damping-off of chickpea with *Penicillium oxalicum*. *Plant Dis.* **68**:806–811.

Kloepper, J. W. 1983. Effect of seed piece inoculation with plant growth-promoting rhizobacteria on populations of *Erwinia carotovora* on potato roots and daughter tubers. *Phytopathology* **73**:217–219.

Kloepper, J. W. and M. N. Schroth. 1978. Plant growth promoting rhizobacteria on radishes. *Proc. IVth Int. Conf. Plant Pathogenic Bacteria (Angers, France)* **2**:879–882.

Kommedahl, T. and C. E. Windels. 1981. Introduction of microbial antagonists to special courts of infection: seeds, seedlings, and wounds. Pages 227–248 in *Biological Control in Crop Production*. G. C. Papavazas, ed. Allenheld, Osmun and Co., Totowa, NJ.

Lang, D. S. and T. Kommedahl. 1976. Factors affecting efficacy of *Bacillus subtilis* and other bacteria as corn seed treatments. *Proc. Am. Phytopathol. Soc.* 3:272 (Abstract).

Ledingham, R. J., B. J. Sallans, and P. M. Simmonds. 1949. The significance of the bacterial flora on wheat seed in inoculation studies with *Helminthosporium sativum*. *Sci. Agric.* **29**:253-262.

Liu, S. and E. K. Vaughan. 1965. Control of *Pythium* infection in table beet seedlings by antagonistic microorganisms. *Phytopathology* **55**:986-989.

Michael, A. H. and P. E. Nelson. 1972. Antagonistic effec of soil bacteria on *Fusarium roseum* 'Culmorum' from carnation. *Phytopathology* **62**:1052-1056.

Mishustin, E. N. and A. N. Naumova. 1962. Bacterial fertilizers, their effectiveness and mode of action. *Mikrobiologiya* **31**:543-555.

Morrow, M. B., J. L. Roberts, J. E. Adams, H. V. Jordan, and P. Guest. 1938. Establishment and spread of molds and bacteria on cotton roots by seed and seedling inoculation. *J. Agric. Res.* **56**:197-207.

Powell, P. E., P. J. Szaniszlo and C. P. P. Reid. 1983. Confirmation of occurrence of hydroximate siderophores in soil by a novel *Escherichia coli* bioassay. *Appl. Environ. Microbiol.* **46**:1080-1083.

Scher, F. M. and R. Baker. 1980. Mechanism of biological control in a Fusarium-suppressive soil. *Phytopathology* **70**:412-417.

Schroth, M. N. and J. G. Hancock. 1982. Disease-suppressive soil and root-colonizing bacteria. *Science* **216**:1376-1381.

Simmonds, P. M. 1947. The influence of antibiosis in the pathogenicity of *Helminthosporium sativum*. *Sci. Agric.* **27**:625-632.

Smiley, R. W. 1978. Colonization of wheat roots by *Gaeumannomyces graminis* inhibited by specific soil microorganisms and ammonium-nitrogen. *Soil Biol. Biochem.* **10**:175-179.

Smith, S. N. and W. C. Snyder. 1971. Relationship of inoculum density and soil type to severity of Fusarium wilt of sweet potato. *Phytopathology* **61**:1049-1051.

Tveit, M. and M. B. Moore. 1954. Isolates of *Chaetomium* that protect oats from *Helminthosporium victoriae*. *Phytopathology* **44**:686-689.

Tveit, M. and R. K. S. Wood. 1955. The control of *Fusarium* blight in oat seedlings with antagonistic species of *Chaetomium*. *Ann. Appl. Biol.* **43**:538-552.

Weller, D. M. 1983. Colonization of wheat roots by a fluorescent pseudomonad suppressive to take-all *Phytopathology* **73**:1548-1553.

Weller, D. M. and R. J. Cook. 1983. Suppression of take-all of wheat by seed treatment with fluorescent pseudomonads. *Phytopathology* **73**:463-469.

Windels, C. E. and T. Kommedahl. 1978. Factors affecting *Penicillium oxalicum* as a seed protectant against seedling blight of pea. *Phytopathology* **68**:1656-1661.

Wright, J. M. 1956. Biological control of a soilborne Pythium infection by seed inoculation. *Plant Soil* **8**:132-140.

Chapter 17

Hyperparasites

The title, hyperparasites, means parasites of parasites. This broad term is used because parasites of fungi and nematodes are included. Up to this point we have discussed antibiotics, siderophores, and bacteriocins, agents of antagonism that are excreted into the soil environment where they may harm other individuals. In parasitism the host is used directly as food. Those interested in pursuing hyperparasitism in depth are referred to Cook and Baker (1983; Chapter 9, *Antagonistae Vitae*) for a detailed treatment and extensive references, and to the book by Barron (1977).

FUNGAL PARASITES OF FUNGI

Trichoderma* spp. Attack Weakened *Armillaria mellea

Armillaria mellea, a higher basidiomycete that forms mushrooms, lives for years in roots and the base of stumps of infected or dead hosts. *A. mellea* is an example of successful saprophytic survival (active possession of substrate) by a tree pathogen. The substrate is substantial (wood), not evanescent like leaves. Part of the infested tissue is deep within the soil, sheltered from extreme variations of temperature or water stress. In addition, this fungus increases its longevity by formation of what Campbell (1934) described and named pseudosclerotia. *A. mellea* develops a dark dense layer (zone line) of thick-walled fungal cells, the pseudosclerotial covering, about the enclosed wood containing live hyphae, mycelial fans, and rhizomorph initials. Several workers (citations in Munnecke et al., 1981) have proved that *A. mellea* enhances its possession of substrate by producing antibiotic.

Fumigation with sublethal levels of carbon disulfide is a practical method of destroying *A. mellea* within wood. Bliss (1951) attributed the success of sublethal rates of fumigants to the action of *Trichoderma viride*. *T. viride* is present in natural soil, yet in natural soil *A. mellea* survives for years. In natural soil after weak fumigation, *T. viride* dominates *A. mellea*. Bliss believed that subjection to weak fumigation weakened *A. mellea*, rendering it susceptible (see Table 13.2).

Studies cited in Munnecke et al. (1981) showed that sublethal fumigation stops growth and antibiotic production by *A. mellea* for up to 4 weeks. It is not only weakened, but it no longer protects itself. Munnecke et al. also suspected that sublethal fumigation stimulated leakage of nutrients from the tissues. The fumigation, sublethal to *A. mellea*, does not kill spores of *Trichoderma* spp. These vigorous fungi respond quickly to opportunity. They invade the weakened pseudosclerotium, and *A. mellea* does not recover.

At this point you might ask, why is this discussion in the chapter on hyperparasites and not under antibiotics? *Trichoderma* spp. are directly parasitic on many fungi. Bliss noted that *T. viride* grew over the top of *A. mellea* cultures, a characteristic of many fungal parasites.

Munnecke et al. referred to *Trichoderma* spp. as a group. The taxonomy of this group is difficult (see Cook and Baker, 1983, p. 374). Webster and Lomas (1964) placed the *T. viride* of early students in *Gliocladium virens*, in the belief that the antibiotic-producers belong in that genus. Munnecke et al. (1981) retained the reference to *Trichoderma* spp., and I have followed their decision.

Use of Both Parasitism and Antagonism

Seeding cotton early in Texas increases yield if good stands result, but *Pythium ultimum* and *Rhizoctonia solani* are serious pathogens of cotton during germination and emergence in cool soil. Howell (1982) inoculated natural soil with sclerotia of *R. solani*, oospores of *P. ultimum*, and spores of *Gliocladium virens*. Germination of sclerotia of *R. solani* was reduced because *G. virens* invaded the sclerotia and killed them. There was no evidence of inhibition of *R. solani* by antibiotics, but *G. virens* hyphae coiled around and penetrated hyphae of *R. solani*. In contrast, *G. virens* did not parasitize *P. ultimum*, but the protoplasm of *P. ultimum* disintegrated in the presence of antibiotic produced by *G. virens*. *G. virens* damaged two different pathogens by two different means.

Effect of Position of Sclerotia on Parasitism

Campbell (1947) described *Coniothyrium minitans,* a fungus that produced pycnidia on sclerotia of *Sclerotinia sclerotiorum.* He suggested it might be useful in biological control. *C. minitans* is known in Australia, New Zealand, North America, and Europe (Ayers and Adams, 1981). *C. minitans* also parasitizes sclerotia of *S. minor, S. trifoliorum,* and some strains of *Botrytis fabae, B. cinerea, B. narcissicola, Sclerotium cepivorum,* and *Clavicips purpurea* (Ayers and Adams, 1981). Hyphae of *C. minitans* penetrate hyphae of *S. sclerotiorum* directly, with no specialized penetration structures. A slight depression in the host cell wall precedes penetration. The parasite grows within host hyphae in either direction, often exiting from the host and growing outside again, (Huang and Hoes, 1976)

Sclerotinia sclerotiorum causes an important "wilt" disease of sunflower. Cankers form at the base of the stalk, and sclerotia of *S. sclerotiorum* form on the surface of the tap root, within the pith of the tap root, and within the stem. The percentage of healthy sclerotia within the root can decline rapidly through the growing season (Huang, 1977), and sclerotia on the outside of the root were heavily parasitized throughout the season (Fig. 17.1). By the end of the growing season Huang estimated that 59% of the sclerotia on the root surface, 76% in the pith cavity of roots, and 29% in the pith cavity of basal stems had been killed by *C. minitans.* Only 4, 9, and 68% of the sclerotia in these locations, respectively, were still viable. Some nonviability was

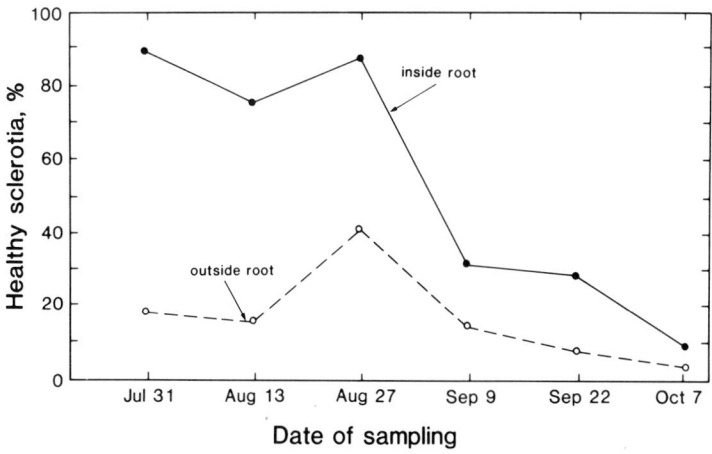

Figure 17.1. Sclerotia of *Sclerotinia sclerotiorum* produced outside the roots of sunflower are heavily and rapidly parasitized by mycoparasitic fungi, mainly by *Coniothyrium minitans.* Most of the sclerotia that remained healthy were produced above the soil line within the stem (not shown). After Huang, 1977.

attributed to attacks by *Trichoderma viride, Gliocladium catenulatum*, and possibly by other organisms, but *C. minitans* was the dominant hyperparasite. Most of the viable sclerotia were produced within the stalk well above the soil surface.

Parasitism Greater than Saprophytism

Uecker et al. (1978) described *Sporidesmium sclerotivorum* as a parasite of *Sclerotinia* spp. *S. sclerotivorum* also produces conidia that fit the genus *Selenosporella*. Conidia form on parasitized sclerotia and on moist filter paper on which sclerotia are incubated. Conidia are produced on live sclerotia more rapidly than on dead sclerotia. The hyperparasite was grown on corn meal agar containing comminuted sclerotia, but growth was more rapid and sporulation greater when living rather than autoclaved sclerotia were used (Ayers and Adams, 1979). Macroconidia of the mycoparasite germinated within 9 mm of sclerotia. Germination of macroconidia was first observed within 3 days, and it was quickest when the spores were close to the sclerotium. When spores were placed on a membrane on soil containing sclerotia, spore germination increased as the concentration of sclerotia increased. Spores of *S. sclerotivorum* remained viable in soils maintained moist or dry for at least 9 months.

Penetration of the sclerotial rind was direct without the aid of appressoria (Adams and Ayers, 1983). Once inside the sclerotium the hyphae became intercellular, taking their shape from the intercellular spaces among sclerotial cells. The parasite lives on the extracellular matrix within sclerotia, aided by enzymes produced by the sclerotia while life remains within them. Parasitized sclerotia remain living for an undetermined period of time, at least in *S. minor*, because parasitized sclerotia sometimes germinate myceliogenically (P. B. Adams; USDA-ARS, Beltsville, MD; personal communication).

Adams and Ayers (1981) surveyed soils for the presence of *S. sclerotivorum* by incubating sclerotia of *S. minor* in field soils. *S. sclerotivorum* was found in soil from 10 widely separated states in the United States. The mycoparasite is widely distributed and not restricted closely by soil type or general climatic conditions.

In the autumn of 1977 a plot of silt loam free of *Sclerotinia minor* (by sieving assay) and of *S. sclerotivorum* (by baiting) was seeded to lettuce and inoculated with laboratory-grown sclerotia of *S. minor* (Adams and Ayers, 1982). The diseased lettuce was left to overwinter to ensure maximum production of sclerotia. In March 1978 the land was again seeded to lettuce. This crop was 85 to 100% infected by May 1978 with an average of 21 sclerotia per 100 g of soil. At that time *S. sclerotivorum* was added to portions of the plot at 0, 1, 10, 100, and 1000 macroconidia per gram of soil. The plots were rototilled in each direction to a depth of 15 cm. The rototilling continued at biweekly intervals until 19 March 1979 when lettuce was again planted. From that time on lettuce was repeatedly grown.

The mycoparasite invaded and increased in the control plots, and by August, 1980 there was little difference between control and inoculated plots. The percentage of bait sclerotia of *S. minor* that became parasitized in the controls was 0 in May 1978 and 90 in August 1980. The mycoparasite had established itself in the control plots in spite of precautions to keep it from spreading. Ayers and Adams (1981) have patented *S. sclerotivorum* as a potential biological control agent.

Pythium spp.

When *Pythium oligandrum, P. pleriplocum,* and *P. acanthicum* are paired in culture with *P. salpingophorum, P. vexans, P. undulatum,* and *P. anandrum,* the former group halts the advance of the latter group (Drechsler, 1946). Hyphae of the former group twine around and invade hyphae of the latter group. *P. vexans* retaliates to some extent against *P. pleriplocum,* so these two species can attack each other. *Pythium oligandrum, P. acanthicum,* and *P. pleriplocum* produce coarse hyphae typical of strong plant pathogens such as *P. ultimum,* and fine delicate branches which initiate the attack on other *Pythium* spp. Dreschler noted that *P. oligandrum* seldom occurred alone, but was usually isolated along with virulent plant pathogens. Drechsler was able to produce a rapid watery rot of cucumber and watermelon fruits with pure cultures of *P. oligandrum* in wounds, but this rarely occurs in nature, and he suspected that its oospores respond to opportunities too slowly for its to compete with successful plant pathogens.

Host-parasite interactions among pairings have many differences, but *P. oligandrum* is relatively unspecialized, especially because young hyphae of many fungi are susceptible (Deacon, 1976). Most host fungi become more resistant with maturity, even those with nonmelanized hyphae. After "maturity," when senescence begins, many fungi increase in susceptibility to hyperparasitism.

Haskins (1963) grew *Pythium acanthicum* (?) and other fungi paired on Difco potato dextrose agar, starting each at opposite sides of the dish. Of 98 fungi, 69 were susceptible and supported oogonia production by the pathogen; 10 were parasitized but no oogonia were produced; 9 were not parasitized (neutral), and 10 inhibited the *Pythium.* *Ustilago maydis* was the strongest antagonist. Haskins illustrated hyphae of *P. acanthicum* twining tightly about those of *Trichoderma viride, Pythium mammillatum, Fusarium sambucinum, Botrytis cinerea,* and about root hairs of wheat seedlings.

Deacon and Henry (1978) seeded potato dextrose agar with *Phialophora radicicola* var. *radicicola* at one side of the dish, incubated the dish 8 to 10 days at 25°C so that the surface was thoroughly colonized, and then three small soil crumbs or small bits of organic matter were placed on the youngest portion of the mycelium. This method enabled them to isolate parasitic *Pythium* spp., *Acremonium, Trichoderma* sp., *Gliocladium roseum* and *Papulaspora* sp. from soil selectively. Soil was a better source of mycoparasitic fungi than organic debris. Debris favored invasion of the cultures by "contaminants" and nematodes.

NEMATODES THAT PARASITIZE FUNGI

Most nematodes that feed on fungi have very short stylets, probably an adaptation to the size of hyphal cells and thinness of the cell walls versus those of higher plants. Nevertheless, some nematodes pathogenic to plants can be reared in culture on fungi. *Aphelenchoides besseyi, Ditylenchus destructor* and to a lessor extent *D. dipsaci* are such nematodes.

Mankau and Mankau (1963) determined the percent of fungivorous nematodes versus the total nematode population in soil under barley (50%), summer fallow (45%), snap beans (45%), alfalfa (7%), olive trees (8%), and lemon trees (8%). *Aphelenchus avenae* was the most abundant fungivorous nematode found. Their abundance is evidence that they are a factor in fungal ecology. *Aphelenchus avenae,* described by Bastian in 1865, is known in Africa, Europe, Central and South America, India, Israel, and the United States.

When maize is inoculated with *A. avenae* and *Pythium arrhenomanes* alone and together simultaneously, top growth of the plants in the nematode and *P. arrhenomanes* + nematodes treatments are equal to that in the controls, but *P. arrhenomanes* alone reduces growth by 75%. Addition of these nematodes with the *Pythium* negates all loss. Many nematodes and eggs can be found in the lesioned roots. There is no evidence that the nematode feeds on healthy roots (Rhoades and Linford, 1959).

A. avenae is associated with and is found in severely diseased tissue, but it lives on fungi in the tissue, not on the plant. *A. avenae* does not congregate around healthy roots (Linford, 1939) as do plant-parasitic nematodes. *A. avenae* increases more rapidly in soil enriched with refuse or manure than with normal roots of healthy plants.

In studies of the rate of reproduction of *A. avenae* on several fungi, six females per Petri dish were used (Mankau and Mankau, 1963). The approximate number after 3 weeks was 71,000 on *Periconia* sp. and 0 on *Pythium ultimum,* the extremes within the test. The taxonomic relationship of the fungi has little bearing on suitabillity as a host for the worms. The most rapid increase in nematodes occurs on fast-growing fungi that are not disorganized by their feeding. Most soil saprophytes (*Penicillium* sp., *Aspergillus terreus, Rhizopus nigricans, Trichoderma viride,* and *Alternaria* sp.) are poor hosts. *A. avenae* feeds on the *Phytophthora* spp. and lays many eggs in the agar, but the eggs usually disintegrate. Hatching usually produces weak larvae which die. *Pythium ultimum* is a very poor host in that the nematodes usually are inactive in the feeding position, and they apparently die.

Possible Relationship with Mycorrhizal Fungi

Mycophagous nematodes can hinder development of ectotrophic mycorrhizal fungi (Sutherland and Fortin, 1968). *Amanita rubescens, Cenococcum graniforme, Rhizopogon roseolus, Russula emetica, Suillus granulatus, S. luteus,* and *S. punctipes* were started on an agar medium for 24 hr, then the nematodes added. *Aphelenchus avenae* thrives on all but *Rhizopogon roseolus,* which apparently produces something toxic to the nematode. *Aphelenchus avenae* prevents *Suillus granulatus* from establishing itself on *Pinus resinosa* in double-inoculation trials.

An *Aphelenchoides* sp. obtained from stands of *Juniperus monosperma* in central New Mexico multiplies on *Suillus granulatus* and on *Mycelium radicis atrovirens* (Riffle, 1967). When the nematode and *S. granulatus* are introduced simultaneously on potato dextrose agar, the fungus fails to grow. When the fungus is given a 4-day start, its growth ceases 12 days after inoculation. The pseudomycorrhizal fungus *Mycelium radicis atrovirens,* is more tolerant to nematode feeding, and on this fungus the increase in nematodes after simultaneous inoculation is 63 times. When the fungus had the 4-day start, the nematode increases 1000 times in 44 days. Riffle had no field evidence that nematode feeding on fungal hyphae reduces mycorrhizal associations, but the laboratory results indicate that it is possible. Many nematodes are favored by light, sandy soils, and in such soils mycorrhizae are important to tree vigor.

AMOEBA THAT PARASITIZE FUNGI AND NEMATODES

Dobell in 1913 (in Old and Darbyshire, 1978) reported that *Arachnula impatiens,* a large amoeba, fed upon bacteria, diatoms, blue green algae, and protozoa. *Theratromyxa weberi* captures, engulfs, and digests soil nematodes. Early work with various amoebae has indicated that few feed on fungi, with the exception of yeasts. Spores of many fungal species are ingested, but they are not digested, nor is their viability reduced. Only spores of *Paecilomyces elegans, Polyscyalatum fecundissimum, Fusarium,* and *Torulopsis fumata,* among filamentous fungi, are digested. The possible importance of amoebae in fungal ecology was not established until Old in 1977 reported perforation and digestion of conidia of *Bipolaris sorokiniana* by a member of the Vampyrellidae (Old and Darbyshire, 1978).

When I first saw pictures of the perfectly round holes in spores of *B. sorokiniana* and *Thielaviopsis basicola* and was told that amoebae cut those holes, I didn't believe it (Fig. 17.2). How could such an amorphous organism "cut" with such precision? The hole begins as a ring, or annulation, which becomes deeper until digestion is complete around the selected spot. When the ring has been digested, a round portion of the fungal spore wall enters the amoeba, followed by much of the spore contents (Anderson and Patrick, 1978).

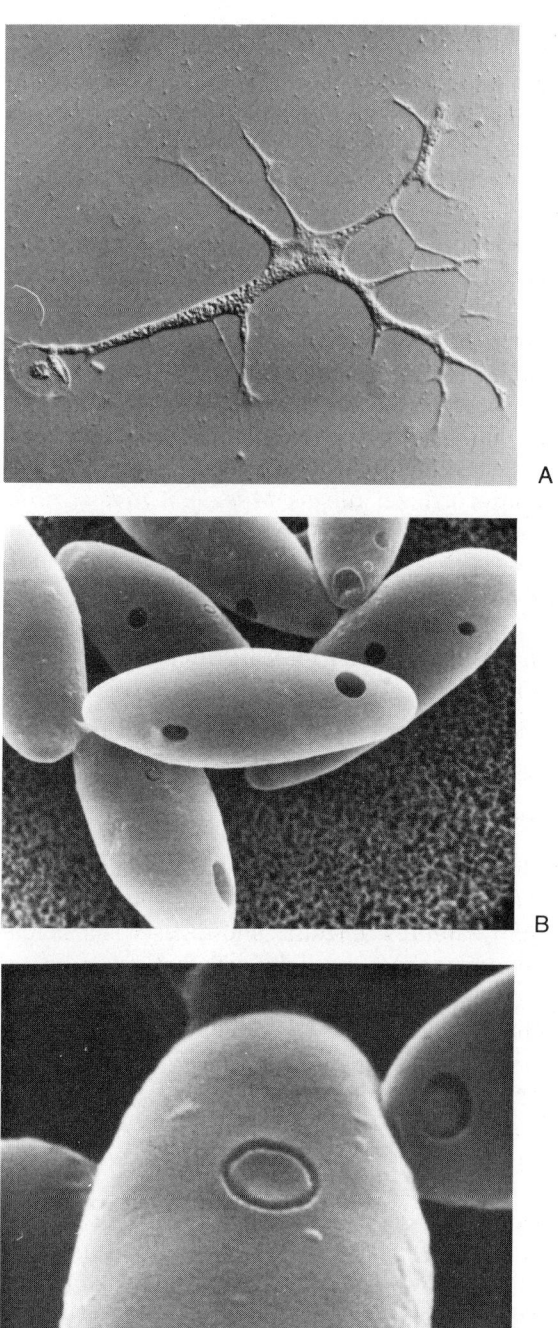

Figure 17.2. Trophic mycophagous amoeba in an active state (*A*), perforated and emptied conidia of *Bipolaris sorokiniana* (*B*), and a conidium partially perforated by an amoeba (*C*). Anderson and Patrick, 1978.

A giant vampyrellid amoeba engulfs and perforates chlamydospores of *Aureobasidium pullulans* and of *Thielaviopsis basicola*, conidia of many species, sporangiospores of *Rhizopus stolonifera*, and pycnidiospores of *Phoma exigua* (Old and Darbyshire, 1978). The amoebae engulf but do not perforate blastospores of *Aureobasidium pullulans*, sclerotia of *Botrytis cinerea*, oospores of *Phytophthora fragariae*, or vegetative cells of two yeasts. The amoebae are unattracted to conidia of *Aspergillus niger* and *Fusarium* sp. or hyphae of *Rhizoctonia solani*. All stages of four nematodes (*Longidorus elongatus*, *Ditylenchus dipsaci*, *Globodera rostochiensis*, and *Pratylenchus penetrans*) are attacked by the amoebae. The amoebae adhere to the cuticle of actively swimming larvae or adults and form discrete holes in them, just as in fungal spores. Pseudopodia enter the nematodes. Digestion is complete, leaving only the empty cuticle. Eggs are likewise susceptible. The giant vampyrellid amoeba digest propagules of many fungi, and melanin offers no protection. A hyaline *Bipolaris sorokiniana* is no more susceptible than the normally pigmented form. Chitin and chitosan, as present in sporangiospores of *Rhizopus stolonifer*, do not prevent penetration, nor do the cuticle or egg shell of the nematodes. The enzyme complement of the amoeba is remarkable.

Earlier descriptions have been of vampyrellid amoebae that lysed large holes (1 to 7 μm) in conidia. Anderson and Patrick (1980) found two amoebae that lyse small holes ($<1\mu$) in the melanized spores of *Thielaviopsis basicola* and *B. sorokiniana*. These small holes were once thought to be caused by bacteria. Two amoebae, one resembling *Vampyrella vorax*, produce small holes.

Duczek (1983), in a survey of amoebae in Saskatchewan soils, assayed the numbers of amoebae that make large holes, using conidia of *B. sorokiniana* as bait. The numbers of amoebae varied from lowest during the warmest part of the season to a peak in early autumn when soil temperatures had declined to 7°C. The autumn peak for amoebae was also observed in Britain and in continental Europe and Russia by other workers. Numbers of amoebae are not correlated with soil water content at the time of the collection, at least not between the permanent wilting point and saturation. Experiments indicate that significant amoebal activity is limited to periods when the soil is wet but in which oxygen is present (Old and Patrick, 1979; Homma and Cook, 1985).

The population density and viability of *B. sorokiniana* conidia in soil decrease with depth (Duczek, 1981). The populations in the 0 to 5 cm profile of Degraded Black, Black, Dark Brown, and Brown soils are 16, 25, 34, and 34, respectively. If we assume the listed sequence goes from moist to relatively dry, conidial populations are greater in drier soils. Duczek also grouped the data as to soil texture, The lowest population (17/g) is in loam and the highest in clay (43/g), but there is no consistent trend in relation to soil texture. There appears to be an inverse relationship between numbers of conidia and of amoebae (Duczek, 1983) (Table 17.1).

Table 17.1. Numbers of conidia of *Bipolaris sorokiniana* and vampyrellid amoebae per gram of Saskatchewan soils

Soil characteristics	Conidia/g	Amoebae/g
Degraded Black	16	2.6
Black	25	2.7
Dark Brown	34	2.1
Brown	34	1.0
Sandy loam	32	0.4
Light loam	27	2.5
Loam	28	2.2
Silty loam	17	3.8
Clay	43	1.1

Source: Duczek, 1981, 1983.

VIRUS IN FUNGI

Mycoviruses have been found in over 100 species of fungi (Hollings, 1978, 1982). Morphologically distinct viruses can simultaneously infect and replicate in the same fungus. Hypovirulence in *Gaeumannomyces graminis* was believed by French workers to be due to viral infection, but British workers could not substantiate this.

Viruses of fungi have been difficult to study because of difficulties in transmission. No vectors are known, and in nature the only known means of transmission are through propagules (serial transmission) or through anastomosis (lateral transmission) (Ghabrial, 1980). Many mycoviruses are avirulent, indicating long association between host and virus. A virulent "seedborne" virus lacking a vector, especially one that kills prior to host reproduction, would be self-eliminating.

Certain strains of *Ustilago maydis* secrete specific toxic proteins (killer factors) that kill susceptible strains in culture. Production of the killer factor is associated with the presence of virus in the smut fungus. At least three specific killer factors are known. Members within each group are not affected by their own toxin but may be susceptible to those produced by members of the other two groups. These phenomena have been observed in culture. Puhalla (1968) found that killer 1 strain would mate with a compatible, susceptible member of killer 2 strain in the host and survive. Hankin and Puhalla (1971) suggested that the pH of maize cell sap is not favorable for production of killer factor, or it is inactivated, translocated away, or otherwise rendered ineffective within the plant.

Rhizoctonia solani is usually stable in culture and retains its virulence for years, but J. R. Parmeter, Jr., noted individual sick cultures within his collection (Castanho and Butler, 1978a). Culture 189, anastomosis group 1, was obtained from Parmeter, and a severely diseased culture was obtained from it during routine transfers (called 189a). Healthy isolates were obtained from 189a by hyphal-tipping, and they remained healthy. The disease-causing agent is apparently not completely systemic but it was necessary to use mass transfers of 189a to keep 189a alive.

Isolate 189a anastomoses freely with healthy offspring, and when the diseased parent and healthy offspring are placed at opposite sides of the same Petri dish and allowed to grow together, all transfers from various parts of both "mates" produce diseased subcultures. The causal agent moves completely through the receptor mycelium, and until transferred, the receptor appears normal. Attempts to transfer the pathogen by anastomosis to other cultures (not offspring) within anastomosis group 1 have failed. The pathogen thus has great specificity, greater than that governing anastomosis groups. Castanho and Butler found over 30 diseased isolates within five anastomosis groups of *R. solani*.

Efforts to isolate virus particles from isolate 189a have failed, but double strands of RNA (ds RNA) of three sizes, 2.2, 1.5, 1.1 $\times$ 10^6 molecular weights have been obtained (Castanho and Butler, 1978c). These ds RNAs are present in minute amounts in the healthy offspring. After pairing the healthy offspring with 189a, the ds RNA strands are abundant throughout. Two other diseased isolates in different anastomosis groups contain ds RNA particles. The ds RNA particles in these isolates differ from each other and from those in 189a. Each sick isolate contains similar but different ds RNA. Castanho and Butler concluded that the healthy offspring of 189a either contained too little ds RNA to become sick (unlikely—if it were the same, why didn't it increase?) or contained a mild, nonpathogenic type.

When Castanho and Butler (1978b) added 189a and its healthy, virulent offspring to soil, damping-off and preemergence blight of susceptible sugar beet seedlings was reduced in comparison to that from the healthy culture alone. Isolate 189a did not persist in soil for 2 months; a healthy offspring survived 2 years in soil tubes and at least 6 months in greenhouse soil. After coinoculation with 189a and a healthy offspring, neither survived longer than 1 month. Anastomosis occurred in soil as well as in Petri dishes. Practical use of such "biological" control depends on a carrier isolate that persists longer in soil and one that will anastomose with unrelated members of the same anastomosis group. At present there is little evidence to indicate that fungal viruses play a significant role within soil.

FUNGAL PARASITES OF NEMATODES

Over 150 species of fungi (Barron, 1977) attack nematodes. The first record was that of Zopf, who in 1888 observed *Arthrobotrys* trapping a nematode. According to Barron, this is the first example of an animal trapped by a fungus. Drechsler in 1933 found that the trapping rings were adhesive, preventing the nematode from escaping. Barron classified fungi that trap nematodes, enveloping them with hyphae, as predators. They invade and digest their prey after they trap them. In contrast, some fungi parasitic on nematodes exist in soil as small propagules that either adhere to the cuticle prior to penetration or are ingested by nematodes and germinate within them. The latter are classed as endoparasitic fungi. Most fungal parasites have some characteristics of both predators and endoparasites (Barron, 1977). Some produce constricting rings (*Arthrobotrys*), some produce spores that inject infective particles into nematodes (*Haptoglossa*), some produce attractants, and some secrete toxins into the host.

Pramer and Stoll (1959) grew *Arthrobotrys conoides*, a nematode-trapping fungus, on maize meal agar for 4 days and then added 1 ml of water in which nematodes had been washed. The wash water induced trap formation. No traps formed in cultures not receiving nematode wash water. They called the unidentified substance that induced trap formation nemin.

Effect of Salts on a Parasitic Relationship

Criconemella (Criconemoides) xenoplax attacks roots of many woody plants, including peach trees. Jaffee and Zehr (1983) studied the effect of solutes and osmotic water potential upon its infection by the fungus *Hirsutella rhossiliensis*. Many *C. xenoplax* extracted from soil were parasitized, but when spores of the fungus and nematodes were placed together in tap or distilled water, only 6% of the adults and 25% of the juveniles were infected. Jaffee and Zehr initially thought that low water potentials would favor infection, because *C. xenoplax* is most common in sandy soils.

Jaffee and Zehr incubated the nematode in KCl solutions to give 0, −6, −12, and −18 bars osmotic potential, each adult female was touched with 20 of the sticky spores and then incubated for 5 days in the water and the KCl solutions; 0, 100, 100, and 100% died. The spores adhered to the cuticle in pure water, but the nematodes were not infected (Fig. 17.3). When nematodes were inoculated and then incubated in water at 0, −0.3, −3 and −6 bars adjusted with KCl, mortality was 0, 95, 100 and 100%, respectively. In sucrose solutions adjusted to the same osmotic potentials mortality was 0, 16, 20 and 29%. In polyethylene glycol solutions similarly adjusted, mortality was 0, 0, 0, and 0%. The effect was not due to water potential. K^+, Ca^{2+}, and Mg^{2+} cations favored infection, but Na^+ did not. Cl^-, NO_3^-, and H_2PO_4 favored infection, but SO_4^{2-} did not. They then tested soil extracts from five orchard soils. The more concentrated the extract, the higher the infection. The osmotic potentials of saturation extracts of five orchard soils, all of which contained large numbers of *C. xenoplax*, ranged from −0.1 to −0.9 bars. Because dilute salt solutions do not stress the nematode, Jaffee and Zehr believe the effect of the salt is upon the fungus. Its spores apparently do not germinate or establish infection when the nematode is within pure water.

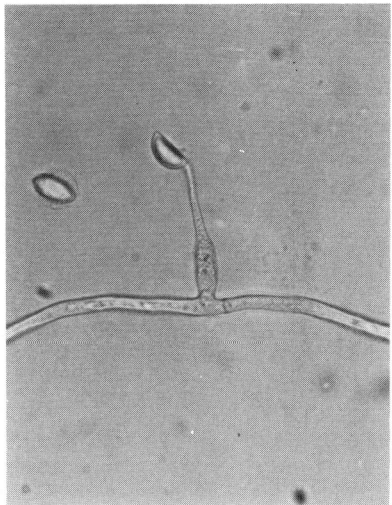 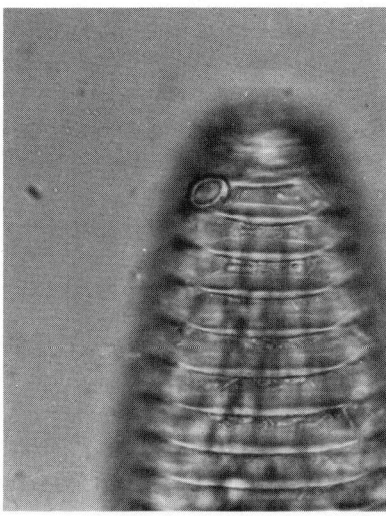

Figure 17.3. One free conidium and one conidium still attached to the conidiophore of *Hirsutella rhossiliensis* (left). A conidium adhering to the cuticle of *Criconemella xenoplax*, anterior end (right). After Jaffee and Zehr, 1982.

A BACTERIAL PARASITE OF NEMATODES

A bacterium with a complex and unusual development, called *Bacillus penetrans* (Mankau, 1975), parasitizes certain nematodes. The same or a similar organism (called *Pasteuria ramosa*) was described by Metchnikoff in 1888 as a parasite of *Daphnia magna* and *D. pulex*. Thorne (1940) observed an organism with spores that adhered to the cuticle of nematodes which he described as *Duboscqia penetrans* (as a protozoan). In 1960 Williams (in Mankau, 1975) in Mauritius found the parasite attacking nematodes in sugarcane fields, and he questioned its supposed protozoan nature. Mankau (1975), after extensive electron microscopy, placed the organism among bacteria. The life stages, beautifully illustrated by Sayre and Wergin (1977), explain why this organism presented so many taxonomic and technical problems. Sayre (1980) compared recent illustrations, after use of the electron microscope, with those of Williams and Metchnikoff. The similarities are testimony to the acute observations of the early workers.

The bacterium forms spores that adhere to the cuticle of host nematodes, primarily on the head or neck. Spores adhere to the second-stage larvae of *Meloidogyne incognita* as they move through infested soil. The spores germinate in about 8 days, after the nematode has penetrated its host plant. Unlike typical bacteria, germination is by a germ tube that penetrates directly through the nematode cuticle. Early growth of the bacterium inside the nematode is at first hyphal, with septations and compound cell walls. The colonies fragment when intercalary hyphal cells lyse. Movements of body fluids within the pseudocoelom may disperse the daughter cells so that a more general infection occurs. The daughter cells develop sporangia within which single endospores form, completing the life cycle. Mankau (1975) found 2.1×10^6 endospores in each parasitized root-knot female. The methods of growth shows affinity to Actinomycetales (Sayre, 1980), but the endospores are typical of true bacteria.

The mode of distribution or the fate of the endospores within the soil is not known. Dutky (in Mankau, 1975) found that endospores of *Bacillus popillae* adhere to soil particles. This could localize the inoculum at sites near the dead female.

The parasite appears to be at least partially host specific (Thorne, 1940), because the soil from which diseased nematodes are found contains 21 species of nematodes that are not attacked. *Dorylaimus obscurus* apparently ingests many of the spores and has many spores inside its intestinal tract with no evidence that it is parasitized. Dutky (in Sayre, 1980) found that spores of one source of *B. penetrans* adhered to *Pratylenchus brachyurus* and not to *Meloidogyne incognita* and that a source from *M. incognita* did not adhere to *P. brachyurus*. Specific strains of the bacterium therefore occur.

COMMENTS

Evidence is strong (this chapter and the next) that hyperparasites are far more important than formerly thought. Some of the fungal parasites of fungi are so parasitic that they develop better on live than on dead host structures, evidence that parasitism is their means of existence. The work of Duczek (1983) in Canada does not prove but it suggests that amoebae affect propagule numbers. Whether nematodes that feed on fungi have significant effects in nature is not well established. As yet, viruses of fungi have not been established as significant enemies of soilborne plant pathogenic fungi.

A myxobacterium, genus *Polyangium,* attacks melanized hyphae of *Rhizoctonia solani* and conidia of *Cochliobolus miyabeanus* (Homma, 1984). The colonies of myxobacteria move slowly on a solid surface in a slimy substance. The cells lack flagella. The slimy colonies encircle *R. solani* hyphae, make holes averaging 0.4 μm in diameter, feed on cell contents, and move away to fresh cells. Conidia of *C. miyabeanus* are often penetrated at one end. The colony moves within the conidium until the conidium is emptied. After 12 to 24 hr the colony abandoned the empty spore. The bacterial polysaccharide slime facilitates movement over solid substances, but the method of movement, without flagella, must be amazing for rod-shaped bacteria. Hocking and Cook (1972) believed they may have some significance in nature.

KEY REFERENCES

Adams, P. B. and W. A. Ayers. 1983. Histological and physiological aspects of infection of sclerotia of *Sclerotinia* species by two mycoparasites. *Phytopathology* **73**:1072–1076.
Anderson, T. R. and Z. A. Patrick. 1978. Mycophagous amoeboid organisms from soil that perforate spores of *Thielaviopsis basicola* and *Cochliobolus sativus*. *Phytopathology* **68**:1618–1626.
Howell, C. R. 1982. Effect of *Gliocladium virens* on *Pythium ultimum*, *Rhizoctonia solani*, and damping-off of cotton seedlings. *Phytopathology* **72**:496–498.
Huang, H. C. and J. A. Hoes, 1976. Penetration and infection of *Sclerotinia sclerotiorum* by *Coniothyrium minitans*. *Can. J. Bot.* **54**:406–410.
Jaffee, B. A. and E. I. Zehr. 1983. Effect of certain solutes, osmotic potential, and soil solutions on parasitism of *Criconemella xenoplax* by *Hirsutella rhossiliensis*. *Phytopathology* **73**:544–546.

REFERENCES

Adams, P. B. and W. A. Ayers. 1981. *Sporodesmium sclerotivorum:* distribution and function in natural biological control of sclerotial fungi. *Phytopathology* **71**:90–93.
Adams, P. B. and W. A. Ayers. 1982. Biological control of sclerotinia lettuce drop in the field by *Sporodesmium sclerotivorum*. *Phytopathology* **72**:485–488.
Anderson, T. R. and Z. A. Patrick. 1980. Soil vampyrellid amoebae that cause small perforations in conidia of *Cochliobolus sativus*. *Soil Biol. Biochem.* **12**:159–167.
Ayers, W. A. and P. B. Adams. 1979. Factors affecting germination, mycoparasitism, and survival of *Sporodesmium sclerotivorum*. *Can. J. Microbiol.* **25**:1021–1026.
Ayers, W. A. and P. B. Adams. 1981. Mycoparasitism and its application to biological control of plant diseases. Pages 91–103 in *Biological Control in Crop Production. Beltsville Symposium in Agricultural Research 5.* G. C. Papavizas, ed. Allanheld, Osmun, London.
Barron, G. L. 1977. *The Nematode-Destroying Fungi. Topics in Microbiology No. 1.* Canadian Biological Publications, Guelph, Ontario.
Bliss, D. E. 1951. The destruction of *Armillaria mellea* in citrus soils. *Phytopathology* **41**:665–683.
Campbell, A. H. 1934. Zone lines in plant tissues. II. The black lines formed by *Armillaria mellea* (Vahl) Quel. *Ann. Appl. Biol.* **21**:1–22.
Campbell, W. A. 1947. A new species of *Coniothyrium* parasitic on sclerotia. *Mycologia* **39**:190–195.
Castanho, B. and E. E. Butler. 1978a. Rhizoctonia decline: a degenerative disease of *Rhizoctonia solani*. *Phytopathology* **68**:1505–1510.
Castanho, B. and E. E. Butler. 1978b. Rhizoctonia decline: Studies on hypovirulence and potential use in biological control. *Phytopathology* **68**:1511–1514.
Castanho, B. and E. E. Butler. 1978c. The association of double-stranded RNA with Rhizoctonia decline. *Phytopathology* **68**:1515–1519.

Cook, R. J. and K. F. Baker. 1983. *The Nature and Practice of Biological Control of Plant Pathogens.* American Phytopathology Society, St. Paul, Minn.

Deacon, J. W. 1976. Studies on *Pythium oligandrum,* an aggressive parasite of other fungi. *Trans. Brit. Mycol. Soc.* **66**:383–391.

Deacon, J. W. and C. M. Henry. 1978. Mycoparasitism by *Pythium oligandrum* and *P. acanthicum. Soil Biol. Biochem.* **10**:409–415.

Drechsler, C. W. 1946. Several species of *Pythium* peculiar in their sexual development. *Phytopathology* **36**:781–864.

Duczek, L. J. 1981. Number and viability of conidia of *Cochliobolus sativus* in soil profiles in summer fallow fields in Saskatchewan. *Can. J. Plant Pathol.* **3**:12–14.

Duczek, L. J. 1983. Populations of mycophagous amoebae in Saskatchewan soils. *Plant Dis.* **67**:606–608.

Ghabrial, S. A. 1980. Effects of fungal viruses on their hosts. *Ann. Rev. Phytopathol.* **18**:441–461.

Hankin, L. and J. E. Puhalla. 1971. Nature of a factor causing interstrain lethality in *Ustilago maydis. Phytopathology* **61**:50–53.

Haskins, R. H. 1963. Morphology, nutrition, and host range of a species of *Pythium. Can. J. Microbiol.* **9**:451–457.

Hocking, D. and F. D. Cook. 1972. Myxobacteria exert partial control of damping-off and root diseases in container-grown tree seedlings. *Can. J. Microbiol.* **18**:1557–1560.

Hollings, M. 1978. Mycoviruses: viruses that infect fungi. *Adv. Virus Res.* **22**:1–53.

Hollings, M. 1982. Mycoviruses and plant pathology. *Plant Dis.* **66**:1106–1112.

Homma, Y. 1984. Perforation and lysis of hyphae of *Rhizoctonia solani* and conidia of *Cochliobolus miyabeanus* by soil myxobacteria. *Phytopathology* **74**:1234–1239.

Homma, Y. and R. J. Cook. 1985. Influence of matric and osmotic water potentials and soil pH on the activity of giant vampyrellid amoebae. *Phytopathology* **75**:243–246.

Huang, H. C. 1977. Importance of *Coniothyrium minitans* on survival of *Sclerotinia sclerotiorum* in wilted sunflower. *Can. J. Bot.* **55**:289–295.

Linford, M. B. 1939. Attractiveness of roots and excised shoot tissue to certain nematodes. *Proc. Helminthol. Soc. Wash.* **6**:11–18.

Mankau, R. 1975. Prokaryote affinities of *Duboscqia penetrans,* Thorne *J. Protozool.* **21**:31–34.

Mankau, R. and S. K. Mankau. 1963. The role of mycophagous nematodes in the soil. I. The relationship of *Aphelenchus avenae* to phytopathogenic soil fungi. Pages 272-280 in *Soil Organisms*: Proceedings of the *Colloquium on Soil Fauna, Soil Microflora and Their Relationships.* Oosterbeek, The Netherlands. Sept. 10–16, 1962. J. Doeksen and J. der Drift, eds. North Holland, Amsterdam.

Munnecke, D. E., M. J. Kolbezen, W. D. Wilbur, and H. D. Ohr. 1981. Interactions involved in controlling *Armillaria mellea. Plant Dis.* **65**:384–389.

Old, K. M. and J. F. Darbyshire. 1978. Soil fungi as food for giant soil amoebae. *Soil Biol. Biochem.* **10**:93–100.

Old, K. M. and Z. A. Patrick. 1979. Giant soil amoebae potential biocontrol agents. Pages 617–628 in *Soilborne Plant Pathogens.* B. Schippers and W. Gams, eds. Academic Press, New York.

Pramer, D. and N. R. Stoll. 1959. Nemin: A morphogenetic substance causing trap formation by predaceous fungi. *Science* **129**:966–967.

Puhalla, J. E. 1968. Compatibility reactions on solid medium and interstrain inhibition in *Ustilago maydis*. *Genetics* **60**:461–474.

Rhoades, H. L. and M. B. Linford. 1959. Control of Pythium root rot by the nematode *Aphelenchus oryzae*. *Plant Dis. Reptr.* **43**:323–328.

Riffle, J. W. 1967. Effect of an *sphelenchoides* species on the growth of a mycorrhizal and a pseudomycorrhizal fungus. *Phytopathology* **57**:541–544.

Sayre, R. M. 1980. Biocontrol: *Bacillus penetrans* and related parasites of nematodes. *J. Nematol.* **12**:260–270.

Sayre, R. M. and W. P. Wergin. 1977. Bacterial parasite of a plant nematode: morphology and ultrastructure. *J. Bacteriol.* **129**:1091–1101.

Sutherland, J. R. and J. A. Foertin. 1968. Effect of the nematode *Aphelenchus avenae* on some ectotrophic, mycorrhizal fungi on a red pine mycorrhizal relationship. *Phytopathology* **58**:519–523.

Thorne, G. 1940. *Dubocqia penetrans* n. sp. (Sporozoa, Microsporidia, Nosematidae), a parasite of the nematode *Pratylenchus pratensis* (de Man) Filipjev. *Proc. Helm. Soc. Wash.* **7**:52–53.

Uecker, F. A., W. A. Ayers and P. B. Adams. 1978. A new hyphomycete on sclerotia of *Sclerotinia sclerotiorum*. *Mycotaxon* **7**:275–282.

Webster, J. and N. Lomas. 1964. Does *Trichoderma viride* produce gliotoxin and viridin? *Trans. Brit. Mycol. Soc.* **47**:535–540

Chapter 18

Disease Suppression

Pathogen-suppressive soils were defined by Baker and Cook (1974, p. 61) as "soils in which the pathogens cannot establish, they establish but fail to produce disease, or they establish and cause disease at first but diminish with continued culture of the crop." This definition is too inclusive to have utility in an ecological sense. *Phymatotrichum omnivorum* cannot survive in cold climates. It could be introduced into all the soils of Wisconsin and would not become established. *Typhula idahoensis,* which increases under snow, could be spread upon all the soil in Louisiana and would not become established. Disease suppressive is real, but the effects of physical and chemical factors (such as temperature, moisture, soil texture, pH, and salinity) should not be included in examples of pathogen-suppressive soils.

In the book *Suppressive Soils and Plant Disease,* edited by Schneider (1982), many examples of "suppressive" soils are discussed. Fusarium wilt of cotton in the Southeast is worse on sandy Coastal Plain soils than on heavier Piedmont soils. This could be due to many things, not least of which would be the prevalence of root-knot nematodes in the lighter soils. The severity of Fusarium wilt of banana in Central America and of peas in Wisconsin is related to soil type. Heavy soils differ from light soils in many respects, and general ecology is probably involved. Baker and Chet (1982) stated that

"Physical factors like soil pH, texture and chemical/geological composition do not reproduce themselves. . . " (Baker and Chet, 1982). It is wise to concentrate on disease suppression attributable to biological entities.

Another type of "disease suppression" should be eliminated, the general suppression that occurs when biological activity within a soil is brought from a low to a high status. If a soil is at a very low level of biological activity, an introduced pathogen develops at first in a relative biologic vacuum, unhindered by effective background microbial activity. Huber and Watson (1970) referred to biological buffering. Inoculations in sterile soil do not often give the same results as inoculations in natural soil. Rovira and Wildermuth (1981) also indicated that the microbiota in a rich soil tends to reduce the severity of attack by many soilborne plant pathogens. These are examples of a general disease suppression. A soilborne pathogen encountering an intense microbiota is in a different environment than one facing sparse background activity.

In this chapter the term suppression is limited to the concept of specific disease suppression. In specific suppression monoculture leads to an initial increase in disease followed by a spontaneous decline in disease. General suppression, in contrast, could reduce many fungal and nematode attacks. Specific suppression is effective against only a single pathogen; it is specific, not affecting the severity of other diseases of the same host. This chapter includes specific declines of potato scab, take-all, oat cyst-nematode, and *Rhizoctonia solani*.

POTATO SCAB DECLINE

Menzies (1959) worked in areas where potatoes had been grown under irrigation for several years and where land had just begun to receive irrigation water and was new to potatoes. He observed that potato scab was of slight importance on land that had been cultivated for many years, but that on new land of similar type under similar management and with the same potato cultivars, scab was serious though sometimes erratic.

Menzies put nine virgin soils (previously uncropped) and three cultivated soils of similar soil type from the Columbia Basin in outdoor beds at Prosser, Washington. Each soil was planted with scab-infected potatoes the first time and with scab-free seed thereafter. In the first year there was little difference between the amount of scab in each soil. In the second year, when scab-free seed was used, scab was light on formerly cultivated land and severe on the former virgin lands. After five successive crops scab was still light in the cultivated soils and heavy in the formerly virgin soils. These soils were treated alike, yet the suppression of scab in the soil from established irrigated farms remained, and scab was serious and sustained in the formerly virgin soils. Something suppressed scab in the old cultivated soil without addition of extra organic matter or any other special treatment.

The soils were similar physically and chemically, low in organic matter, and neutral or slightly alkaline; all should have been favorable to scab. The scab-suppressing factor was destroyed by sterilization, and it could be introduced into scab-conducive soil by mixing 10% by volume of suppressive soil with conducive soil, or with a 1% addition of suppressive soil plus alfalfa meal. Menzies concluded that biological factors, not physical or chemical factors, accounted for the differences among these soils, and he referred to this phenomenon as biological suppression. Menzies reviewed the early literature on control of potato scab resulting from incorporation of massive amounts of organic matter and concluded that the resulting burst of microbial activity would be effective for only a short time.

Potatoes grown under irrigation in Nebraska suffered less from *Streptomyces scabies* after several years of continuous potatoes than in rotation with sugar beets, oats, and maize (Goss and Afanasiev, 1938). A decline of potato scab under monoculture was documented by Weinhold et al. (1964) at Shafter, California. In monoculture scab increased from 1949 to 1956 and then declined. Decline was slow to develop.

TAKE-ALL DECLINE

Take-all of wheat and barley, incited by *Gaeumannomyces graminis* var. *tritici*, is the best documented, most widespread, most studied decline of an important disease that occurs with continuous, unbroken production of a suscept (Hornby, 1979). Observations of a decline in take-all have been recorded since the end of the last century (Shipton, 1975). In 1917 Perkins reported decline of take-all on new lands in the Eyre Peninsula of South Australia. This decline occurred over a 25-year period after the land was brought under cultivation. Fellows and Ficke also observed a decline in Kansas in 1934, and Mary Glynne, in 1935, recorded a decline of take-all at Rothamsted, England. Zogg in 1951 reported that barley grown in monoculture in Switzerland was seriously diseased for several years, followed by a decline of take-all to relative insignificance. Shipton credits Slope with first appreciating the significance of this phenomenon, and Slope and Cox (1964) coined the term *take-all decline*. Shipton recorded the incidence of take-all of spring wheat at Hampshire, England from 1963 to 1969. He documented an increase in disease with monoculture for 4 years, a plateau during year 5, a moderate decline during year 6, and severe decline in disease severity in year 7.

G. graminis var. *tritici* has historically been controlled by short rotations to nonsuscepts, because the fungus has no effective dormant propagule and dies when infested host refuse is decomposed. Rotations of sufficient duration are effective, provided weed grasses that are hosts of the pathogen are not prevalent among the nonhost plants. One of the outstanding features of take-all decline is that the decline effect is lost or weakened when an effective rotation (break) crop is grown.

DISEASE SUPPRESSION

Take-all decline is transferrable using decline soil to inoculate conducive soil. It is destroyed by heating moist soil 30 min at 60°C. It develops only in the presence of diseased roots or active *G. graminis* var. *tritici* and is specific for *G. graminis* var. *tritici*, so far as is known. Cook and Baker (1983) ruled out hypovirulence due to virus infections, cross-protection by infection of roots by *Phialophora graminicola*, and vampyrellid amoebae as the cause of the widespread decline that develops in widely differing soils and environments over the world. They conclude that possession of the root surface by antagonistic *Pseudomonas* sp. is the most likely cause.

Repeated additions of living but not dead mycelium of *G. graminis* var. *tritici* to soil induces a specific suppression (Zogg and Amiet, 1980), and the number of hyphae emerging from infested host debris is reduced in take-all decline soil (Wildermuth et al., 1979). These effects have been observed in the absence of a living host, so in these observations rhizosphere phenomena are not involved. Something pathogenic or highly inhibitory to *G. graminis* must increase in the soil in the presence of living hyphae of that fungus. The cause of take-all decline may still be unknown. See Hornby (1979) for a thorough discussion of take-all decline.

OAT CYST-NEMATODE DECLINE

The oat- or cereal-cyst nematode, *Heterodera avenae*, is common in northern Europe, and in some situations it causes serious losses. Evidence of an important decline was obtained in northwest Nottinghamshire, England by Gair et al. (1969) (Fig. 18.1). These workers studied spring oats and barley, both of which are hosts, in monoculture and in 1- and 2-year rotations on a rather infertile, coarse, sandy soil that is subject to drought and is low in potash and nitrogen. They recorded grain yields in the various crop-

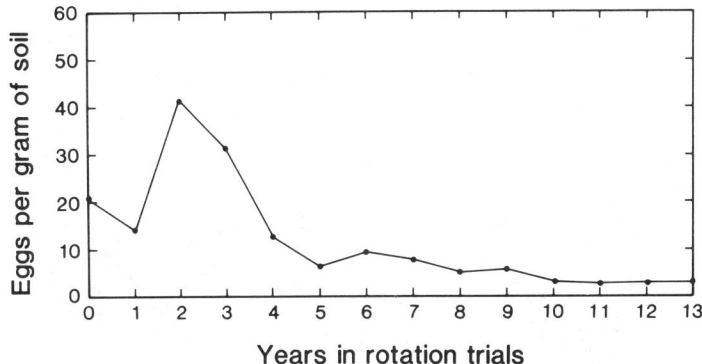

Figure 18.1 The decline in number of eggs of *Heterodera avenae* per gram of soil infested by *Nematophthora gynophilia*, even though hosts of the nematode were frequently or continuously grown. After Gair et al., 1969.

ping systems and monitored the number of eggs of *H. avenae* in the soil. The initial population of *H. avenae* (20 to 30 eggs per gram of soil in all plots) was sufficient to cause economic losses. The egg count increased moderately during the first 3 years and then began to decline. On the fourth year the level was below 10 eggs per gram of soil in most plots, and by the seventh year it did not reach 10 eggs/g in any plot. On the thirteenth year all plots contained fewer than 3 eggs/g, even those in continuous host.

This decline, unlike that of take-all, was not affected by break crops; it persisted after rotation with plants not hosts of *H. avenae*. The decline occurred in all rotations, even in those with a host every third year. The agent of decline, therefore, persisted in effective levels at least 2 years in the absence of *H. avenae* in an active state. Monoculture is not required to sustain the decline so that suscepts can be grown in rotation. Numbers of predatory nematodes (chiefly monochids) did not increase during these trials. The clover cyst nematode (*H. trifolii*) increased on clover during some of the rotation cycles. Ectoparasitic nematodes were numerous. The authors could not explain this specific decline of *H. avenae*, which occurred only on some soils.

By 1975, after studying parasites of *H. avenae* in decline soil, Kerry found an unknown fungus, which he and Crump named *Nematophthora gynophila* (Kerry and Crump, 1980). In inoculation trials *N. gynophila* attacked *H. avenae, H. carotae, H. cruciferae, H. schachtii,* and *H. trifolii*. It did not attack *Globodera rostochiensis*. It has been found in nature only where *H. avenae* is present. Evidence so far indicates a special relationship between *N. gynophila* and *H. avenae*. *N. gynophila* produces oospores with thick walls, up to 3000 per diseased *H. avenae* female (Fig. 18.2). In addition, laterally biflagellate zoospores are liberated from about 40 sporangia per female. The zoospores are strong swimmers. They may encyst and become motile again (Kerry and Crump, 1980). The long-lived oospores,

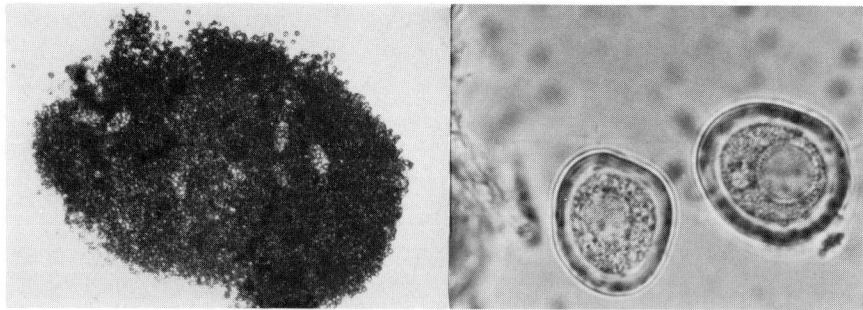

Figure 18.2. A crushed mature female of *Heterodera avenae* with its insides replaced by oospores of *Nematophthora gynophila* (left) and individual, thick walled oospores (right). Photographs courtesy of B. R. Kerry, Rothamsted Station, England.

which account for the failure of break crops to destroy the decline, have not been germinated, and the fungus has not been grown in culture.

Soil surveys for the presence of *N. gynophila* are conducted by adding healthy cysts to the test soils, growing a suscept, and examining female cyst-nematodes for this hyperparasite. The females are attacked only after their bodies erupt through the cortex of roots, exposing them to the soil solution. They become flaccid and gray. Hyphae ramify the female, the body wall and cuticle are disrupted, and the fungal tubes protrude into the soil, enabling zoospores to emerge. In England the female erupts from the root between May 21 and June 4. *N. gynophila* infects 15% of the females by June 4. Females are most numerous June 18, and within 2 weeks 50% are dead, before embryonated eggs have formed. In one soil *H. goettingiana* and *H. trifolii* are attacked, but not in another. *H. avenae* is the most susceptible host. Death can occur within 7 days, and rapid disintegration makes counting diseased or dead females difficult.

Oospores initiate the primary infections (Kerry, 1980). To be effective, oospores must be numerous and well distributed. Zoospores then infect other females. Zoospores can move 1 to 3 cm in soil, and they function even in rather heavy soils. Kerry et al. (1980) grew barley in soil infested with both *N. gynophila* and *H. avenae* for 5 weeks, by which time many females had erupted through the cortex. Heavily watered plants were larger, and at 12 weeks 36% of the females in their roots were diseased. In contrast, only 13% of the females were diseased in pots receiving less water. This decline can fail to control the nematode population in a drought, in that fewer than 1% of the females may become diseased. Wet soil is essential for zoospore movement and for the success of this decline phenomenon. *H. avenae* decline has been reported in Great Britain, Germany, Holland, and Denmark on a wide range of soils and on both winter and spring cereals.

Nematode-trapping fungi must trap large numbers of larvae to have a significant effect, because only about 8% of nematode larvae mature even under favorable conditions. In contrast, *N. gynophila* is efficient, in that it attacks established females, not juveniles, within the soil that may fail to establish in a host. The ability of *N. gynophila* to persist over 2 years is a great advantage for crops benefiting from rotation. Also, once established in the soil, no manipulation by the grower is necessary. This is the only documented case of a natural agent giving long-term control of a cyst-nematode.

RHIZOCTONIA SOLANI DECLINE

Rhizoctonia blight of radishes in a clay loam increases for three successive plantings and then declines (Henis and associates, 1978, 1979). The decline occurs only when *R. solani* is present and active. *R. solani* decline does not occur if the soil contains the pathogen but not the host, or the host without the pathogen. Henis and co-workers explored the changes in microflora associated with the decline and observed an increase in *Trichoderma* spp. As decline intensifies, the number of *Trichoderma* spp. propagules increases. Chet and Baker (1980) and Liu and Baker (1980) continued these studies and established *T. harzianum* as the cause of the decline.

Chet and Baker (1980) infested soil with *R. solani* initially, after which no new inoculum was added. Disease of radish was assessed after 7 days, and the process was repeated each week. At each sampling the soil was assayed for *R. solani* and for *Trichoderma* spp. If disease decreased with time, the soil was considered to have acquired suppressiveness. If disease did not decrease, the soil was considered conducive to disease. They compared the development of disease suppression in a clay loam at pH levels 8.1, 6.5, and 5.7. At pH 8.1, disease increased with successive croppings of the three hosts, each nearing 100% disease after six cycles. At the lower pH's, disease decreased.

T. harzianum grows well in culture over a pH range of 4 to 8. The conidia of *T. harzianum*, however, germinate best on water agar at low pH (4 to 5), and poorly above pH 7. Conidiophores form most abundantly between pH 4 and 6. Spore germination and production are favored by acidity. Chet and Baker (1980) saw no evidence that antibiotics are involved in the suppression of *R. solani*, but direct parasitism is important. Hyphae of *T. harzianum* coil around hyphae of *R. solani*. They release β-(1,3) glucanase and chitinase, lysing walls of many fungi, including those of *R. solani* (Hadar et al., 1979). The pH optimum for β-(1,3) glucanase from *T. viride* is 4.5 and for its chitinase, 5.3 (Jones and Watson, 1969). This additionally supports the importance of acidity to the effectiveness of *Trichoderma* spp. in antagonistic interactions.

The soil in which radishes inoculated with *R. solani* had been grown has a 7- to 8-fold increase in *Trichoderma. Streptomyces* and bacteria are unchanged (Liu and Baker, 1980). The increase of *T. harzianum* is attributed to the presence of active *R. solani*. *T. harzianum* hyphae encircle hyphae of *R. solani* and lyse individual cells but do not enter them. Host cells separate at the septa, so hyphae are essentially cut. The cutting of hyphae weakens *R. solani* to the extent that it cannot mobilize reserves, leading to its decline.

PROPER USE OF SUPPRESSIVE

Big vein of lettuce, caused by an infectious agent transmitted by *Olpidium brassicae*, is prevalent on certain soils in the Salinas Valley, California, and is of little importance in adjacent soils, even though the Salinas Valley is an intensive lettuce-growing area and cultivars susceptible to the malady have been grown repeatedly. This is an ideal situation in which to investigate edaphic factors affecting disease development—a widespread suscept in a small area, so climatic differences should be minimal. Big vein is more prevalent on heavy soils than on light soils (Westerlund et al., 1978). They classified soils big-vein-prone and big-vein-suppressive on the basis of repeated observation of disease incidence and severity. Big vein is favored by wet soil and moderately low temperature.

Resting spores of *O. brassicae* require 6 days to germinate at 0 matric water potential, the optimum wetness. Germination is reduced at -0.04 bar and slight at -0.06 bar, and no germination is detected at -0.1 bar. Infection is correlated with the percent germination of resting spores. Prolonged wetness is required for spore germination. Some type of endogenous dormancy prevents all resting spores from germinating during single wet periods alternated with drying cycles.

Zoospores are released at 0 to -0.01 bar in sand, wetter than is required for zoospore movement. Zoospore movement is greatest at 0 matric water potential and less at -0.02 bars and -0.04 bars. In solutions adjusted with osmotica, zoospores are released from sporangia down to -5.5 bars. Zoospores sense water content more than water activity. Westerlund et al. concluded that the main difference between "conducive" and "suppressive" in this case is due to differences in the rapidity with which the two soils drained, either from rain or irrigation. Big vein is strongly associated with wet spots in the field.

In my opinion the term disease suppressive should not be applied to the coarser-textured soil. Well-drained soil does not favor disease, poorly drained soil does. A coarse-textured soil may favor root-knot nematode, an acid soil may favor club root, snow may favor snow mold. Suppressive soils should be reserved for phenomena documented as being associated with microbial interactions, even if poorly understood, and not for physical-chemical differences. The work of Westerlund et al. is a classic study of water relations, and it has been included in this chapter to protest the use of suppressive soil terminology in this manner. I agree with Baker and Chet (1982), that soil factors that do not reproduce themselves should not be called suppressive. The term *unfavorable soil* would depict physical-chemical factors more precisely.

COMMENTS

Menzies (1959) pioneered the study of disease-suppressive soils and stressed their potential for disease control. The causes of the potato scab and take-all declines are still in doubt, but bacteria are probably responsible. The most exciting advance has been elucidation of the cereal cyst-nematode decline in Britain. Cyst-nematodes are among the most difficult nematodes to control. Maybe "search gardens" should be established, planting various hosts year after year without rotation, where the various nematodes exist. If done in enough places, similar pathogens of these nematodes might be found. *Nematophthora gynophila,* with its tough resting spore, is particularly valuable, in that the decline factor is not lost in crop rotation as is the factor(s) responsible for take-all decline.

A disease that declines in severity with monoculture defies the principle that repeated growth of the host would increase inoculum and allow the pathogen to become even better adapted to the host, and that disease should increase. In the cases in which the cause of the decline is known (cyst-nematode, *Rhizoctonia solani*), parasites of the pathogen itself increase to bring about control.

If take-all decline is caused by antagonistic microflora that responds directly to and increases in the presence of the active pathogen, these antagonists must be very poor competitors in soil. They depend on continual support by the presence of the active pathogen and rapidly decline in numbers or activity when rotation breaks that support, requiring several cycles of the disease to reestablish decline to effective levels.

KEY REFERENCES

Hornby, D. 1979. Take-all decline: A theorist's paradise. Pages 133–156. in *Soilborne Plant Pathogens*. B. Schippers and W. Gams, eds. Academic Press, New York.

Kerry, B. 1981. Fungal parasites: A weapon against cyst-nematodes. *Plant Dis.* **65**:390–393.

Shipton, P. J. 1975. Take-all decline during cereal monoculture. Pages 137–144 in *Biology and Control of Soilborne Pathogens*. G. W. Bruehl, ed. American Phytopathology Society., St. Paul, Minn.

REFERENCES

Baker, K. F. and R. J. Cook. 1974. *Biological Control of Plant Pathogens*. H. Freeman, San Francisco.

Baker, R. and I. Chet. 1982. Induction of suppressiveness. Pages 35–50 in *Suppressive Soils and Plant Disease*. R. W. Schneider, ed. American Phytopathology Society, St. Paul, Minn.

Chet, J. and R. Baker. 1980. Induction of suppressiveness to *Rhizoctonia solani* in soil. *Phytopathology* **70**:994–998.

Cook, R. J. and K. F. Baker. 1983. *The Nature and Practice of Biological Control of Plant Pathogens*. American Phytopathology Society., St. Paul, Minn.

Gair, R., P. L. Mathias, and P. N. Harvey. 1969. Studies on the cereal nematode populations and cereal yields under continuous or intensive culture. *Ann. Appl. Biol.* **63**:503–512.

Goss, R. W. and M. M. Afanasiev. 1938. Influence of rotations under irrigation on potato scab, Rhizoctonia, and Fusarium wilt. *Nebr. Agric. Exp. Sta. Bull.* **317**:1–18.

Hadar, Y., I. Chet, and Y. Henis. 1979. Biological control of *Rhizoctonia solani* damping-off with wheat bran culture of *Trichoderma harzianum*. *Phytopathology* **69**:64–68.

Henis, Y., A. Ghaffar, and R. Baker. 1978. Integrated control of *Rhizoctonia solani* damping-off of radish: effect of successive plantings, PCNB, and *Trichoderma harzianum* on pathogen and disease. *Phytopathology* **68**:900–907.

Henis, Y., A. Ghaffar, and R. Baker. 1979. Factors affecting suppressiveness to *Rhizoctonia solani* in soil. *Phytopathology* **69:**1164–1169.

Huber, D. M. and R. D. Watson. 1970. Effect of organic amendment on soilborne plant pathogens. *Phytopathology* **60:**22–26.

Jones, D. and D. Watson. 1969. Parasitism and lysis by soil fungi of *Sclerotinia sclerotiorum* (Lib.) de Bary, a phytopathogenic fungus. *Nature* **244:**287–288.

Kerry, B. 1980. Biocontrol: fungal parasites of female cyst-nematodes. *J. Nematol.* **12:**253–259.

Kerry, B. R. and D. H. Crump. 1980. Two fungi parasitic on females of cyst-nematodes (*Heterodera* spp.). *Trans. Brit. Mycol. Soc.* **74:**119–125.

Kerry, B. R., D. H. Crump and L. A. Mullen. 1980. Parasitic fungi, soil moisture, and the multiplication of the cereal cyst-nematode. *Nematologica* **26:**57–68.

Liu, S. and R. Baker. 1980. Mechanism of biological control in soil suppressive to *Rhizoctonia solani*. *Phytopathology* **70:**404–412.

Menzies, J. D. 1959. Occurrence and transfer of a biological factor in soil that suppresses potato scab. *Phytopathology* **49:**648–652.

Rovira, A. D. and G. B. Wildermuth. 1981. The nature and mechanisms of suppression. Pages 385–415 in *Biology and Control of Take-All*. M. J. C. Asher and P. J. Shipton, eds. Academic Press, New York.

Schneider, R. W. 1982. Suppressive Soils and Plant Disease. R. W. Schneider, ed. American Phytopathology Society., St. Paul, Minn.

Slope, D. B. and J. Cox. 1964. Continuous wheat growing and the decline of take-all. *Rep. Rothamsted Exp. Sta.* 1963, Pt. 1, 108.

Weinhold, A. R., J. W. Oswald, T. Bowman, J. Bishop, and D. Wright. 1964. Influence of green manures and crop rotation on common scab of potatoes. *Am. Potato J.* **41:**265–273.

Westerlund, F. V., R. N. Campbell, R. G. Grogan, and J. M. Duniway. 1978. Soil factors affecting the reproduction and survival of *Olpidium brassicae* and its transmission of the big-vein agent to lettuce. *Phytopathology* **68:**927–935.

Widermuth, G. B., A. D. Rovira, and J. H. Warcup. 1979. Mechanism and site of suppression of *Gaeumannomyces graminis* var. *tritici* in soil. in *Soilborne Plant Pathogens*. B. Schippers and W. Gams, eds. Academic Press, New York. Pages 157–164.

Zogg, H. and J. A. Amiet. 1980. Laboratory studies on decline with different foot and root rot pathogens of wheat. *Phytopath. Z.* **97:**193–213.

Chapter 19

Mycorrhizae

Frank in 1885 coined the term *fungus-root* (mycorrhiza) to describe the intimate association of fungi with tree roots that he observed in Prussia The term implies a mutualistic symbiosis, the fungus obtaining elaborated nutrients from the host and the host receiving improved mineral nutrition from the fungus. Frank believed the fungus, by hyphal threads within the soil, acted as enlarged root hairs.

According to the review of Hetrick (1984), no one knows the rate at which mycorrhizal fungi grow though soil, but it is thought that the growth is not rapid. Spores of several mycorrhizal fungi pass through rodents unharmed, so even though the fungus may sporulate below ground, it can have an "aerial" vector. Spores have been found in nests of mud daubers, robins, swallows, and sparrows.

ECTOTROPHIC MYCORRHIZAE

In ectotrophic mycorrhizae a dense fungal sheath several cells thick forms around the rootlets (Fig. 19.1). The rootlets are greatly deformed and appear to be diseased. Within the mycorrhizal root living epidermal and cortical cells are surrounded by hyphae, forming the Hartig net which may extend to

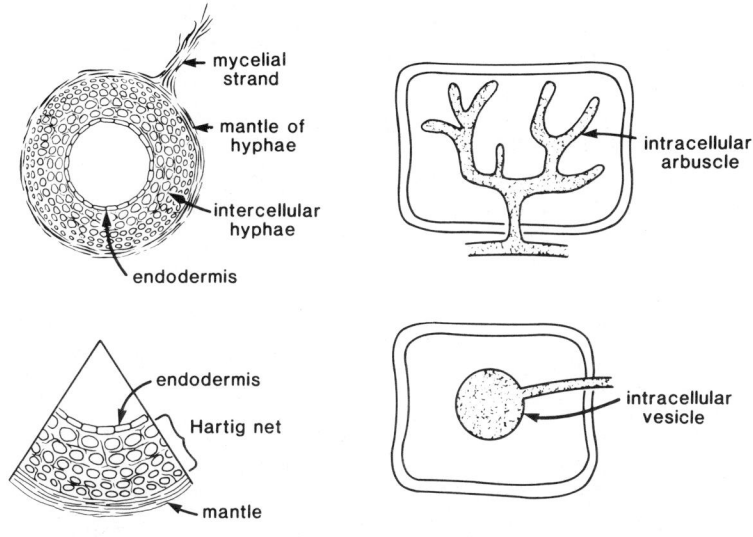

Figure 19.1. In ectotrophic mycorrhizae (left) the rootlet is encased by a hyphal mantle, with intercellular hyphae (the Hartig-net) encircling living host cells inward to the endodermis. In vesicular-arbuscular (right) mycorrhizae there is no fungal mantle, but fungal hyphae ramify the cortex, mostly intercellularly, with intracellular vesicles or arbuscles.

the endodermis. Ectotrophic mycorrhizae generally form only on roots totally derived from a terminal meristem, i.e., those without secondary thickening.

Ecto refers to the external mantle of fungus surrounding the root. Assimilative hyphae, mainly intercellular, are within the root. Ectomycorrhizal is not comparable to ectoparasitic as applied to nematodes that do not enter the host but feed from the outside. Ectomycorrhizae occur on Pinaceae, Fagaceae, Betulaceae, and Myrtaceae, and the fungi are usually basidiomycetes or ascomycetes (Schenk, 1981).

Zak (1964) described ways that mycorrhizal fungi could protect roots from pathogens: (1) by utilizing exudates, thus reducing attraction or support of pathogens in the rhizosphere, (2) with ectotrophic mycorrhizae, excluding them by means of the fungal mantle, (3) by producing antibiotics. (4) by being conducive to beneficial organisms within the rhizosphere. Marx (1972) also reviewed ectomycorrhizae.

Marx and Davey (1969) scraped the mantle from some roots and found that the Hartig net protected the exposed cortex from *Phytophthora cinnamomi*. Whether protection is due to the presence of the fungus or to induced host resistance in response to the mycorrhizal fungus could not be determined. The surface of ectotrophic mycorrhizae differs from that of "normal" roots, and this could lead some pathogens sensitive to surfaces (thigmotropic responses) to fail to identify a host. Douglas fir mycorrhizal roots (with *Laccaria laccata*) accumulate phenolic compounds that made the roots resistant to *Fusarium oxysporum*, a pathogen of seedlings in forest tree nurseries (Silvia and Sinclair, 1983). According to Marx (1972), mycorrhizae should improve host nutrition, resulting in increased resistance or tolerance to some pathogens. In some cases protection is highly localized, limited to the mantle area. Microflora of mycorrhizal roots differs quantitatively and qualitatively from that of nonmycorrhizal roots.

Phytophthora cinnamomi

Little leaf of pines is a major disease of shortleaf and loblolly pines. It was found in Alabama in 1934 and was soon recognized in nearby states. It is most severe on heavy soils with poor internal drainage. The disease usually attacks trees 30 to 50 years old, resulting in yellowing and stunting. *Phytophthora cinnamomi*, the pathogen responsible for this problem, attacks fine feeder roots, and nitrogen deficiency is believed to be a major cause of the symptoms (Campbell et al., 1953).

Culture filtrates of *Leucopaxillus cerealis* var. *piceina* inhibit mycelial growth and zoospore germination in *P. cinnamomi* (Marx, 1969a). An unusual characteristic of this mycorrhizal fungus is that in culture the antibiotic, diatretyne nitrile, is produced in greatest quantity during the period of most rapid fungal growth and incipient autolysis stages. After the growth phase, much of the primary antibiotic is reduced to diatretyne amide and diatretyne 3, which are antibacterial but not antifungal (Marx, 1969b). The primary antibiotic totally inhibits zoospore germination at 2 ppm and is partially effective at 50 to 70 ppb. Very small amounts would be important when produced in the root zone. Most nonmycorrhizal roots in water in Petri dishes near mycorrhizal roots (*L. cerealis* var. *piceina*) are protected from zoospores of *P. cinnamomi* under conditions resulting in 100% infection of nonmycorrhizal roots exposed separately. This study provides evidence that the antibiotic diffuses in the water and can therefore be effective in nature (Marx and Davey, 1969). Seedlings are damaged by 10 ppm of diatretyne nitrile, so the range between fungitoxicity and phytotoxicity is not great.

The fungal mantle of fully developed mycorrhizae with *L. laccata* on pine protects the root from infection. The fungal mantle acts as a mechanical barrier, not by producing antibiotic. These studies provide direct evidence for two of Zak's hypotheses, that antibiotics and physical exclusion of the pathogen could protect trees from invasion by pathogens.

VESICULAR-ARBUSCULAR MYCORRHIZAE (VAM)

Vesicular-arbuscular mycorrhizae (VAM) do not deform the roots. In VAM roots, vesicles (enlarged hyphal swellings) form intra- or intercellularly, and arbuscules (highly branched haustoria) form intracellularly. They are connected to hyphae external to the host, and no fungal sheath or mantle forms on the root. Variations and integrations of the two major types of mycorrhizae exist.

In contrast to fully developed ectotrophic mycorrhizae, the VAM root is not surrounded by a fungal mantle, and physical exclusion of a pathogen is not possible. The hyphae of these fungi ramify cortical tissues, enter cells, and alter the host's physiology. The interactions of vesicular-arbuscular mycorrhizae with nematodes are complex and have been discussed by Hussey and Roncadori (1982).

Mycorrhizae are generally more abundant and more important in infertile soil, but germination of VAM spores is not inhibited by the presence of phosphorus, nitrogen or potassium. Strzemska (1975) noted decreased colonization of rye, wheat, barley, and oats when fertilizers are generously used, but infection of bean roots is not altered by fertilization. Menge et al. (1978) grew sudangrass in split-root technique. Fertilizing half of the root system with phosphorus reduced colonization by fungi in the unfertilized part of the root system, demonstrating that the host controls colonization of its roots by VAM fungi. Leakage of nutrients from roots is increased by low phosphorus levels, but the extent to which these fungi respond to root exudates is not known. Drought also increases colonization, but this may result indirectly from reduced phosphorus uptake in droughty soils or from the additional exudate that emanates from drought-damaged root cells.

Mosse et al. (1981) state that benomyl is toxic to some mycorrhizae, that some nematicides such as aldicarb increase mycorrhizal development, probably by reducing nematodes, and that methyl bromide or dibromochloropropane can be used in forest nurseries without danger to mycorrhizal establishment (Hacskaylo and Palmer, 1957). Mycorrhizae are abundant among pineapples in Hawaii even though nematicides are commonly used. Several fungicides used repeatedly reduce establishment of *Glomus fasiculatus* on creeping bent grass golf turf (Rhodes and Larsen, 1981).

VAM fungi that sporulate little are favored in native bushlands, while heavy sporulators are favored in cultivated or grasslands (Baylis, 1969). When growing roots are constantly present, nonsporulators are favored; when such roots are frequently absent, sporulators dominate (see Chapter 24 for discussion of fungal colonists of cultivated soil).

Phytophthora megasperma

When Ross (1972) inoculated Lee soybeans with an *Endogone* sp. and with *Phytophthora megasperma,* the doubly-inoculated plants weighed 204 g/plant, the *Phytophthora*-infected plants weighed 153 g, and the controls weighed 181 g. A numbered soybean line weighed 126 g with *Phytophthora* alone, the survivors of plants inoculated with both fungi weighed 126 g each, and healthy plants weighed 284 g. In the double infection, 33% of the soybeans of the numbered line died, while no plants were killed by *Phytophthora* alone. *Endogone* made the numbered line more susceptible to *P. megasperma.*

Davis et al. (1978) did a similar experiment with citrus sweet orange, avocado, and alfalfa and *Phytophthora parasitica, P. cinnamomi,* and *P. megasperma,* respectively, along with *Glomus fasiculatus.* In their experiments the reactions of citrus and alfalfa seedlings to the pathogens was not altered, but avocado seedlings with mycorrhizae were more susceptible to disease.

COMMENTS

Our knowledge of the effects of mycorrhizae on soilborne plant diseases is limited, but we know that some interactions, as with *Phytophthora cinnamomi,* are beneficial, and some, as with *P. megasperma,* are not. Also, mycorrhizae are not always beneficial, with or without the presence of plant pathogens (see Buwalda and Goh, 1982). *Glomus macrocarpum* causes a severe stunt of tobacco in Kentucky (Modjo and Hendrix, 1986).

KEY REFERENCES

Marx, D. H. 1969b. The influence of ectotrophic mycorrhizal fungi on the resistance of pine roots to pathogenic infections. II. Production, identification, and biological activity of antibiotics produced by *Leucopaxillus cerealis* var. *piceina. Phytopathology* **59:**411–417.

Ross, J. P. 1972. Influence of *Endogone* mycorrhiza on Phytophthora rot of soybean. *Phytopathology* **62:**896–897.

Schenck, N. C. 1981. Can mycorrhizae control root disease? *Plant Dis.* **65:**230–234.

REFERENCES

Baylis, G. T. S. 1969. Host treatment and spore production by *Endogone. N. Z. J. Bot.* **7:**173–174.

Buwalda, J. G. and K. M. Goh. 1982. Host fungus competition for carbon as a cause of growth depression in vesicular-arbuscular mycorrhizal ryegrass. *Soil Biol. Biochem.* **14:**103–106.

Campbell, W. A., O. L. Copeland, Jr., and G. H. Hepting. 1953. Littleleaf in pines in the southeast. Pages 855–857 in *USDA Yearbook of Agric. 1953. Plant Diseases.* A. Stefferud, ed. U. S. GPO, Washington, D.C.

Davis, R. M., J. A. Menge, and G. A. Zentmyer. 1978. Influence of vesicular-arbuscular mycorrhizae on Phytophthora root rot of three crop plants. *Phytopathology* **68:**1614–1617.

Hacskaylo, F. and J. G. Palmer. 1957. Effects of several biocides on growth of seedling pines and incidence of mycorrhizae in field plots. *Plant Dis. Reptr.* **41:**354–358.

Hetrick, B. A. Daniels, 1984. Ecology of vesicular-arbuscular mycorrhizal fungi.

Pages 35–57 in *Vesicular Arbuscular Mycorrhizae*. C. U. Powell and D. J. Bagyaraj, eds. CRC Press, Boca Raton, Fla.

Hussey, R. S. and R. W. Roncadori. 1982. Vesicular-arbuscular mycorrhizae may limit nematode activity and improve plant growth. *Plant Dis.* **66**:9–14.

Marx, D. H. 1969a. The influence of ectotrophic mycoorrhizal fungi on the resistance of pine roots to pathogenic infections. I. Antagonism of mycorrhizal fungi to root pathogenic fungi and soil bacteria. *Phytopathology* **59**:153–163.

Marx, D. H. 1972. Ectomycorrhizae as biological deterents to phytopathogenic root infections. *Ann. Rev. Phytopathol.* **10**:429–454.

Marx, D. H. and C. B. Davey. 1969. The influence of ectotrophic mycorrhizal fungi on the resistance of pine roots to pathogenic infections. IV. Resistance of naturally occurring mycorrhizae to infections by *Phytophthora cinnamomi*. *Phytopathology* **59**:559–565.

Menge, J. W., D. Steirle, D. J. Bogyaraj, E. L. V. Johnson, and R. T. Leonard. 1978. Phosphorus concentrations in plants responsible for inhibition of mycorrhizal infection. *New Phytol.* **80**:575–578.

Modjo, H. S. and J. W. Hendrix. 1986. The mycorrhizal fungus *Glomus macrocarpum* as a cause of tobacco stunt disease. *Phytopathology* **76**:686–691.

Mosse, B., D. B. Stribley, and F. LeTacon. 1981. Ecology of mycorrhizae and mycorrhizal fungi. Pages 137–210 in *Advances in Microbial Ecology*. Vol. 5. M. Alexander, ed. Plenum Press, New York.

Rhodes, L. H. and P. O. Larsen. 1981. Effects of fungicides on mycorrhizal development of creeping bent grass. *Plant Dis.* **65**:145–147.

Silvia, D. M. and W. A. Sinclair. 1983. Phenolic compounds and resistance to fungal pathogens induced in primary roots of Douglas-fir seedlings by the ectomycorrhizal fungus *Laccaria laccata*. *Phytopathology* **73**:390–397.

Strzemska, J. 1975. Mycorrhiza in farm crops grown in monoculture. Pages 527–535 in *Endomycorrhizas*. F. E. Sanders, B. Mosse, and P. B. Tinker, eds. Academic Press, London.

Zak, B. 1964. Role of mycorrhizae in root disease. *Ann. Rev. Phytopathol.* **2**:377–394.

Chapter 20

Soilborne Viruses

The realization that several important viruses are intimately associated with soil is a relatively recent development. Mayer in 1886 and Beijerinck in 1898 believed that tobacco mosaic virus was soilborne to some extent, but the ease of mechanical transmission of the virus and the relative unimportance of the soil as a source of infection resulted in these observations having little effect on plant pathology. Behrens, as early as 1889, described tobacco rattle and concluded that the pathogen was in the soil (in Raski and Hewitt, 1963). It was not until McKinney (1923, 1925) established that wheat soilborne mosaic virus was soilborne that a relationship was really proved. The fungus *Polymyxa graminis* was associated with soilborne wheat mosaic in 1954, but it was not until several years later that substantial proof of the fungus-virus relationship was established. Big-vein of lettuce was observed in the Imperial Valley of California in 1922, and by 1929 it was recognized as being generally distributed in that state. Jagger observed the fungus *Olpidium brassicae* in lettuce roots, but it was not until 1957 that the fungus was associated with big-vein by Fry in New Zealand and Grogan et al. (1958) in California. In the same year Hewitt et al. (1958) proved that Grapevine fanleaf virus was transmitted by the nematode *Xiphinema index*.

By 1962 Teakle had established the role of *O. brassicae* as a vector of tobacco necrosis virus. The subject of soilborne viruses has expanded rapidly since.

Harrison (1960) defined soilborne viruses as "viruses with an underground natural method of spread which does not depend simply on contact between tissues of infected and healthy plants."

VIRUS FREE IN THE SOIL

Hoggan and Johnson (1936) found that tobacco mosaic virus was released from decaying infested host debris into soil and that it persisted in moist soil but was quickly inactivated in dry soil. They considered infested soil a hazard during transplanting of susceptible plants. When soil is dried, a physical disruption of capsid and ribonucleic acid of tobacco mosaic particles occurs (Allen, 1984).

Many plant viruses are quickly inactivated in soil, but some are sufficiently stable to be of consequence. Allen (1981) found tobacco mosaic virus in greenhouse soil. Infectivity was not appreciably reduced when potting mix containing naturally infested soil was steamed 30 min after reaching 85 to 88°C. Virus splashed from the potting mix onto tomato leaves caused infection when the plants were handled. Allen suspected that cucumber necrosis, tobacco necrosis, and tomato bushy stunt viruses posed similar dangers. Steaming for 30 min at 82°C, the recommended soil treatment, is not adequate for destroying tobacco mosaic virus.

Roots of *Cleome spinosa* infected with tobacco necrosis and tobacco mosaic viruses release virus into drainage water (Yarwood, 1960). Tobacco necrosis virus is seldom important except in greenhouses, and it will be discussed under soilborne viruses with fungal vectors. Yarwood did not believe that virus is released by living cells, but rather from dead epidermal or cortical cells during their decomposition. Diseased plants in sterilized quartz sand with nutrient solution release tobacco necrosis, cucumber necrosis, petunia asteroid mosaic, tomato bushy stunt, tobacco mosaic, and southern bean mosaic viruses into the drainage water (Smith et al. 1969). Most of the viruses retained infectivity in moist soil for at least 25 days but were inactivated within 2 days in dry soil. It is obvious that some viruses are present free in soil and that most of this virus is harmless. Exceptions include cases of contamination during transplanting, or when a virus has a vector like *Olpidium brassicae* that accumulates virus as its zoospores swim in a virus suspension.

VIRUS IN ROOTS

According to Zeyen (1979), Allard in 1916 was probably the first to recover a virus from roots. Fulton (1941) studied several virus-host combinations and found that virus was rapidly translocated to roots, sometimes unidirectionally, and that translocation from roots upward was generally slow. Virus often spreads laterally within roots faster than nematode or fungal vectors can move through soil.

Grapevine fanleaf virus and *Xiphinema index* were introduced from Europe on *Vitis rupestris* 'St. George' rootstocks. In California the virus and these rootstocks have the same distribution (Hewitt et al., 1958). *X. index* multiplies well on most grape roots, but it does not maintain itself on *V. rotundifolia*. Boubals and Pistre (in Bouquet, 1981) reported failure to get symptoms of grapevine fanleaf in *V. rotundifolia* within 5 years after exposure to viruliferous nematodes. Bouquet (1981) corroborated this finding. *V. rotundifolia* shoots become infected when they are grafted to diseased *V. rupestris* rootstocks. When susceptible *V. rupestris* scions were grafted onto *V. rotundifolia* rootstocks and viruliferous *X. index* were introduced, the susceptible scions remained healthy. Whether the resistance of *V. rotundifolia* in nature is due to resistance to the nematode or to the virus in roots is not known. Unfortunately, this species has been a poor rootstock for grapes in France.

IS THE VECTOR A NEMATODE OR A FUNGUS?

Several viruses transmitted by a fungus persist within the vector in air-dried soil. In contrast, viruses transmitted by nematodes are not transmitted from soil that has been air-dried for very long. The simplest way to determine if a suspected soilborne virus has a fungus or nematode vector is to collect soil from a heavily infested area, then store some in a cool, moist state and store another portion in an air-dry condition for 2 weeks. Next, plant suscepts in the soils and incubate them under cool, moist conditions. Allow plenty of time for expression of symptoms or translocation of virus to the shoots. If incubation conditions are suitable and appropriate test plants are used the following diagnoses can be made.

Air-dried soil, positive transmission = fungal vector
Air-dried soil, negative transmission = probably not a fungal vector
Moist soil, positive transmission = either
Moist soil, negative transmission probably not soilborne

Even if the natural host is a woody plant, it is wise to include herbaceous plants in the trials. Herbaceous plants have low concentrations of tannins and acids, substances that cause viruses to be more difficult to transmit mechanically or to purify. In addition, the symptoms will usually appear more quickly on herbaceous plants. *Xiphinema index* multiplies well on grape and poorly on herbaceous plants, yet grapevine fanleaf virus is transmitted to *Chenopodium amaranticolor* by the nematode. Nematode-transmitted viruses and most of those with fungal vectors are mechanically transmissible.

Field assays for nematode-transmitted viruses have often employed indicator plants. Most nematodes feed on many plants, and most of these viruses have wide host ranges. Harrison and Cadman (1959) collected soil around strawberry, raspberry, and white clover plants and used peas as test plants for arabis mosaic virus transmitted by *Xiphinema diversicaudatum*. Jha and Posnette (1959) used virus-free strawberries and petunias as indicators. Teliz et al. (1966) used cucumber plants to test for *X. americanum* and tomato ringspot, peach yellow bud mosaic, and grape yellow vein viruses. Raski and Hewitt (1963) present a detailed early history of nematode-virus associations with many clues as to how to study them.

NEMATODE-TRANSMITTED VIRUSES

Maggenti (1981, pp. 160–161) listed several nematode vectors of plant viruses. The list includes *Longidorus attenuatus* as a vector of tomato black ring virus, *L. elongatus* as a vector of tomato black ring virus and raspberry ringspot virus, *L. macrosoma* as a vector of raspberry ring spot virus. *Xiphinema diversicaudatum* transmits cherry leaf roll, hop strain, strawberry latent ringspot, raspberry ringspot, and arabis mosaic viruses. *X. index* transmits grapevine fanleaf virus, and *X. coxi* transmits arabis mosaic, cherry leaf roll, strawberry latent ringspot, and tobacco ringspot viruses. *X. vuittenezi* transmits cherry leaf roll virus.

Paratrichodorus anemones transmits pea early browning virus. Five *Paratrichodorus* spp. transmit tobacco rattle virus, and three *Trichodorus* spp. transmit pea early browning virus. Some viruses are transmitted by only one nematode, some viruses by several. Harris (1981) listed several nematode-transmitted viruses not in the list of Maggenti. He commented that usually the vectors of a virus belong to the same nematode genus, but he listed raspberry ringspot virus as transmitted by *L. elongatus, L. macrosoma,* and *X. diversicaudatum* and brome mosaic virus transmitted by *L. macrosoma* and *X. diversicaudatum* as exceptions. The vectors all belong to the other Dorylaimida. The family Longidoridae contains *Longidorus* and *Xiphinema,* and the family Trichodoridae contains *Trichodorus* and *Paratrichodorus.*

The vectors belong to a small, closely related group. All are ectoparasitic in feeding habit. They, in contrast to the vast majority of plant-parasitic nematodes, have a two-part esophagus and some form of odontostylet that originates in the wall of the esophagus. This contrasts with the three-part esophagus and a stomatostylet found in the majority of plant-parasitic nematodes. The two-part esophagus has the general shape of a flask with a thick neck.

The stylets of *Longidorus* are up to 200 μm long, and they can reach any tissue of a rootlet. However, even though the vectors have relatively long stylets, they feed mostly shallowly. These nematodes may feed from a few minutes to 4 days in one site (Wyss, 1982). They are easily dislodged and many fall off when roots are removed from soil (Raski and Hewitt, 1963).

The stylets of *Trichodorus* spp. are relatively short, about 50 μm long. After each molt a new tooth moves from the esophagus wall into the tip of the odontophore. They typically feed on epidermal cells and root hairs 1 to 3 mm behind a root tip, and when their numbers are high, root elongation ceases. They ingest cell contents, including nuclei, and some species kill the cells they feed on (Wyss, 1982).

Transmission

There is no known special structure that enables nematodes to transmit viruses, but virus transmission depends on adsorption and release of virus. Electron micrography reveals virus on the surface of the cuticle of the stylet-guiding sheath in *Longidorus elongatus*. Virus is retained on the odontophore, anterior, and posterior esophagus of *Xiphinema index*. In *Trichodorus pachydermus* tobacco rattle virus adheres to the cuticular lining of the entire esophagus and buccal cavity. Sufficient virus is retained for serial transmission (transmission to more than one plant successively). No plant virus is circulative in a nematode, none is transmitted transovarially, and all virus is lost during molts. A vector can remain viruliferous for weeks or months if it does not molt. Members of all vector genera are polyphagous, making them more effective as vectors.

Individuals of some species of *Xiphinema* and *Longidorus* live for 3 to 5 years in Britain. In these species long periods elapse between molts. *X. index* kept in moist sand without a host for 8 months is still infectious—if it does not molt. Murant and Taylor (1965) found that *L. elongatus* lost raspberry ringspot virus during overwintering in fallow soil, and Lister and Murant (1967) found that *L. elongatus* lost raspberry ringspot virus in about 9 weeks in fallow soil. In most cases all active stages can acquire virus and inoculate a plant, except for *L. elongatus*, in which adults do not transmit the Scottish strain of tomato black ring virus. Most workers report access feeding periods rather than acquisition and test feeding periods because of difficulties in determining when a nematode is actually feeding, particularly when tests are conducted in soil or sand. *X. americanum* can acquire tomato ringspot virus in 1 hr, *T. allius* can acquire tobacco rattle virus in 1 hr, and *X. index* acquires grapevine fanleaf virus in 1 day. *X. americanum* can inoculate a plant in 1 hr, and all motile stages are vectors (Telez et al., 1966). There is no evidence of a latent period or of cross-protection in the vector (Fulton, 1967; Sauer, 1966). All these observations support the simplest of virus-vector relationships; i.e., they are noncirculative and nonpropagative.

This is, however, not equatable with stylet-borne vector relations, as in aphids, because of the length of time a nematode can remain infective and of the potential for serial transmission.

The salivary ducts of nematodes are in positions that expose virus on cuticular surfaces to saliva. It is possible that pH changes associated with ejected saliva and ingested food regulate adsorption or release of virus from the vector. The internal cuticular surfaces have slightly different staining properties and different origins, and these differences may account for retention of viruses in different parts of the feeding apparatus (Maggenti, 1981). The retention of virus, by all evidence, depends on physical characteristics (adsorption) of the internal nematode cuticle and of the virus capsid.

The natural vector of the Scottish strain of raspberry ringspot virus (RRV) is *Longidorus elongatus* and that of the English strain is *L. macrosoma*. Harrison et al. (1974) wondered if the distribution was an "accident" of geography or if it were related to species of nematode. *L. macrosoma* transmitted the English strain but not the Scottish strain. They constructed pseudorecombinants of the English and Scottish RRV strains and determined that vector-specificity is governed by capsid properties. Taylor and Robertson (1975) found that adsorption of the virus is not the only requirement. *L. macrosoma* adsorbs both the English and the Scottish strain, but transmits only the English strain. The Scottish strain is not released by *L. macrosoma* during feeding.

Cadman (1963) grouped viruses transmitted by nematodes into two groups: NETU viruses—*NE*matode-transmitted *TU*bular viruses, and NEPO viruses—*NE*matode-transmitted *PO*lyhedral viruses. The NETU-group name has since been abandoned, with the Tobravirus (*TOB*acco *RA*ttle) group name taking its place. *Longidorus* and *Xiphinema* spp. transmit nepoviruses, and *Trichodorus* spp. transmit tobraviruses. Taylor (1980) reviewed vector-specificity of nematode-transmitted plant viruses.

Patchiness and Speed of Spread

If the virus is present in propagating materials, as grapevine fanleaf virus in grape rootstocks, infection can be uniform and complete at inception of the planting. In most cases disease resulting from nematode-transmitted viruses occurs in patches within fields, and spread is slow. Illustrations of patches of diseased strawberries in Great Britain indicate that cultivation does not spread viruliferous nematodes to any extent. The patches are relatively stable, expanding at the periphery.

McNamara (1980) collected virus-free *Xiphinema diversicaudatum* near Kent, England, from 1967 to 1976. In doing so he obtained evidence of the spread of arabis mosaic virus at a rate of about 11m/year in uncultivated soil. The nematode was present before the sampling began, so lack of a vector did not account for the initial lack of virus. Arabis mosaic virus is seedborne in many weeds, and McNamara believed dissemination of infected seeds initiated the spread. Patchiness, if appropriate nematodes are present but the virus is lacking, could originate around establishment of a source plant arising from an infected seed. Harrison and Winslow (1961) and Fritzsche (1968) each estimated the natural spread of *X. diversicaudatum* under field conditions to be about 0.3m/year. The observed 11m/year recorded by McNamara is not the result of nematode movement alone.

If the virus is readily translocated within the plant, the root system extensive, and the vector preexisting in the soil, then the rate of spread would be influenced largely by the size of the root systems. Bergeson et al. (1964) calculated that tobacco ringspot virus spread 9 cm/week in soybean roots. After considering weed seed, nematode movements, and movement of virus through roots, McNamara concluded that the 11m/year he observed must include all three methods of spread.

Vertical Distribution in Soil

In general, except in stress conditions, nematodes are found in greatest abundance in the soil zones containing the most roots. Flegg (1968) studied the vertical distribution of *Xiphinema vuittenezi* and *Longidorus profundorum* on quince in a light sandy loam in England. The vertical distribution of the two species differed markedly. *X. vuittenezi* was most abundant in the upper 20 cm, and *L. profundorum* was most abundant at about 70 cm on the same host in the same soil (Figure 20.1). There was little evidence of vertical migration by either species during winter or summer. Why the two species preferred different depths within the soil was not discovered.

Vertical migration of nematodes explains the failure of standard soil fumigation to control potato corky ringspot in deep sandy soil in Florida (Weingartner et al., 1983). Potato corky ringspot virus is transmitted by *Trichodorus* and *Paratrichodorus* spp. These ectoparasites are killed by ordinary soil fumigation, but in the problem soil they occur in the entire soil profile to the depth of the groundwater. Nematicides effective in the surface soils, but which dissipate rapidly, increase potato yields but do not reduce the incidence of corky ringspot. Viruliferous nematodes that migrate from deep in the soil inoculate young tubers, destroying the marketability of the product. Control is by slowly volatile nematicides (aldicarb) that give lasting control. Control is enhanced by combining aldicarb within the soil with foliar application of a systemic nematicide (oxamyl). Weingartner et al. stated, however, that this experience is unusual. Standard fumigations are effective in Europe and in other parts of the United States. The coarse soil texture to deep profiles in the Florida soil favors a wide vertical distribution of the vector and facilitates movement, accounting for the severe local problem.

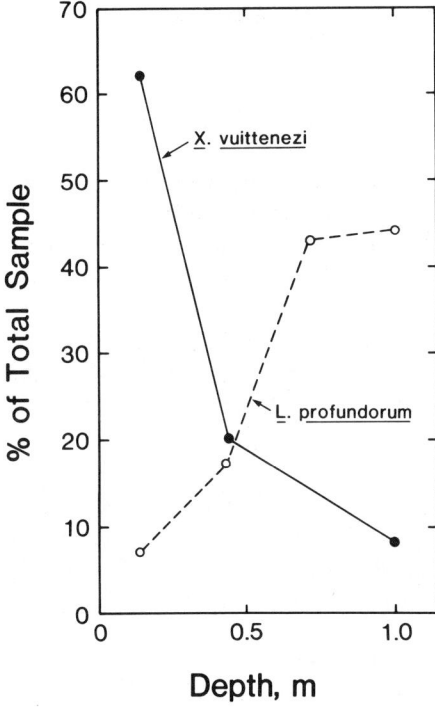

Figure 20.1, For reasons nor understood, the vertical distribution of *Xiphinema vuittenezi* and *Longidorus profundorum* under the same conditions differs greatly. After Flegg, 1968.

A Problem Vector

Xiphinema americanum is difficult to study in the greenhouse. It apparently overwinters largely as eggs, because soil sievings result in few motile stages. One generation per year reduces its reproductive capacity, and it does not produce high populations in frequently cultivated soils (Dropkin, 1980). Dropkin stated that this species can live in moist soil at 10°C for 49 months without a host. Miller (1980) attempted to produce populations of *X. americanum* for greenhouse studies. He inoculated 37 species of plants with 36 nematodes/50 g of soil and estimated populations after 169 days. In fallow soil, 4 nematodes/50 g of soil were alive. The most favorable hosts were *Potentilla canadensis,* Chicory, dandelion, Red osier dogwood, red fescue, *Stellaria media,* and *Quercus borealis.* After 169 days no live nematodes were found on watermelon, cucumber, and oats. Most workers use cucumbers extensively, and cucumber is a poor host of *X. americanum.* Fewer nematodes than in the original inoculum were found on cabbage, muskmelon, tobacco, and several other plants. It is little wonder that workers have experienced difficulties increasing this nematode. *X. americanum* was given special attention because of its wide distribution, the difficulty that many workers have had in rearing it, and that it transmits cherry rasp leaf, blueberry necrotic ringspot, peach yellow bud mosaic, grape yellow vein, tobacco ringspot, and tomato ringspot viruses.

Long-Lived Nematodes

Some *Longidorus* spp. have life cycle of 3 to 5 years, with extended periods between molts. Sturhan (1963) found that *L. maximus* had very large eggs and a slow rate of reproduction and suggested that the life cycle was long. Flegg (1968) found that eggs of *L. profundorum* were produced from May through July in a pear orchard near Kent, England. One crop of eggs was produced each year. Adults made up a relatively constant proportion of the population, not exceeding 20% at any time during the year. It is possible that *L. elongatus* likewise has an extended life cycle with much overlapping of life stages.

Loss of Infectivity Murant and Lister (1967) observed that those viruses most often seedborne in weeds are lost in the vectors in about 9 weeks if the vector has no access to infected plants. *L. elongatus* was highly infective when under strawberries suffering from raspberry ringspot in October, 1960. This crop was removed and planted to grass. In December, 1961, the nematode was still present in significant numbers, but it was noninfective. When allowed to become infective and then kept in fallow soil, it lost infectivity in 10 weeks. Murant and Lister concluded that access to virus at least within every 2 or 3 months is necessary for transmission. Virus is apparently lost even without molting. This loss of infectivity is fortunate, because if these nematodes are as long-lived as some species within the genus, they could remain viruliferous from one season to the next if recharging were not essential. For fallowing to be effective, adequate control of weeds that carry the virus is essential. See Taylor (1972, 1980) for more extensive treatment.

Importance of Weeds

Weeds on which *Longidorus elongatus* will reproduce are important in the spread of raspberry ringspot virus among raspberries. *L. elongatus* transmits raspberry ringspot virus to raspberry, but it declines in number when confined to raspberry (Taylor, 1967). Aqueous extracts of raspberry canes and roots are toxic to *L. elongatus,* so it is not unexpected that this plant is a poor host (see Chapter 7). *L. elongatus* multiplies on strawberry, so it transmits virus in this crop in a sustained way in the relative absence of weeds. Murant and Lister (1967) found more infected weed seeds among strawberries than among raspberries, largely because strawberries contributed to the increase of the vector. In strawberries, a good host of *L. elongatus,* weeds are not so critical for spread of the virus.

By the time Lister and Murant (1967) wrote their paper, tomato black ring virus was known to be seedborne in 19 species in 13 plant families, arabis mosaic virus in 13 species in 11 families, raspberry ringspot virus in 6 species in 5 families, and tomato ringspot, cherry leaf roll, and tobacco rattle virus were known to be seedborne in some plants. They noted that most of the hosts that carried virus in seed are symptomless, with infections ranging from 10 to 100% of the seed, and that virus transmission through seed is persistent through two or three weed generations. In Britain, *Capsella bursa-pastoris, Chenopodium album, Polygonum persicaria, Myosotis arvensis, Lamium amplexicaule,* and *Senecio vulgaris* are important hosts of several viruses. Some transmission occurs in raspberry and strawberry seeds. Virus remains infective in dormant seeds for long periods.

Cherry rasp-leaf virus developed in a few cherry trees when virus-free cherry trees were set in land previously in apples for 43 years (McElroy, 1975). Symptoms appeared 2 years after transplanting. The virus was recovered from symptomless dandelions and plantains within the orchard. *Xiphinema americanum* was present about the diseased trees to a depth of 60 cm. McElroy recommended good weed control as a means of reducing the amount of spread. Rosenberger et al. (1983) found that 25% of the dandelions in four orchards in New York were infected with tomato ringspot virus. Tomato ringspot virus, transmitted by *X. americanum* and/or *X. rivesi,* causes prunus stem pitting and apple union necrosis, both of which are important in peach or apple orchards in Pennsylvania. Even with clean nursery trees, significant new infections continue. Mountain et al. (1983) found that about 24% of the seed of infected dandelions produced infected dandelion seedlings. Infected and virus-free dandelion seed germinated equally and had equal longevity. Dandelion seeds are certainly adapted for dissemination.

With the exception of strawberry latent ringspot virus, for which no results are available, all NEPO viruses known at present can be transmitted through seed of at least some of their hosts. Less is known about Tobraviruses. To be important as a source of virus, the carrier plant should be a suitable host of the vector. Some discrepancies in frequency of seed-transmission in nature and in experiments are due to feeding preferences of nematodes. A plant may be susceptible to the virus but unattractive to the nematode.

FUNGAL VECTORS

Infectious Agent within the Vector

Teakle (1980) reviewed transmission of viruses by fungi and listed tobacco necrosis, its satellite, and cucumber necrosis viruses as being transmitted in a nonpersistent manner, and lettuce big-vein agent, wheat soilborne mosaic, and tobacco stunt viruses as being transmitted in a persistent manner.

Teakle treats many other virus-fungal-host relationships, and the reader is referred to his article for them.

Wheat Soilborne Mosaic

A high proportion of winter wheat plants remained in the rosette stage in the spring of 1919 in fields near Granite City, Illinois, on a heavy, black alluvial "gumbo" soil near the Mississippi River. The same disease was found in a few other localities between the Mississippi River and the Atlantic Coast. It was at first diagnosed erroneously, but McKinney (1923) suspected that the rosette disease was caused by a soilborne virus. The virus was reported in Egypt in 1931, and it was known in Japan in 1937. It has since been found in Italy, Brazil, and Florida. It was not until 1944 that soilborne wheat mosaic was found west of the Mississippi River, widely distributed throughout the central Great Plains. Nykaza et al. (1979) reported 810,000 ha of infested land in Kansas alone.

Symptoms Symptoms are influenced by the age of the host when infected, the cultivar, the weather, and the virus strain. In Harvest Queen winter wheat a slight yellowish cast is evident in early spring when growth is resumed. Close observation reveals a mosaic, but this is obscured as the leaves formed the previous fall become deep bluish-green (McKinney et al., 1925). Diseased plants have more tillers than healthy ones. After an extreme rosette stage, mosaic symptoms are again visible within 7 to 10 days in new leaves formed in the spring. If the spring remains cool, rosette-susceptible cultivars remain in the rosette stage. With warm weather, some recovery will occur.

McKinney tested about 150 wheats, and 6% developed rosette; the rest had the milder, mosaic reaction. Webb et al. (1923) tested 200 wheats, and 4% developed rosette. The wheat cultivar Currell, in the same nurseries with Harvest Queen, never rosetted, but it developed a striking yellow mosaic. Yellowish-brown flecks were observed in the parenchyma of basal leaf sheaths and stem tissues. McKinney et al. (1923) found X-bodies (viral inclusions) similar to those of tobacco mosaic virus in cells in these tissues.

Etiology Wheat mosaic virus is difficult to transmit mechanically, but McKinney (1925) used juice of yellowed Currell to inoculate healthy Currell and Harvest Queen in the 2 to 3 leaf stage. Leaves of Harvest Queen became dark green and brittle, like rosette, and leaves of Currell developed the yellow mosaic but no rosette reaction. Some strains of the virus incited the rosette reaction in Harvest Queen, and some incited only a mosaic reaction. Both rosette and nonrosette strains occurred in all nurseries east of the Mississippi River (McKinney, 1948), but Sill (1958), using rosette-susceptible wheats, detected no rosette strains in Kansas. Wada and Fukano (1937) reported different kinds of inclusions in nonrosette strains (Type A), rosette strains (Type B), and in mixed infections (Type M). Tsuchizaki et al. (1973) also studied strains from the United States and Japan and described three types of virus inclusions. All efforts to transmit the virus mechanically using soil extracts have failed. No transmission occurs when manually inoculated and healthy plants are grown together in sterilized soil, and no infection occurs when leaf material from diseased plants is added to noninfective soil. All evidence points toward the existence of a soilborne vector. Brakke et al. (1965) found virus in the roots of naturally infected plants 2 to 3 weeks after planting, several weeks before any symptoms appeared. The virus remained localized in the roots of most plants during most of the autumn. The above reports are from natural infections in the field, and they differ considerably from some laboratory, in vitro results.

The Vector Ledingham (1939) published the morphology, cytology, and host-parasite relations of *Polymyxa graminis* and placed it in the Plasmodiophorales. It was known only in southeastern Ontario until Linford and McKinney (1954) suspected it might be the vector of soilborne wheat mosaic virus. They found it in roots of wheat from five states where wheat soilborne mosaic occurs. *Polymyxa graminis* was associated with the disease, but it occurred in healthy wheat also. No proof of the role of *P. graminis* as a vector was obtained, but Linford and McKinney must have had strong convictions to write the report, naming only one fungus.

Individual resting spores of *P. graminis* are spherical or polyhedral, 5 to 7 μm in diameter, with yellow-brown outer walls. They resist drying and remain alive in dry soil for many years. Field soils are more infective in autumn than in spring (Brakke et al., 1965). Rao (1968) conditioned roots of diseased plants at 28°C for 2 months, with occasional watering to simulate passage through a Nebraska summer, after which they were dried and powdered. The powder was added to water at 17°C for 4 days, at which time wheat seedlings were placed in the water for 5 more days. Biflagellate zoospores were swimming in the water, and infection ensued. Fungal penetration occurred mainly through root hairs. The fungus developed sporangia which released zoospores into water through evacuation tubes; 7 to 12 days were required from the time of exposure of healthy roots to infective zoospores for the production of a new generation of infective zoospores. Shortly after zoospore production, the multinucleate plasmodia cleaved into cysts to complete the life cycle, usually within 2 to 4 weeks of initial infection.

Host Range Wheat is the only crop to suffer losses sufficient to require control measures. McKinney (1930) listed seven *Triticum* spp., *Hordeum sativum*, and *Secale cereale* as hosts. Since then two *Bromus* spp., winter annual weed grasses, have been added. By manual transmission, mostly local lesions of some type were produced on several other plants, and some differences in host range among Japanese and American isolates have been detected (Tsuchizaki et al., 1973).

Polymyxa graminis was reported on *Agropyron repens*, barley, wheat, and rye by Ledingham (1939). He did not find it on oats. Barr (1979) extended this list to include additional grasses, and Thouvenel and Fauquet (1980) reported *P. graminis* on *Sorghum arundinaceum* in Upper Volta, Africa. Eversmeyer et al. (1983) mentioned that susceptible grain sorghums may have increased the range of WSBMV in Kansas. *Polymyxa graminis* is the vector of wheat spindle streak mosaic virus, and possibly of barley yellow mosaic and rice necrosis mosaic viruses. *P. betae* transmits beet necrotic yellow vein virus (in Barr, 1979).

Environmental Factors

Soil. Wheat mosaic can occur in whole fields, in spots without relation to topography, or on slopes in any direction, but the disease is often prevalent in areas receiving drainage water (Sill, 1958). The agent persists longer in silts than in sandy soils (McKinney, 1930). The disease is not found in wheat in light-colored silt loams between the Wabash and Illinois Rivers in Illinois, even though both river valleys are infested (Koehler et al., 1952). Though common on heavy upland or river bottom soils in Kansas, wheat mosaic has not been found in Kansas on light or sandy soils (Sill, 1958). The report by Nykaza et al. (1979) that 810,000 ha are known to be infested in Kansas is evidence of some tolerance to different soil types. Heavy soils rich in organic matter are favorable, and high infectivity has been found to a depth of 65 cm in soil (Koehler et al. 1952).

Temperature. Early planting of winter wheat exposes it to warmer soil for seedling establishment than is true for later plantings. Webb (1927) studied the effect of soil temperature, planting date, and plant size on rosette development. Wheat was planted 4 August, 18 August, 1 September. The drop in soil temperature from 22.9°C on 4 August to 15.8 on 1 September resulted in a change in rosette development from 9 to 82%. Much rosette developed when soil temperatures during the first week after planting were between 10 and 16°C, and less developed at 6°C. Ikata and Kawai (1937), also working in naturally infested soil, reported that the optimum for infection was near 15°C, that 10°C was favorable, 20°C was unfavorable, and 25°C inhibited disease development.

A cool, prolonged spring favors symptom expression and maximum yield loss. Warm weather in early spring results in remarkable recovery. Nykaza et al. (1979) observed that even though symptoms were obvious in early spring, losses varied widely. In some seasons mosaic symptoms disappear in 10 to 14 days, whereas in others, mosaic persists to the time of bloom. They stated that 17°C is the critical temperature. Below that, symptoms persist. If the temperature rises above 17°C quickly in the spring, losses are slight. Eversmeyer et al. (1983) believe that at temperatures below

15°C the virus multiplies faster than the host grows, and that above 15°C, the host develops faster than the virus.

Water. Webb (1927) adjusted the infested heavy black gumbo soil from Granite City, Illinois, to a water content of 52 to 58% of its water-holding capacity (WHC) (wet but still workable), to 42 to 45% WHC (ideal tilth), and to 30 to 32% WHC (too dry to support good germination of wheat seed). Kernels of Harvest Queen were soaked overnight in water and planted in the soils on 23 September. After 30 days the plants were dug, the roots cleaned and soaked in aqueous $HgCl_2$ (1:1000) for 5 min, rinsed, and transplanted to noninfested soil. At the 52% water level, 100% infection occurred. At the 42% water level, 2 to 15% infection developed. No disease developed in the dry soil series. For the disease to be prevalent, soil moisture must exceed that present at the stage of good tilth. In Kansas (Sill, 1958) epidemics are accompanied by adequate to abundant moisture during long, cool autumns and springs.

W. Willis (Kansas State Univ., personal communication) states that in eastern Kansas wheat mosaic is general, that the further west you go (from humid to drier), the effect of low spots and drainage ways become more pronounced, and that there is a transitional zone about 150 km wide in which the disease is noted in the west during wet years, but in dry years the disease retreats eastward.

Control

Seeding Date. Seeding early in warm soil controls the disease, but seeding winter wheat when the soil is still warm subjects it to many hazards. Seeding late reduces disease but results in lower yields. Changing the seeding date drastically from the optimum agronomic time of planting is not a practical control.

Rotation. Heavily infested soil was seeded for 4 consecutive years to a highly susceptible wheat, an immune wheat, maize, oats, and soybeans. After 4 years, susceptible wheat was seeded to assess the effect of the various crops upon the infestation (Koehler et al., 1952). The average yield was 25 bu/a following the susceptible wheat, 53 bu/a after immune wheat, and 58 to 59 bu/a after oats, soybeans, or maize. Four years without a host greatly reduced the loss but did not eliminate the infestation. Rotation was deemed impractical.

Koehler et al. (1952) tried to determine how quickly a soil infestation could become significant. They determined, as had others, that drying the soil did not destroy infectivity, so infestation could occur through wind-disseminated dust. They collected soils from permanent pasture, an old orchard, and grass strips adjacent to infested fields but not subject to water runoff from the infested lands. Susceptible wheat planted in these soils remained disease-free. The level of infestation was too low to detect with a single planting of suscept. They then added 0, 0.01, 0.1, 1.0, and 10.0% infested soil by weight to noninfested soil and grew four crops of susceptible wheat. The control wheat (0 infested soil added) remained healthy. The 10% level gave almost maximum infection in the first crop, reaching essentially 100% diseased plants in year 2. In the 0.01% infestation, no disease (rosette) was detected the first year (0.01% — approximately 224 kg of soil per 15 cm of top soil per ha) but by the fourth year mosaic reached 90% (Figure 20.2). In nature, starting from wind-blown dust, it would require several crops of susceptible winter wheat to develop a significant infestation.

Resistance. Resistance is the only practical control in the areas in which wheat mosaic is a problem. Greatest losses result from the rosette reaction. In the beginning, before the effect of mosaic without rosette was appreciated, a wheat that did not rosette was considered resistant. Because most wheats do not rosette, "resistance" was easily achieved. McKinney and others selected nonrosetting individuals from highly rosette-susceptible cultivars, and some valuable wheats were obtained in this way (Koehler et al., 1952). The 10 to 30% losses from mosaic without rosette are significant. Nykaza et al. (1979) calculated a 22% loss from mosaic on infested soil for some adapted wheats by use of modified isogenic lines. McKinney (1948) reported that some wheats resistant in the field were susceptible when mechanically inoculated. Bockus and Niblett (1984) developed an effective, quick laboratory test for resistance using the vector.

Miyake (1938) reported that resistance was governed by a single partially dominant gene in some Japanese wheats, and Modawi et al. (1982) confirmed a single locus for resistance but the background host genome influenced resistance.

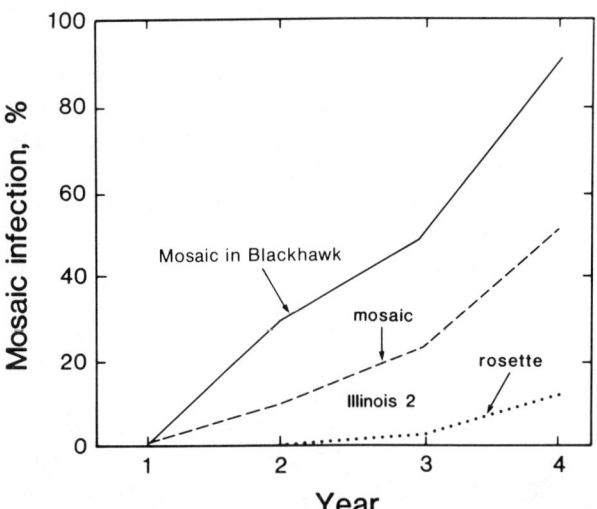

Figure 20.2. Development of an economic infestation of soil by viruliferous *Polymyxa graminis* following wind-blown dust would require years. The percentage of plants with mosaic symptoms in Blackhawk winter wheat (susceptible) and in Illinois 2 (more resistant) following incorporation of 0.01% infested soil (224 kg/ha). After Koehler et al., 1952.

Big-Vein of Lettuce

The veins of diseased plants appear thickened and light in color, and the leaf lamina does not expand normally. Vein clearing gives the illusion of vein thickening. Jagger and Chandler (in Grogan et al., 1958) believed big-vein to be akin to soilborne wheat mosaic, but neither they nor anyone else could transmit the agent mechanically, with aphids, with nematodes or through seed. Air-dried soil stored 8 years still contained the causal agent. Allen (in Grogan et al., 1958) found transmissibility resistant to several soil fumigants.

Superficial root scrapings from diseased plants make good inoculum, tissues deep in the root do not (Grogan et al., 1958). Resting spores of *Olpidium brassicae* are produced in superficial tissues, but not in the stele. When a zoospore suspension is filtered to remove the zoospores, no disease results. Plants with big-vein from California, Arizona, Utah, and New Zealand all contain *O. brassicae*. Campbell et al. (1961) transmitted an agent from a big-vein plant to a healthy plant by grafting, thereby producing the disease without the fungus. Since *O. brassicae* does not by itself produce big-vein, it is a vector.

Campbell (1962) treated resting spores of the fungus with hydrochloric acid or with trisodium phosphate buffers to inactivate virus on the spore surface. The resting spores survived the treatments, and zoospores obtained from them initiated disease, proving that the infectious agent is borne within the spore. *O. brassicae* from lettuce infects sugar beet, and Campbell inoculated sugar beets with infectious zoospores. Zoospores obtained from the sugar beet did not produce big-vein of lettuce. The fungus lost the infectious entity within the beet, and the "virus" did not multiply in sugar beet or in the vector. Tomlinson and Garrett (1964) cycled *O. brassicae* from diseased lettuce to *Plantago major* or *Veronica persica* and then back to healthy lettuce. The unknown agent was lost in the transfer. The fungus from healthy lettuce was cycled through diseased lettuce to healthy lettuce, and transmission occurred. They confirmed that *O. brassicae* is a vector and that the agent does not multiply in the vector. Growing lettuce from transplants results in less disease than from direct seeding, which suggests that wounding is not a factor (Campbell et al., 1980). See Chap. 18 for affects of water relations on *O. brassicae*.

Big-vein of lettuce is caused by double-stranded ribonucleic acids (Mirkov and Dodds, 1985).

Vector, Virus, Host Interactions

Hiruki (1967) inoculated 71 species of plants in 24 families with tobacco stunt virus (TSV) using viruliferous zoospores of *O. brassicae*. TSV is mechanically transmissible, so he was able to test for complex interactions. He inoculated plants by pouring zoospore suspensions on the soil or by submerging roots in a zoospore suspension for 30 to 60 min. Five cultivars of tobacco, three other *Nicotiana* spp., soybean, and *Tetragonia expansa* were systemically infected by TSV. In 20 species susceptible to *O. brassicae* the virus was confined to the roots. In 11 species, *O. brassicae* multiplied in the roots, but the virus did not. In five species susceptible to *O. brassicae* the plants were susceptible to TSV by mechanical inoculation of leaves but not by inoculation with viruliferous zoospores. Some plants were susceptible to TSV by mechanical inoculation but not to *O. brassicae,* and some were susceptible to neither. Most interesting to me is that the virus was not translocated from the roots of some plants and that in some plants the leaves accepted virus mechanically but the virus did not establish in roots by means of a vector. *O. brassicae* contains several strains, some with very broad host ranges, and some with more restricted host ranges (Lin, 1979).

Infectious Agents Outside the Vector

Tobacco Necrosis Tobacco necrosis virus (TNV) is easy to transmit mechanically and is easy to work with, so progress with it has been rapid. Teakle (1962) proved that zoospores of *O. brassicae* transmit tobacco necrosis virus. *O. brassicae* zoospores alone or uninjured roots exposed to virus suspension alone result in little disease, whereas zoospores swimming in a virus suspension result in much disease. Zoospores transmit tobacco necrosis virus to cowpea, mung bean, tomato, and tobacco.

TNV is not retained in an infectious state in resting spores. It is external on zoospores. Virus escapes from roots of disease plants and zoospores in the soil suspension become viruliferous external to the host by adsorption of the virus. Smith et al. (1969) demonstrated that zoospores acquire TNV from soil extracts too low in virus content for detection by mechanical transmission. Transmission is prevented when zoospores produced in diseased plants swim in a preparation of TNV-antiserum prior to exposure to test plants, evidence that TNV is on the outside of the zoospores (Kassanis and MacFarlane, 1964; Campbell and Fry, 1966). Resting spores treated with acid, as done in experiments with big-vein agent, results in no transmission. sion.

There are no reports of soils retaining infectivity after long drying (years) as in big-vein, but Hiruki (1965) obtained TNV from air-dried soil up to 4 months. TNV may remain infectious adsorbed to soil colloids or in plant debris for a limited time. There is no evidence that it is preserved within the fungal resting spore. Smith et al. (1969) recovered TNV from roots infected with both TNV and *O. brassicae* that were dried for 130 days.

Transmission

Temmink et al. (1970), in a classic use of electron micrography, demonstrated adsorption of TNV to the surface membranes of zoospores and the flagellum when transmission occurs and lack of transmission when the zoospore does not adsorb virus while swimming in a virus suspension. As in virus transmission by nematodes, vector surface: virus particle physical interactions (adsorption) are critical. Two *O. brassicae* accessions adsorb TNV and transmit the virus, but an isolate of the fungus from mustard does not adsorb the virus and does not transmit it, which is proof that this vector can acquire TNV outside the host. Nematodes, in contrast, acquire virus only from a host by feeding.

Temmink et al. (1970) referred to acquisition of virus from a virus suspension as in vitro acquisition. *O. radicale* (*O. cucurbitacearum*) apparently acquires cucumber necrosis virus in vitro (Dias, 1970), and Kassanis and MacFarlane obtained in vitro acquisition of tobacco necrosis and its satellite virus with *O. brassicae*. The zoospore flagellum is retracted during encystment (Sahtiyanci, 1962; Koch, 1968), and cucumber necrosis virus adsorbed to the surface of the *O. radicale* flagellum is introduced into fungal cytoplasm when the flagellum is retracted into the spore body (Stobbs et al., 1982). There is no evidence for multiplication of the virus within the vector. The fungus may expel the virus particles as foreign substances when within the host.

COMMENTS

Students of soilborne viruses have added greatly to our knowledge of the role of virus capsid properties in virus transmission by vectors. The surfaces of the feeding apparatus of dorylaimid nematodes permit adsorption of viruses, a property apparently lacking in all other plant-parasitic nematodes. Also, both retention of virus and release from the surface during feeding are important, because *Longidorus macrosoma* adsorbs both the English and Scottish strains of raspberry ringspot virus and transmits only the English strain. The ability of surfaces of zoospores of *Olpidium brassicae* to attract and retain some viruses acquired while swimming in a virus suspension also proves the importance of adsorption. If the virus is not adsorbed, it is not transmitted.

The "piercing-sucking" mouthparts of nematodes should be ideal for virus transmission. If some factors do not operating to limit development of nematode-virus interactions, we would be overwhelmed by virus diseases. The details of the life cycles and feeding habits of a few vector nematodes have been presented because, in some respects, they differ markedly from the life cycles most of us know best (cyst- and root-knot nematodes).

Mathews (1970) stressed that a virus such as tobacco necrosis virus, which has a wide host range and is adsorbed superficially on the surface of zoospores of *O. brassicae*, would require frequent transmission among annual plants if it is to survive, because it does not persist within the resting spore of the vector. In contrast, soilborne wheat mosaic virus and the lettuce big-vein agent persist for years within the resting spore of their fungal vectors. These infectious entities have narrower host ranges, but persistence within the resting spore more than compensates for the relatively narrow host ranges.

Resting spores of *Polymyxa graminis* and *O. brassicae* survive extreme desiccation and cold, and infectious entities within them also survive these physical conditions. In contrast, viruliferous nematodes cannot survive extreme conditions. Nematodes compensate for this weakness to some extent by vertical migrations, escaping a dry surface by migrating deeper into the soil. The importance of vertical migration is illustrated by the problem of controlling potato corky ringspot on deep sandy soils in Florida. After standard fumigation, which kills vectors in the surface soils, infective nematodes from deep in the soil migrate upward, and the potatoes develop corky ringspot.

Viruses transmitted by nematodes spread slowly when compared to those spread by leafhoppers, aphids, or mites. With an active aerial vector, a field of plants can be healthy one week and be endangered the next. Soilborne viruses lack this epidemic quality. The widespread nature of soilborne wheat mosaic in the central Great Plains must be the result of wind-disseminated viruliferous *P. graminis* spores. The work of Koehler et al. (1952) indicates that starting from infestation by wind-blown dust, several years would be required to develop a significant soil infestation (Figure 20.2), but winter wheat (and sorghum?) has been widely and repeatedly grown on soils in that region. To my knowledge, no virus transmitted by nematodes has established a soil infestation as thorough or extensive as has been demonstrated for *P. graminis* and soilborne wheat mosaic virus.

The persistence of patchiness in strawberry and raspberry plantings is evidence that cultivation is less effective in spreading infective nematodes than one would expect.

The ingenuity of researchers in the area of soilborne viruses deserves admiration.

KEY REFERENCES

Campbell, R. N. and P. R. Fry. 1966. The nature and associations between *Olpidium brassicae* and lettuce big-vein and tobacco necrosis viruses. *Virology* **29**:222–233.

McKinney, H. H. 1930. A mosaic of wheat transmissible to all cereal species in the tribe Hordeae. *J. Agric. Res.* **40**:547–556.

Murant, A. F. and R. M. Lister. 1967. Seed-transmission in the ecology of nematode-borne viruses. *Ann. Appl. Biol.* **59**:63–76.

Raski, D. J. and W. B. Hewitt. 1963. Plant parasitic nematodes as vectors of plant viruses. *Phytopathology* **53**:539–41.

Smith, P. R., R. N. Campbell and P. R. Fry. 1969. Root discharge and soil survival of viruses. *Phytopathology* **59**:1678–1687.

Taylor, C. E. 1980. Nematodes. Pages 375–416 in *Vectors of Plant Pathogens*. K. F. Harris and K. Maramorosch, eds. Academic Press, New York.

Teakle, D. S. 1980. Fungi. Pages 417–438 in *Vectors of Plant Pathogens*. K. F. Harris and K. Maramorosch, eds. Academic Press, New York.

REFERENCES

Allen, W. R. 1981. Dissemination of tobacco mosaic virus from soil to plant leaves under glass house conditions. *Can. J. Plant Pathol.* **3**:163–168.

Allen, W. R. 1984. Mode of inactivation of TMV in soil under dehydrating conditions. *Can. J. Plant Pathol.* **6**:9–16.

Barr, D. J. S. 1979. Morphology and host range of *Polymyxa graminis*, *Polymyxa betae*, and *Ligniera pilorum* from Ontario and some other areas. *Can. J. Plant Pathol.* **1**:85–94.

Bergeson, G. B., K. L. Athow, F. A. Laviolette, and M. Thomasine. 1964. Transmission, movement, and vector relationships of tobacco ringspot virus in soybean. *Phytopathology* **49**:332–334.

Bockus, W. W. and C. L. Niblett. 1984. A procedure to identify resistance to wheat soilborne mosaic in wheat seedlings. *Plant Dis.* **68**:123–124.

Bouquet, A. 1981. Resistance to grape fanleaf virus in Muscadine grape inoculated with *Xiphenema index*. *Plant Dis.* **65**:791–793.

Brakke, M. K., A. P. Estes and M. L. Schuster. 1965. Transmission of soilborne wheat mosaic virus. *Phytopathology* **55**:79–86.

Cadman, C. H. 1963. Biology of soil-borne viruses. *Ann. Rev. Phytopath.* **1**:143–172.

Campbell, R. N. 1962. Relationship between the lettuce big-vein virus and its vector, *Olpidium brassicae*. *Nature* **195**:675–677.

Campbell, R. N., A. Greathead, and F. Westerlund. 1980. Big vein of lettuce: infection and methods of control. *Phytopathology* **70**:741–746.

Campbell, R. N., R. G. Grogan, and D. E. Purcifull. 1961. Graft transmission of big-vein of lettuce. *Virology* **15**:82–85.

Dias, H. 1970. Transmission of cucumber necrosis virus by *Olpidium cucurbitacearum* Barr and Dias. *Virology* **40**:828–839.

Dropkin, V. H. 1980. *Introduction to Plant Nematology*. Wiley, New York.

Eversmeyer, M. G., W. G. Willis, and C. L. Kramer. 1983. Effect of soil fumigation on occurrence and damage caused by soilborne wheat mosaic. *Plant Dis.* **67**:1000–1002.

Flegg, J. J. M. 1968. The occurrence and depth distribution of *Xiphenema* and *Longidorus* spp. in southeastern England. *Nematologica* **14**:189–196.

Fritzsche, R. 1968. Beitrag zum Wanderungsverhalten von *Xiphenema diversicaudatum* (Mikoletzky) Thorne, *X. coxi* Tarjan und *Longidorus macrosoma* Hooper sowie der Ausbreitung des Rhabarbermosaik-Virus im Feldbestand. *Biol. Zentralbl.* **87**:481–488.

Fulton, J. P. 1967. Dual transmission of tobacco ringspot virus and tomato ringspot virus by *Xiphenema americanum*. *Phytopathology* **57**:535–537.

Fulton, R. W. 1941. The behavior of certain viruses in plant roots. *Phytopathology* **31**:575–598.

Grogan, R. G., F. W. Zink, W. B. Hewitt, and K. A. Kimble. 1958. the association of *Olpidium* with the big-vein disease of lettuce. *Phytopathology* **48**:292–297.

Harris, K. F. 1981. Anthropod and nematode vectors of plant viruses. *Ann. Rev. Phytopathol.* **19**:391–426.

Harrison, B. D. 1960. The biology of soilborne plant viruses. *Adv. Virus Res.* **7**:131–161.

Harrison, B. D. and C. H. Cadman. 1959. Role of dagger nematode (*Xiphenema* sp.) in outbreaks of plant diseases caused by arabis mosaic virus. *Nature* **184**:1624–1626.

Harrison, B. D., A. F. Murant, M. A. Mayo, and I. M. Roberts. 1974. Distribution of determinants for symptom production, host range and nematode transmissibility between two RNA components of raspberry ringspot virus. *J. Gen. Virol.* **22**:233–247.

Harrison, B. D. and R. D. Winslow. 1961. Laboratory and field studies on the relation of arabis mosaic virus to its nematode vector *Xiphenema diversicaudatum* (Micoletzky). *Ann. Appl. Biol.* **49**:621–633.

Hewitt, W. B., D. J. Raski and A. C. Goheen. 1958. Nematode vector of soilborne fanleaf virus of grapevine. *Phytopathology* **48**:586–595.

Hiruki, C. 1965. Transmission of tobacco stunt virus by *Olpidium brassicae*. *Virology* **25**:541–549.

Hiruki, C. 1967. Host specificity in transmission of tobacco stunt virus by *Olpidium brassicae*. *Virology* **33**:131–136.

Hoggan, I. A. and J. Johnson. 1936. Behavior of the ordinary tobacco mosaic virus in the soil. *J. Agric. Res.* **52**:271–294.

Ikata, S. and I. Kawai. 1937. Some experiments concerning the development of yellow mosaic disease (white streak) of wheat. Relation between the development of yellow mosaic disease of wheat and soil temperature. *J. Plant Prot. (Tokyo)* **24**:491–501, 847–854 (in Japanese). Abstract in *Rev. Appl. Mycol.* **18**:98. 1939.

Jha, A. and A. F. Posnette. 1959. Transmission of a virus to strawberry plants by a nematode (*Xiphenema* sp.). *Nature* **184**:962–963.

Kassanis, B. and I. MacFarlane. 1964. Transmission of tobacco necrosis virus by zoospores of *Olpidium brassicae*. *J. Gen. Microbiol.* **36**:79–93.

Loch, W. J. 1968. Studies of the motile cells of chytrids. Part 5. Flagellar retraction in posterly uniflagellate fungi. *Am J. Bot.* **55**:841–859.

Koehler, B., W. M. Bever, and O. T. Bonnett. 1952. Soilborne wheat mosaic. *Univ. of Ill., Agric. Exp. Sta. Bull.* **556**:567–599.

Ledingham, G. A. 1939. Studies on *Polymyxa graminis*, n. gen., n. sp., a plasmodiophoraceous root parasite of wheat. *Can. J. Res.* **17**:38–51.

Lin, M. T. 1979. Occurrence and host range of *Olpidium brassicae* in central Brazil. *Plant Dis. Reptr.* **63**:10–12.

Linford, M. B. and H. H. McKinney. 1954. Occurrence of *Polymyxa graminis* in roots of small grains in the United States. *Plant Dis. Reptr.* **38**:711–713.

Lister, R. M. and A. F. Murant. 1967. Seed-transmission of nematode-borne viruses. *Ann. Appl. Biol.* **59**:49–62.

Mathews, R. E. F. 1970. *Plant Virology*. Academic Press, New York.

Maggenti, A. 1981. *General Nematology*. Springer-Verlag, New York.

McElroy, F. D. 1975. Nematode-transmitted viruses in British Columbia, Canada. Pages 287–288 in *Nematode Vectors of Plant Viruses*. F. Lamberti, C. E. Taylor, and J. W. Seinhorst, eds. Plenum Press, London and New York.

McKinney, H. H. 1923. Investigation on the rosette disease of wheat and its control. *J. Agric. Res.* **23**:771–800.

McKinney, H. H. 1925. A mosaic disease of winter wheat and winter rye. *U.S.D.A. Bull.* 1361.
McKinney, H. H. 1948. Wheats immune from soilborne mosaic viruses in the field, susceptible when inoculated manually. *Phytopathology* **38**:1003-1013.
McKinney, H. H., S. H. Eckerson, and R. W. Webb. 1923. The intracellular bodies associated with rosette disease and a mosaic-like leaf mottling of wheat. *J. Agric. Res.* **26**:605-608.
McKinney, H. H., R. W. Webb, and G. H. Dungan. 1925. Wheat rosette and its control. *Ill. Agric. Exp. Sta. Bull.* **264**:275-296.
McNamara, D. G. 1980. The spread of arabis mosaic virus through noncultivated vegetation. *Plant. Pathol.* **1980**:173-176.
Miller, P. M. 1980. Reproduction and survival of *Xiphenema americanum* on selected woody plants, crops and weeds. *Plant Dis.* **64**:174-175.
Mirkov, T. E. and J. A. Dodds. 1985. Association of double-stranded ribonucleic acids with lettuce big-vein disease. *Phytopathology* **75**:631-635.
Miyake, N. 1938. Mendelian inheritance of resistance against the virus disease in wheat strains. *Jap. J. Genet.* **14**:239-242 (*Plant Breed. Abstr.* 1939, **8**:800).
Modawi, R. S., E. G. Heyne, D. Brunetta, and W. G. Willis. 1982. Genetic studies of field reaction to wheat soilborne mosaic virus. *Plant Dis.* **66**:1183-1184.
Mountain, W. L., C. A. Powell, L. B. Forer and R. F. Stouffer. 1983. Transmission of tomato ringspot virus from dandelion via seed and dagger nematodes. *Plant Dis.* **67**:867-868.
Murant, A. F. and C. E. Taylor. 1965. Treatment of soil with chemicals to prevent transmission of tomato black ring and raspberry ringspot viruses of *Longidorus elongatus* (de Man). *Ann. Appl. Biol.* **55**:227-237.
Nykaza, S. M., E. G. Heyne, and C. L. Niblett. 1979. Effects of wheat soilborne mosaic on several plant characters of winter wheat. *Plant Dis. Reptr.* **63**:594-598.
Rao, A. S. 1968. Biology of *Polymyxa graminis* in relation to soilborne wheat mosaic virus. *Phytopathology* **58**:1516-1521.
Rosenberger, D. A., M. B. Harrison, and D. Gonsalves. 1983. Incidence of apple union necrosis and decline, tomato ringspot virus, and *Xiphenema* vector species in Hudson Valley orchards. *Plant. Dis.* **67**:356-360.
Sahtiyanci, S. 1962. Studien uber einige wurzelparasitare Olpidiaceen. *Arch. Mikrobiol.* **41**:187-228.
Sauer, N. I. 1966. Simultaneous association of strains of tobacco ringspot virus within *Xiphenema americanum*. *Phytopathology* **56**:862-863.
Sill, W. H., Jr. 1958. A comparison of some characteristics of soilborne wheat mosaic viruses in the Great Plains and elsewhere. *Plant Dis. Reptr.* **42**:912-924.
Stobbs, L. W., G. W. Cross, and M. S. Manocha. 1982. Specificity and methods of transmission of cucumber necrosis virus by *Olpidium radicale* zoospores. *Can. J. Plant Pathol.* **4**:134-142.
Sturhan, D. 1963. Der pflanzenparasitische Nematode *Longidorus maximus*, seine Biologie und Okologie, mit Untersuchungen an *L. elongatus* und *Xiphenema diversicaudatum*. *Z. Angew. Zool.* **50**:129-193.
Taylor, C. E. 1967. The multiplication of *Longidorus elongatus* (de Man) on different host plants with reference to virus transmission. *Ann. Appl. Biol.* **59**:275-281.
Taylor, C. E. 1972. Transmission of viruses by nematodes. Pages 226-247 in *Principles and Techniques in Plant Virology*. C. I. Kado and H. O. Agrawal, eds. Van Nostrand-Reinhold, New York.

Taylor, C. E. and W. M. Robertson. 1975. Acquisition, retention and transmission of viruses by nematodes. Pates 253–275 in *Nematode Vectors of Plant Viruses*. F. Lamberti, C. E. Taylor, and J. W. Seinhorst, eds. Plenum Press, London and New York.

Teakle, D. S. 1962. Transmission of tobacco necrosis virus by a fungus, *Olpidium brassicae*. *Virology* **18**:224–231.

Teliz, D., R. G. Grogan, and B. F. Lownsbery. 1966. Transmission of tomato ringspot, peach yellow bud mosaic, and grape yellow vein viruses by *Xiphenema americanum*. *Phytopathology* **56**:658–663.

Temmink, J. H. M., R. N. Campbell, and P. R. Smith. 1970. Specificity and site of in vitro acquisition of tobacco necrosis virus by zoospores of *Olpidium brassicae*. *J. Gen. Virol.* **9**:201–213.

Thouvenel, J. C. and C. Fauquet. 1980. *Polymyxa graminis* on new *Sorghum* species in Africa. *Plant. Dis.* **64**:957–958.

Tomlinson, J. A. and R. G. Garrett. 1964. Studies on the lettuce big-vein virus and its vector *Olpidium brassicae* (Wor.) Dang. *Ann. Appl. Biol.* **54**:45–61.

Tsuchizaki, T., H. Hibino, and Y. Saito. 1973. Comparisons of soilborne wheat mosaic virus isolates from Japan and the United States. *Phytopathlogy* **63**:634–639.

Wada, E. and H. Fukano. 1937. On the difference and discrimination of wheat mosaics in Japan. *J. Imp. Agric. Exp. Sta. Tokyo* **3**:8–15, 93-128. (Abstr. in *Rev. Appl. Mycol.* **16**:665, 1937).

Webb, R. W. 1927. Soil factors influencing the development of the mosaic disease in winter wheat. *J. Agric. Res.* **35**:587–614.

Webb, R. W., C. E. Leighty, G. H. Dungan and J. B. Kendrick. 1923. Varietal resistance in winter wheat to the rosette disease. *J. Agric. Res.* **26**:261–270.

Weingartner, D. P., J. R. Shumaker, and G. C. Smart, Jr. 1983. Why soil fumigation fails to control potato corky ringspot disease in Florida. *Plant Dis.* **67**:130–134.

Wyss, U. 1982. Virus-transmitting nematodes: feeding behavior and effect on root cells. *Plant. Dis.* **66**:639–644.

Yarwood, C. E. 1960. Release and preservation of virus by roots. *Phytopathology* **50**:111–114.

Zeyen, R. J. 1979. Viruses. Pages 179–205 in *Ecology of Root Pathogens*. S. V. Krupa and Y. R. Dommergues, eds. Elsevier, Amsterdam.

Chapter 21

Historical Epidemiology

This chapter portrays the spread of some major pathogens and the successful efforts of pioneer pathologists to save important plants. Some fusarium wilts follow their hosts so relentlessly that, without control, the hosts would be of minor importance today. Conventional plant breeders receive little publicity in scientific circles, primarily because their advances are slow, continuous, and seldom spectacular. This chapter emphasizes this important facet of plant pathology.

FUSARIUM WILT

Among the many serious diseases of plants caused by *Fusarium* spp. are the Fusarium wilts. Suscept specificity led most early workers to describe the *Fusarium* of each wilt fungus of each host as a new species, and the list became long. *F. lycopersici* was described from tomato by Saccardo in 1886, *F. vasinfectum* from cotton by Atkinson in 1892, *F. niveum* from watermelon by E. F. Smith in 1894, *F. lini* from flax by Bolley in 1901, *F. cubense* from banana by Smith in 1910, *F. conglutinans* from cabbage by Wollenweber in 1913, and *F. oxysporum* var. *nicotianae* from tobacco by J. Johnson in 1921, and *F. orthoceras* var. *pisi* from peas by Linford in 1928.

Linford described the pea wilt fungus as a variety of *F. orthoceras*, a species listed in the Section Elegans of Wollenweber (1913). Johnson described the tobacco fungus as a variety of *F. oxysporum*, and Carpenter, in 1918, saw the affinity of the okra pathogen to the cotton fungus and also to *F. oxysporum*. The wilt fusaria all fall within the Elegans section of the genus *Fusarium*. Snyder and Hansen (1940) classified all these fungi as formae speciales of *F. oxysporum*, differentiated by pathogenicity rather than by morphology. The list of pathogenically identified formae speciales (f. sp.) is now long. Booth (1971) listed 72, and Armstrong and Armstrong (1981) listed 70.

Fusarium oxysporum, the probable progenitor, is one of the most widely distributed *Fusarium* spp. (Messiaen and Cassini, 1981), making up 51 to 66% of the isolates of Gordon in Canada, and 40 to 79% of the *Fusarium* isolates of Guillemat and Montegut in France. It is an efficient competitive saprophyte and a semi-parasite of roots of many species. The pathogenic formae speciales surely arise from this soil inhabitant. No morphologic changes in the fungus are required, and it retains an adaptation to moderate to high temperatures in all its formae speciales.

Flax Wilt

Linum angustifolium, a wild flax with which cultivated flax, (*L. usitatissimum*) will hybridize, is native of the Mediterranean region, and flax probably originated there. The ancient Egyptians made cloth from flax, and the white togas of wealthy Romans were made of flax (linen). The Romans thought flax exhausted the soil, because flax did poorly following flax.

Flax production moved ever westward in North America as yields declined in established regions. In Russia flax was grown mostly on land cleared of scrub timber and in the Netherlands mostly on newly reclaimed polder. In 1892 Miyabe and in 1893 Hiratsuka in Japan believed it subject to attack by a *Fusarium* sp. (in Bolley and Manns, 1932). Bolley, not knowing of the Japanese work, discovered the cause of flax failure in North Dakota and described *Fusarium lini* in 1901. Bolley found the fungus on flax seed and in flax trash in threshed seed (Bolley, 1901) and Bolley and Manns (1932) obtained this fungus from seed lots from Russia, Austria, Canada, Holland, and North America.

Luggar of Minnesota told Bolley that flax in some way brought about its own destruction. At Bolley's instigation, J. H. Shepperd analyzed wheat, barley, corn, oats, peas, mangels, potatoes, and flax and found that flax did not remove any more nutrients from the soil than other crops. If flax produced a toxic condition in soil, it did not affect other plants, because other crops thrived after sick flax, as did weeds among the sick flax. Bolley convinced himself that flax sick soil was caused by *F. lini* (*F. oxysporum* f. sp. *lini*).

Knowing that a fungus caused the disease, that the fungus was disseminated with seed, and that North Dakota at that time had much land on which flax had never been grown, Bolley made strenuous efforts to halt the spread of flax wilt, and thereby preserve the ability of North Dakota soils to produce this valuable plant. North Dakota became a state in 1889, and much of it was unsettled at that time. All his efforts failed. The spread of the fungus was relentless. By 1903 Bolley knew that no rotation would rid the soil of the pathogen once it was established.

A piece or virgin Red River Valley soil near Fargo, North Dakota was seeded to flax in 1893 and to flax every year since. Yields were good until 1895, when they began to diminish. In 1900 all plants were dead by July 4. Bolley began seeding flaxes from various sources on this land in search of resistance. He, and others using this method, succeeded. Resistance in flax is not complete (Flor, 1953) because all flaxes, even the most resistant, wilt to some extent on severely infested soil in warm, dry seasons.

Through resistance, flax wilt has been controlled, even though essentially all the land in the flax region of the northern Great Plains is now infested. Resistant cultivars removed flax from the refugee status. There would now be little new land to flee to.

Cotton Wilt

Plants belonging to the genus *Gossypium* are native in India, Australia, Hawaii, Africa, Arabia, and from Peru to Arizona in the Americas. Most wild species have 13 pairs of chromosomes, as do *G. herbaceum* and *G. arboretum* of India. *Gossypium herbaceum*, grown in India, Iran, China and Japan, was the main cotton of Asia before introduction of *G. hirsutum*. Cottons with 26 pairs of chromosomes, including upland cotton (*G. hirsutum*) and sea island or Egyptian cotton (*G. barbadense*), are from the New World. Cotton fabric was found in pueblo Indian ruins in Arizona. The natives of the West Indies used cotton before Columbus arrived in the Americas, and the Mexicans were wearing cotton when conquered by Cortez. *Gossypium barbadense* probably originated in South America (Hill, 1952). Cultivated Egyptian cotton of the Nile Basin was introduced from Central America. *Gossypium hirsutum* probably originated in Guatemala or southern Mexico, and, according to Wilhelm (1981), about 95% of the cotton of the world is now produced by *G. hirsutum*.

The American Cotton Belt Sea island cotton was introduced into Georgia from the Bahama Islands in 1785. The long high quality fibers of this cotton (a strain of *G. barbadense*) are relatively easy to remove from the seed, and beautiful fabric is woven from it. It is produced mainly near the coast of South Carolina and Georgia. It was the only profitable cotton in the United States prior to the invention of the cotton gin by Eli Whitney in 1793 (Todd and Curti, 1961).

Upland cotton (*G. hirsutum*) has shorter fibers, and it takes one man a day to free about 0.45 kg of fibers from the seed. With a hand-powered gin a man can obtain 22 kg in a day, and with power a gin can separate 450 kg a day. Cotton has thus been transformed from an expensive oddity to a practical substance. For the previous 4000 years India had been the worlds' leading producer of cotton, but now the American South expanded its production.

In 1791 the United States produced 400 bales (227 kg of compressed cotton fibers = 1 bale); by 1810, 171,000; by 1830, 731,000; by 1860, 5,000,000; by 1900, 11,236,383 bales. By 1860 the United States produced two-thirds of the cotton of the world, and cotton was the largest export of this country. In 1900 Texas, Mississippi, Georgia, Alabama, South Carolina, and Arkansas together equalled the rest of the world. For many years the United States produced at least one-half the cotton of the world in the old Cotton Belt. The purpose of this history of the old Cotton Belt is to establish the extent and the length of time that cotton dominated the region (Figure 21.1).

Discovery of Cotton Wilt The history of cotton in the Southern states proves that cotton wilt did not flame across the South. Unlike flax, cotton was grown for years before wilt became serious. The area of longest cotton culture was the sea island cotton area, but wilt was widespread and serious there in 1895. E. F. Smith began his studies on the wilts of cotton, watermelon, and cowpeas at that time. Cotton wilt was also severe in southern Alabama, because it was there that Atkinson (1892) described the disease and named the pathogen *F. vasinfectum*. It was worst on sandy soils,

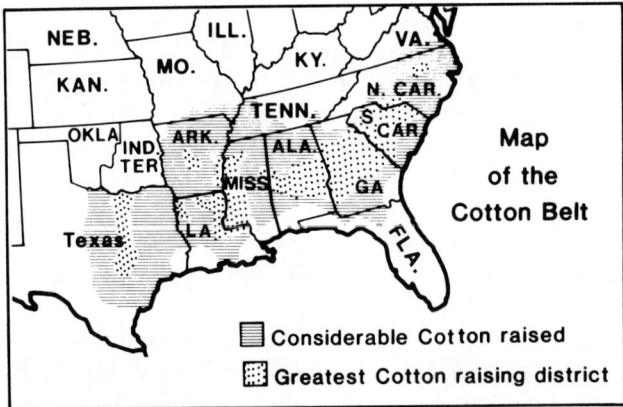

Figure 21.1. The cotton belt in the United States about 1900 (approximate). Tarr and McMurray, 1904.

aggravated by the root-knot nematode. E. F. Smith named the fungus from cowpea *F. tracheiphilum.* Judged by the specific epithets (*vasinfectum*, infector of vascular tissue; *tracheiphilum,* lover of tracheae), both Atkinson and Smith were impressed by invasion of the xylem.

Orton (1908) drew a map of the known distribution of wilt. Wilt was most common on the lighter soils of the Coastal Plain and less common on the heavier soils of the Piedmont. He estimated that the disease had probably been present for 25 to 30 years in a few localities at that time. In most places the wilt was of limited distribution, absent on some farms, and on many present only in spots. But the disease was steadily spreading. By 1914 Gilbert reported wilt in every cotton state from North Carolina to Texas.

Spread E. F. Smith found microconidia of *F. oxysporum* f. sp. *vasinfectum* in xylem vessels, sporodochia with macroconidia on the stems of dead hosts, and chlamydospores in diseased tissue (Higgins, 1911). Chlamydospores in soil and host debris can blow with wind or run with water. Smith (1953) stated that pathologists believed the pathogen originated in Mexico or Central America in the same vicinity as upland cotton. If it did, how did it get to the southeastern Gulf States? Elliott (1923) observed individual wilted plants in otherwise healthy fields, and often in practically virgin land in Arkansas. Farmers suspected that the fungus was introduced on seed.

Elliott observed cotton with wilt in all stages of severity, from plants dead before any bolls formed, to lightly infected plants with essentially normal bolls, to mixtures of plants variously diseased. During wet weather the fungus sporulates profusely on dead branches and stems, and macroconidia can splash onto the open bolls of plants sufficiently vigorous to produce seed. Conidia can live on such seed to the following planting time.

He disinfested the surface of seed and then isolated the pathogen from them. Taubenhaus and Ezekiel (1939) confirmed these studies. Several workers failed to obtain *F. oxysporium* f. sp. *vasinfectum* from seed from diseased plants. We now know that seed is most apt to be infected in humid climates and that few seeds are infected through the vascular system, but rather by inoculum splashed by rain onto the flowers and fruits (Smith et al., 1981). Seed infection seldom occurs in climates with little rain. In general, many true vascular wilt pathogens lacking a leaf-spot phase are rarely transmitted in true seed, but the rare transmissions are important.

Races Armstrong and Armstrong (1978) recognized six races of *F. oxysporum* f. sp. *vasinfectum.* Races 1, 2, and 6 are very similar. They attack *G. hirsutum* and some lines of *G. barbadense,* or 26-chromosome cottons. Race 1 is prevalent in the old Cotton Belt, especially on acid, sandy soils. It is also known in South America, Asia, Europe, East Africa, the USSR, and India, presumably following American upland cottons to these regions. Race

1 isolates differentially pathogenic to *G. hirsutum* cultivars are unknown in the United States (Armstrong and Armstrong, 1978). It is associated with nematode injury, particularly with *Meloidogyne incognita*.

Race 3 is prevalent on some lines of *G. barbadense* on heavy, black clay loams in Egypt. It became serious on Sakel, a high-quality, long-staple cotton in Lower Egypt. Resistance was found among survivors in some of the fields (Fahmy, 1927). Ashmouni, a shorter staple cotton popular in Upper Egypt, was not susceptible. *Meloidogyne* spp. are not associated with cotton wilt in these soils, but the reniform nematode, *Rotylenchus reniformis*, increases susceptibility of some cottons (Khadr et al., 1972). Race 3 is known in the USSR and possibly in Israel. Race 5 was reported from the Sudan by Ibrahim in 1966. It is closely related to Race 3 of Egypt Because it wilts Ashmouni cotton, it exhibits host-specificity on an intraspecies basis. Race 3 attacks Sakel but not Ashmouni. Race 5 attacks both. Race 4 is prevalent on susceptible lines of *G. arboreum* and *G. herbaceum* in India on heavy, black alkaline or neutral soils (usually pH 7.8 to 8.3). It does not attack *G. hirsutum* or *G. barbadense*.

The races of f. sp. *vasinfectum* are broad, based more on differential virulence on species than on intraspecies (cultivar) differentials, and they differ in adaptation to soils. It is tempting to believe that race 1 gave rise to races 2 and 6, that race 3 gave rise to race 5, and that race 4 evolved independently in India. When the length of time and the extent of cotton culture is considered, the number of races is small.

From this point on, only work with race 1 of f. sp. *vasinfectum* in the former Cotton Belt of the United States is considered.

Resistance Fusarium wilt of cotton can be ameliorated somewhat by proper host nutrition, especially ample potassium, but in general two strategies are important: genetic resistance and nematode control. Resistance in cotton was noted by a sea island (*G. barbadense*) planter, E. L. Rivers, who selected a resistant plant in 1895. It was resistant, not an escape, but it had poor quality. In 1899 he found a plant that combined resistance and quality, and it became known as Rivers Sea Island cotton. In 1900 W. A. Orton selected a sea island cotton which he named Centerville, after the Rivers' plantation. Orton's selection was grown on the mainland rather than on the islands (Ware, 1936).

The first resistant upland cotton (*G. hirsutum*) was selected by Orton at Dillon, South Carolina, in 1900. In 1902 he tested a selection later named Dillon and many other cottons at Troy, Alabama. In 1902 he selected Dixie. These cottons were resistant at all test sites and they were used for several years where wilt was devastating. Dillon had poor agronomic type, Dixie had better type, and Dixie × Triumph resulted in the first really valuable wilt-resistant cotton, Dixie Triumph. A variety called Cook was selected by a farmer, J. R. Cook, at Ellaville, Georgia. H. B. Tisdale, the developer of black-shank-resistant tobacco, selected Cook 307-6 from Cook. Cook 307-6, released in 1913, was widely grown in Alabama, and it has contributed resistance to several modern cultivars.

Mr. T. S. Williams, a farmer from Monetta, South Carolina, wrote Orton 25 August 1900, stating that he grew a cowpea called Iron that was practically impervious to the pea wilt, or the pea sickness. Orton (1902) investigated and found that Iron not only resisted f. sp. *tracheiphilum* but also root-knot. Being healthy, it was better nodulated also. He strongly recommended this cowpea for use in rotations with cotton.

Orton (1908) was convinced that the root-knot nematode played a role in cotton wilt. Root-knot was favored by the same conditions favoring cotton wilt, being worse in the light, acid sandy loams. At that time the cowpea was widely used and highly recommended to maintain fertility in these light soils of the coastal plain. He recommended the use of the Iron cowpea, not regular cowpeas. Gilbert (1914) recommended crop rotation to control the nematodes, not the fungus. By this time it was realized that, once f. sp. *vasinfectum* was thoroughly established, it could not be eradicated or even substantially reduced in rotations, but the nematodes could be reduced. He recommended rotations with the Brabham and Iron cowpeas, grasses, cereals, and peanuts.

Nematodes reduce resistance and complicate field studies, especially efforts to select resistant cottons or to study inheritance of resistance. Smith (1953) went so far as to report that in Alabama control of nematodes controlled wilt. Smith also stated that until about 1940 all wilt-resistant upland cottons were sufficiently inferior agronomically that they could not compete on wilt-free soils.

The inheritance of resistance to f. *vasinfectum* race 1 has been difficult to study. Kappelman (1971) concluded that resistance is governed by an undetermined number of genes with additive effects. Complex inheritance explains why so many have worked so long to develop highly resistant, commercially acceptable cottons, as does the lack of pathogenic specialization within race 1. Kappelman also noted some dominance and some epistasis. Kappelman (1975) compared the results obtained in the greenhouse using stem punctures with 2×10^6 spores of mixed virulent cultures in the absence of nematodes with field results where nematodes were present. Although there were some discrepancies between field and greenhouse trials, he concluded that the greenhouse trial has great value because it permits testing more material than can be handled properly in field plots.

Meloidogyne incognita multiplies little on Bayou, a cotton resistant to both wilt and root-knot (Jones and Birchfield, 1967). Root-knot–resistant cottons, like resistant cowpeas, benefit succeeding crops by reducing the nematode population. Tamcot SP 21 and Tamcot 23, developed at Texas A & M, are also resistant to both. It may be difficult to work with the complex resistance in upland cotton, but the results should be dependable.

Nematode Interactions When ethylene dibromide became available, Smith (1948) treated a deep, fine, sandy loam that was so heavily infested that wilt-susceptible cotton could not be grown. Both the meadow and root-knot nematodes were abundant. With solid fumigation (Deltapine 14), a susceptible cotton produced 1012 kg of lint per ha; the control produced 72 kg. The yields of the most wilt-resistant cottons were increased 50% by fumigant applied only under the row.

After Smith demonstrated the value of ethylene dibromide, which is not a strong fungicide, in controlling Fusarium wilt of cotton in Alabama, Martin et al. (1956) investigated the relationship of nematodes to wilt in Louisiana. They used Deltapine 15 (wilt-susceptible) and Coker 100 Wilt (wilt-resistant), and *Meloidogyne, Trichodorus, Tylenchorhynchus,* and *Heliocotylenchus* nematodes. The nematodes thrived on both cottons. The fungus and the nematodes were introduced into the soil prior to planting. Plants in pots with the fungus alone were undamaged. The root-knot nematodes, added in numbers equal to a severe infestation, reduced growth about 50%, and these nematodes (*M. incognita* and *M. acrita*) plus the fungus reduced cotton 75%. The other nematodes did not increase wilt.

The reniform nematode, *Rotylenchulus reniformis*, heavily infested experimental plots near Baton Rouge, Louisiana (Neal, 1954). When a highly wilt susceptible cotton was grown in the presence of *F. oysporium* f. sp. *vasinfectum* alone, little wilt developed. With the fungus and heavy nematode infestation, plants were severely wilted. Several resistant varieties withstood the reniform nematode plus f. sp. *vasinfectum* in these field plots. When the 22 strains resistant to the reniform nematode-wilt complex at Baton Rouge were tested at Natchitoches, Louisiana, on a fine sandy loam with a heavy root-knot + wilt infestation, most were susceptible. Root-knot nematodes exert a stronger predisposing effect than reniform nematodes.

Resistance of cotton to root-knot nematodes is not based on prevention of entrance by the larvae, but on restricted development of larvae after penetration (Minton, 1962). Nematodes penetrate the resistant cottons; thus they wound them. Breaking or reducing resistance to f. sp. *vasinfectum* is due to more than penetration wounds. Brodie et al. (1960) reported that *M. incognita acrita* entered the roots of Auburn 56 (wilt resistant) as readily as it did the roots of other cottons in their trials.

Potassium Deficiency Potassium applications reduces Fusarium wilt in cotton, but when root-knot nematodes are abundant, they negate the beneficial effect of the potash (Young, 1938). Oteifa (1952) noted that *Meloidogyne incognita* reduced the K content of lima beans and that beans in the absence of nematodes did not respond to K, but when nematode infection occurred they responded to heavy K applications. In pot tests in which no nematodes are present, adequate K reduces Fusarium wilt of tomatoes (Walker and Foster, 1946). There is a relationship between host nutrition, nematode injury, and Fusarium wilt.

BLACK SHANK OF TOBACCO

Black shank of tobacco and *Phytophthora nicotianae* were described by Breda de Haan in Java in 1896. Black shank was observed in southern Georgia, in 1915, and by 1922 it was important in cigar wrapper production areas of adjacent southwest Georgia and northwest Florida. Cook reported it in Puerto Rico in 1924. Hansford reported it in Uganda in 1927. It was first observed in North Carolina in 1931 and in Kentucky and Tennessee in 1935. Tobacco is the only natural host of the pathogen, and its origin is puzzling (Lucas 1975). Tucker reduced *P. nicotianae* to a variety, *P. parasitica* var. *nicotianae*. Lucas accepted Tucker's designation, mainly because it is so well established in the literature, even though Waterhouse prefers *P. nicotianae* (Lucas, 1975).

The Spaniards valued *Nicotiana tabacum*, grown by the Indians of Yucatan, and introduced its seed to Santo Domingo about 1531 and then to Trinidad. Commercial production had begun in Cuba by 1580. The Indians in North America east of the Mississippi River grew *N. rustica*. *N. tabacum* was introduced by the settlers at Jamestown, Virginia in 1612, and by 1619, 20,000 pounds were shipped to England. Jean Nicot, French ambassador to Portugal, grew tobacco in 1559 and introduced its use to the royal court of Paris. *Nicotiana* was named after him. By 1639 Maryland and Virginia exported 1,500,000 pounds of tobacco, and these two states dominated production until the close of the American Revolution, after which it extended into North Carolina, Kentucky, Tennessee, Ohio, and Missouri.

The Orinoco variety has given rise to most of the flue-cured, dark air-cured, and fire-cured leaf types. Maryland Broadleaf gave rise to northern cigar leaf and probably White Burley. The Cuban variety gave rise to some cigar filler and wrapper tobaccos. These seminal "varieties" were susceptible to many diseases, including black shank. How could so much tobacco be grown for so long without devastation by black shank?

In the United States occurrences of black shank are well documented. Tobacco has been grown for long periods, and a disease as devastating as black shank could not have been overlooked. Early tobaccos were susceptible. It is unlikely that *P. parasitica* var. *nicotianae* is a recent introduction just establishing itself. It is more likely that its limited host range (for practical purposes, only tobacco?) and its chlamydospores do not equip it for long survival without the frequent production of the suscept on the same land. It must have been introduced and died out repeatedly, except under extraordinarily favorable situations.

Recent Spread

In Georgia black shank has been present constantly since 1915 in two counties adjacent to Gadsen County, Florida, where cigar wrapper tobacco was intensively grown under shade. The disease appeared sporadically in the extensive flue-cured tobacco area of Georgia only to disappear. About 1956 (Gaines, 1960), the disease began a more aggressive spread, and by 1959 it was present in 26 counties outside the wrapper area. Gaines attributed this rise to greater than normal rainfall early in the growing seasons of 1955–1959, increased sprinkler irrigation from streams and ponds, and to the very susceptible variety 402.

Observations made on the Georgia Coastal Plain Experiment Station provided evidence of aerial spread. Black shank had been unknown at this location, but it appeared in a transplant bed, even though the soil had been fumigated. The disease also developed in plots set with the transplants, and it spread to plots planted with healthy transplants. A few scattered leaf lesions appeared on cultivar 402 on 20 May. Rain fell 16 days between 20 May and 11 June, with only two clear, sunny days among 23. Sporangia formed on the under side of leaf lesions, especially on mid-ribs. Lesions were abundant on the tops of the plants with no evidence of infection through the roots. In addition, spread was greatest in the direction of the prevailing winds. The variety 402 favored aerial spread, because leaf lesions developed on it before they did on other susceptible tobaccos at the experiment station.

Black shank was first found in two fields in Kentucky in 1935. Both fields were seeded to grass, and the disease disappeared. In 1940 it was found on two farms in two counties. Both fields were seeded to grass, and the disease was not reported from those counties for many more years. One case spread along a creek to the Kentucky River and eventually along the Ohio River. In 1942 a farmer swept out the soil in the bottom of a truck after hauling material from a Southeastern state. Black shank resulted, and it spread both in a creek and to higher land. In 1951, 600 farms in that county had black shank. By 1952 the disease was established in 60 counties, mostly in small plots along creek bottoms. Black shank became established in Kentucky when farmers ceased to rotate and grew tobacco repeatedly on the same land. A 3-year rotation is adequate for controlling this disease.

Lucas (1975) attributes much spread in North Carolina to the efforts of farmers to get transplants when the bluemold (*Peronospora tabacina*) epidemic of 1949 invaded many transplant beds. Plants were brought from many outside nurseries to farms that had grown their own plants for years.

Resistance

W. B. Tisdale (1931) tested tobacco varieties for resistance to black shank, and by 1931 he had developed resistant lines by selection and inbreeding within Big Cuba, Little Cuba, Dubek from Russia, and from Santiago from Java. In contrast, when some varieties such as Connecticut Round Tip were grown on heavily infested soil, no plants survived, so progress in them was not possible. Resistance within *N. tabacum* was rare. The first selections from Big Cuba (*N. tabacum*) were made in 1922 in severely diseased commercial fields. Big Cuba had mediocre quality but the resistant selections were equal in quality to it. With inbreeding, the resistance of the resistant selections gradually increased.

Tisdale crossed the most resistant selections of Big Cuba (poor quality) × Little Cuba (excellent quality but poor yield) in 1924. A certain family, 301, of this cross had good resistance. Florida 301 was the only source of resistance used until 1964, so his work had widespread application (Lucas, 1975). The resistance in Tisdale's 301 is complex, and subsequent workers have been unable to analyze it genetically. Its resistance is governed by a major recessive gene enhanced by several modifiers (Apple, 1962). The resistance of Florida 301 is multigenic with linkage to undesirable factors (Wallace and Wilkinson, 1975). Only about 60% of its resistance can be incorporated into agronomically useful varieties (Clayton, 1953).

The difficulties of working with Florida 301 stimulated efforts to transfer the high resistance of *N. longiflora* and *N. plumbaginifolia* into commercial types. This resistance is monogenic and dominant (Wallace and Wilkinson, 1975). The resistance in these species is far greater than in any *N. tabacum* sources, and by 1965, NC 2326 with resistance from *N. plumbaginifolia* was released (Lucas, 1975). Valleau et al, 1960 (in Apple, 1962), predicted that, though the resistance in these two species was simply inherited, it would be of little value, because isolates attacking them had already been found in Kentucky.

Apple (1962) surveyed isolates of *Phytophthora parasitica* var. *nicotianae* for virulence on *N. plumbaginifolia* in 1957. He found no isolates virulent on that tobacco in Tennessee, Georgia, Florida, North Carolina, Kentucky, and from other countries. Isolates sent to him later from Kentucky devastated this species. Apple designated isolates that attack the exotic resistance as race 1. The unselected fungus was race 0. Rapid multiplication of race 1 in fields heavily infested with race 0 limits the value of the monogenic resistance of *N. plumbaginifolia* and *N. longiflora* (Wallace and Wilkinson, 1975). Nonspecific, multigenic resistance is difficult to work with, but it has been highly rewarding and lasting.

Nematode Interactions

Tisdale (1931) observed that any plant that resisted black shank to maturity had little root-knot. Sasser et al. (1955) inoculated Dixie Bright 101, moderately resistant to black shank, with the cotton nematode (*M. incognita acrita*) and with *P. parasitica* var. *nicotianae*. When Dixie Bright 101 plants that had grown 30 days in the nematode-infested soil were inoculated with the fungus, black shank began developing within 1 week. When plants of the same age but without nematodes were inoculated with the fungus, they remained relatively healthy for 30 days, even though by then 80% of the plants inoculated with nematodes and fungus had black shank. Wounding the roots with a knife or exposing them to toxic substances from decaying plant residue did not increase the severity of black shank. The predisposing effects caused by the root-knot nematodes were more complex than wounding. Powell and Nusbaum (1960) found that the tissue of nematode galls was extremely favorable for growth of *P. parasitica* var. *nicotianae.* Hyphae ramified the galls profusely and killed the giant cells within 72 hr. Cells at the margin of the galls, those just undergoing hyperplasia, were tolerant. Root-knot nematodes predispose normally resistant tobacco to black shank. Resistance to both pathogens is required on soils infested with both.

Inoculum

Chlamydospores in diseased tissues are released into the soil as the tissues decompose (Lucas, 1975), and they are the main long-term survival structures. They germinate directly or indirectly. Sporangia can form on wet, decaying stems or leaves during warm weather, and a cycle from zoospore to zoospore can be complete in as little as 72 hr. *P. parasitica* var. *nicotianae* is heterothallic, and the role of oospores in the life cycle is unknown.

Zoospore inoculum facilitates a standardized inoculation technique not influenced by the many variables of field plots. Gooding and Lucas (1959) inoculated individual tobacco seedlings of 402 (highly susceptible) and Dixie Bright 101 (moderately resistant) when they were 10 cm tall by injecting zoospores into the root zone. With 10^3 zoospores per plant, the disease index of 402 was 7; with 10^4 zoospores, 31; with 10^5 zoospores, 86; and with 10^6 zoospores, 91. With moderately resistant Dixie Bright, the disease indexes were 0, 0.1, 1, and 12 at the above inoculum levels, respectively. The disease indexes of six cultivars obtained with 10^6 zoospores per plant 20 days after inoculation corresponded to their known field reactions when nematodes are not a factor. Susceptibility was influenced by frequency of fertilization. In sand culture, with 10^5 zoospores per plant, Dixie Bright 101 plants receiving nutrient solution every day were all dead within 2 weeks. Those receiving nutrients on alternate days were more resistant.

Virulence of isolates is correlated with ability to produce vigorous zoospores, evidence that the vigor and health of the isolate are important (Dukes and Apple, 1962). Weakly virulent cultures had been maintained in the laboratory for considerable periods. Zoospores are the most important means of spread in the summer months (Dukes and Apple, 1961). Zoospore accumulate en masse on wounds in roots of tobacco, potato, pepper, and eggplant, but not on tomato roots. Roots of *N. plumbaginifolia,* which is highly resistant, and roots of resistant cultivars of *N. tabacum* attract zoospores as readily as roots of susceptible tobacco. Zoospores placed in the root zone of the very susceptible cultivar 402 encyst, germinate, and penetrate directly within 3 hr (Nusbaum, 1952). Hyphae reach the stele within 6 hr with little host response. In the more resistant Dixie Bright 101 and 102, infection is usually followed by cell collapse, with host cells dying in advance of the fungus. If the fungus reaches the stele, however, invasion proceeds as in a susceptible tobacco.

Water

Field observations indicate that black shank is favored by water. Abundant water favors saprophytic growth of the pathogen in infested refuse, sporangium production, zoospore production, and dissemination (McCarter, 1967). With zoospore inoculum, the susceptible tobacco (Burley 21) was severely diseased at all water levels tried (19, 59, 79, 102% of WHC). The resistant Burley 11B was damaged most in the driest soil (19% WHC) and least at 59 to 79% WHC. Either sustained very dry (19% WHC) and wet (99 to 102% WHC) conditions exceeds what could exist in economic tobacco production, but it shows that if infection occurs, the disease will develop to some extent over the entire range of possible water relations.

Temperature

In northern Florida black shank seldom developed in the plant beds because it was too cool at that time, but in Puerto Rico the disease was often severe in plant beds (Tisdale, 1931). The fungus does not become aggressive until the soil temperature reaches 20°C or above. Size of the plants when inoculated, however, influences the temperature response (Gratz and Kincaid in Lucas, 1975). Susceptible plants, transplanted when about 4 cm tall, are infected at 16°C, but if inoculation is delayed for several weeks after transplanting, the minimum temperature for disease development is 24°C. McCarter (1967) inoculated seedlings 6 cm tall and grew them at temperatures from 16 to 30°C. Disease developed at all temperatures, but was favored by the highest temperatures tested (28 to 30°C).

Under the shade in the Florida cigar wrapper area, temperatures reach 35 to 40°C for a few hours daily. Rainfall is usually light during these periods, resulting in curtailed growth by the tobacco and probable lessening of any natural resistance in the tobacco (Tisdale, 1931). Black shank is primarily a warm-weather disease.

PHYTOPHTHORA ROOT AND STEM ROT OF SOYBEAN

Soybeans were cultivated in China before written records. It is one of the five sacred grains of China, along with rice, wheat, barley, and millet. The greatest concentration of soybeans was in Manchuria, where they occupied 25% of the cultivated land. Korea and Japan have also grown soybeans for many years.

The soybean was introduced into the United States in 1804 (Morse and Carter, 1937) but it remained a curiosity for years. In 1920, 81×10^6 kg were grown in 14 states. North Carolina produced 55% of the crop. By 1931 production had increased to 418×10^6 kg, with Illinois and Indiana assuming importance. By 1935, 92% of production was in the Corn Belt states of Illinois, Indiana, Iowa, Missouri, and Ohio. Southwestern Ontario, Canada also produced 6000 ha of soybeans. By 1980, industrial, food, and feed uses had expanded greatly, with 16,964,000 harvested ha in the United States. Production in North America and Brazil surpassed that in China.

Why this history? Phytophthora root and stem rot (and seedling blight) is a severe disease not easy to overlook, yet it was first observed in 1948 in northeastern Indiana, in 1951 in northwestern Ohio, and in 1954 in southeastern Ontario. The first publications regarding this disease appeared in 1955. It is not mentioned in Dickson's *Diseases of Field Crops*, second edition, published in 1956. The soybean encountered a "new" pathogen in the New World only after intensive production. The early history is reviewed briefly by Kittle and Gray (1979).

Damping-off of soybean was reported by Skotland (1955) in North Carolina and root rot by Suhovecky and Schmitthenner (1955) in Ohio. The latter authors proved the pathogen could attack soybeans of any age, from seedlings to nearly mature plants, and Herr (1957) noted that young plants were more susceptible than older plants. By 1957 this root rot of soybean was known in every important soybean-producing county in Ohio, with an estimated loss of $1,500,000. Kaufmann and Gerdemann (1958) in Illinois described *Phytophthora sojae* as a new species and stated that the disease was favored by heavy, poorly drained soils, though not restricted to them. Hildebrand (1959) in Ontario described the pathogen as *P. megasperma* var. *sojae*. Kuan and Erwin (1980) reduced the pathogen from a variety to f. sp. *glycinea*. Skotland (1955) observed oospores in diseased tissues. The pathogen is homothallic (Savage, et al., 1968). Leaf lesions have been observed (Morgan, 1963), proving that the disease has a limited aerial phase.

Hildebrand (1959) attempted to discover the origin of *P. megasperma* f. sp. *glycinea*. He inoculated several weeds in Ontario soybean fields, and all were nonhosts, as were 14 diverse crop species. Only garden beans were somewhat susceptible. Suhovecky and Schmitthenner (1955) and Kaufmann and Gerdemann (1958) confirmed these results, but they found also that in plants grown in steamed, infested soil, appreciable damping-off also occurred in alfalfa and a trace could be found in sweet clover. Sinclair and Shurtleff (1975), stated that three native species of *Lupinus* are hosts. In addition, they list tomato, alfalfa, garden pea, snap bean, subterranean clover, and white clover. What constitutes a host? A plant in nature? A plant in an extreme test? Only soybeans are important hosts in nature. *P. megasperma* f. sp. *glycinea* is known only in Canada and the United States. It exists in nature in an obscure, unimportant way to become prominent on intensive cultivation of soybeans.

Spread

The fungus moves from the roots up the stem and from the stem downward (Hildebrand, 1959). In one experiment Hildebrand grew Harosoy (susceptible) plants 7.5 cm apart on clean soil. He inoculated the stems of alternate plants in rows at a point 5 to 7.5 cm above the soil surface. When inoculated at 23 days of age, 92% of the inoculated plants died, and 60% of the noninoculated intervening plants also became diseased. When inoculated at 31 days of age, 4% of the noninoculated plants became diseased. When inoculated early, the disease moved down the roots to infect roots of the adjacent plant, or it spread from plant-to-plant in some other way. Hildebrand spread soil from an infested field on clean land. Soybeans became diseased where the infested soil was worked into the clean soil, so the pathogen can be spread by movement of infested soil (animals, man, machine, water, wind).

Klein (1959) proved spread during the growing season by means of zoospores. He flooded 1-month-old diseased plants with tapwater. Sporangia developed on the diseased hypocotyls, and zoospores were present in the water within 48 hr. Sporangia continued to increase in number during the 120-hr flood period. This flood water was used as inoculum, and disease ensued, so zoospores in water spread the pathogen. Morgan (1963) inoculated leaves with zoospores and sporangia. The fungus grew down the petioles to cause stem cankers. If resistant beans were inoculated, only local lesions on leaflets developed.

Klein (1959) found seeds infected via lesions on the pods. He could isolate the fungus only from immature seeds. Oospores formed on the seed coats, but he could not germinate them. He planted seeds, and the pathogen grew from two of them. The pathogen is seedborne to a limited extent, sufficient to introduce the pathogen to a new area.

Soil

Heavy soils favor disease, and the disease is more abundant in low areas subject to water-logging and drainage water (Hildebrand, 1959). Kittle and Gray (1979) grew soybeans in sand, a 1:1 soil:sand mix, and in a soil. These soils were steamed, inoculated with mycelial inoculum, and watered daily to saturation. Disease was greatest in the soil and least in the sand. They concluded that low total porosity (sand), characterized by large individual pores and rapid draining, was not conducive to disease development. High total porosity, characterized by the many small pores and slow drainage of clay, favors this disease. Suhovecky and Schmitthenner (1955) first found the disease in Ohio on old lake bed soils that were heavy clays. Plants on compacted soils, with reduced pore size and root aeration, have higher disease levels.

In a poorly drained field, root rot increases with distance from drainage tiles at low and normal fertility levels (Dirks et al., 1980). At high fertility (3 times normal) disease is uniformly severe over the entire area. Canady and Schmitthenner (1979) added nitrogen in the forms of urea and ammonium nitrate solution (at 4, 16, 40, and 100 ppm N) 10 days before planting. At 40 and 100 ppm damping-off was increased. Root rot and stunting of plants 30 days old were increased by as little as 16 ppm N. These results were obtained in a growth chamber, but they, in conjunction with the field tests of Dirks et al. (1980) indicate that increased succulence from high fertility increases susceptibility.

Herbicide Interaction

The herbicide, 2, 4-DB [4 (2, 4-dichlorophenoxy) butyric acid], applied at flowering time to control *Xanthium saccharatum* (cocklebur), increases the disease in susceptible soybeans. The interaction between herbicide and disease is not observed when disease-resistant cultivars are sprayed (Walters and Caviness, 1968).

Age of Plant and Compensation

Hildebrand (1959) inoculated alternate plants in the stem 5 to 7.5 cm above the soil surface at different ages from planting. The plants were thinned to 7.5 cm spacings in the row. Data were collected from pairs of plants selected for uniformity at the time of inoculation. The results were spectacular (Table 21.1). *P. megasperma* f. sp. *glycinea* damage to soybeans is more severe from early infections than late infections. The diseased-healthy pairs illustrate the compensating ability of the plant. The total yields of pairs ranged only from 1044 to 1241 grams of beans.

Data of this type should not be used to imply that 40% dead or nonproductive plants gives only a small yield loss. If healthy and diseased plants are perfectly alternated, this would be so, but they seldom are. When dead plants occur in patches in the field, there is little or no compensation.

Adaptability of the pathogen

The interactions between pathogen, host, and predisposing factors [drought, fertilizer, drainage, northern root-knot (*Meloidogyne hapla*), herbicides] have been obtained with soybean cultivars lacking specific genes for resistance. Specific resistance is relatively stable, little affected by environmental

Table 21.1. Effect of age when inoculated or noninoculated with *Phytophthora megasperma* var. *sojae* on the yield of pairs of Harosoy soybean plants

Age when inoculated (days after planting)	Yield (g)			Yield loss by inoculated plants (%)
	Inoculated	Noninoculated	Total per pair	
23	22	1022	1044	98
38	116	1015	1131	98
51	10	1043	1053	99
66	361	665	1026	46
79	583	658	1241	11
94	587	594	1181	1

Source: Hildebrand, 1959.

stresses. But *P. megasperma* f. sp. *glycinea* has proved to be a formidable adversary of specific resistances. By 1983, 23 pathogenic races were known (White et al., 1983). While Athow and Laviolette (1982) studied the inheritance of one resistance gene, "Eleven new physiologic races of *P. megasperma* f. sp. *glycinea* were reported while this work was in progress." The progenitor, what ever it was, certainly has remarkable ability to adapt to this plant new to it.

COMMENTS

Fusarium oxysporum f. sp. *lini* follows flax relentlessly. It quickly becomes a limiting factor in production on a wide variety of soils. *F. oxysporum* f. sp. *vasinfectum* race 1, in contrast, does not spread with rapidity on cotton; it spread most rapidly on the sandy, acid soils of the southeastern part of the Cotton Belt, associated with *Meloidogyne incognita*. Rosen (1928) reviewed the efforts of many workers to complete Koch's postulates with the cotton wilt fungus and found many failures. Atkinson (1892) believed the fungus too avirulent to cause disease without assistance. E. F. Smith in 1899 failed to complete Koch's postulates with the cotton wilt fungus but succeeded with the watermelon fungus under the same conditions. Rosen stated that f. sp. *lycopersici* was more virulent on tomato than f. sp. *vasinfectum* on cotton. He commented that susceptible cotton was often grown for many years but susceptible tomatoes could not be. If the cotton wilt fungus were as aggressive as some other forms, "cotton growing by this time would be a lost art." It is obvious the various forms differ in virulence.

For sound research, differences in resistance must first be firmly established by field trials. Once field trials have ranked a few cultivars with significant differences, greenhouse or laboratory tests should place the cultivars in the same approximate order of resistance. This is not as easy as it seems. Reliable laboratory tests require research, but they advance progress greatly.

In testing for nonspecific resistance, with shades of gray rather than black and white as occurs with specific resistance, considerable precision is required. High levels of resistance may be unavailable, but each increment of advance is important. In black shank of tobacco, nonspecific resistance, though complex and difficult to work with, has been useful and stable. In root and stem rot of soybean, with race-specific resistance, the breeders have faced a volatile problem. *P. megasperma* f. sp. *glycinea* has great ability to overcome specific resistances, and *P. parasitica* var. *nicotianae* has overcome the specific resistance it encountered. The rapid advance of the soybean pathogen, both geographically and pathogenically (at least 23 known races by 1983) was accelerated by the tendency toward less and less rotation, made possibly largely by chemical fertilizers and herbicides. The next chapter is entitled "Activities of Man," but man, like so many things, appears in all the chapters.

KEY REFERENCES

Athow, K. L. and F. A. Laviolette. 1982. $Rpse_6$, a major gene for resistance to *Phytophthora megasperma* f. sp. *glycinea* in soybean. *Phytopathology* **72**:1564–1567.

Gooding, G. V. and G. B. Lucas. 1959. Effect of inoculum level on the severity of tobacco black shank. *Phytopathology* **49**:274–276.

Hildebrand, A. A. 1959. A root and stalk rot of soybeans caused by *Phytophthora megasperma* Drechsler var. *sojae* var. nov. *Can. J. Bot.* **37**:927–957.

Kappelman, A. J., Jr. 1975. Correlation of Fusarium wilt of cotton in the field and greenhouse. *Crop Sci.* **15**:270–272.

Minton, N. A. 1962. Factors influencing resistance of cotton to root-knot nematodes (Meloidogyne spp.). *Phytopathology* **52**:272–279.

Smith, A. L. 1948. Control of cotton wilt and nematodes with a soil fumigant. *Phytopathology* **38**:943–947.

Smith, S. N., D. L. Ebbels, R. H. Garber and A. J. Kappelman, Jr. 1981. Fusarium wilt of cotton. Pages 29–38 in *Fusarium Diseases, Biology, and Taxonomy.* P. E. Nelson, T. A. Toussoun, and R. J. Cook, eds. Penn. State Univ. Press, University Park.

Wills, W. H. 1964. Autumn weather in relation to subsequent occurrence of tobacco black shank in Virginia. *Plant Dis. Reptr.* **48**:32–36.

REFERENCES

Apple, J. L. 1962. Physiologic specialization within *Phytophthora parasitica* var. *nicotianae*. *Phytopathology* **52**:351–354.

Armstrong, G. M. and J. K. Armstrong. 1978. A new race (race 6) of the cotton-wilt fusarium from Brazil. *Plant Dis. Reptr.* **62**:421–423.

Armstrong, G. M. and J. K. Armstrong. 1981. Formae speciales and races of *Fusarium oxysporum* causing wilt diseases. Pages 391–399 in *Fusarium. Diseases, Biology, and Taxonomy*. P. E. Nelson, T. A. Toussoun, and R. J. Cook, eds. Penn. State Univ. Press, University Park.

Atkinson, G. F. 1892. Some diseases of cotton. III. Frenching. *Ala. Agric. Exp. Sta. Bull* **41**:19–29.

Bolley, H. L. 1901. Flax wilt and flax sick soil. *N. D. Agric. Exp. Sta. Bull.* 50.

Bolley, H. L. and T. F. Manns. 1932. Fungi of flaxseed and of flaxsick soil. *N. D. Agric. Exp. Sta. Tech. Bull.* 259.

Booth, C. 1971. *The Genus Fusarium*. Commonwealth Mycological Institute, Kew, England.

Brodie, B. B., L. A. Brinkerhoff, and F. B. Struble. 1960. Resistance to the root knot nematode, *Meloidogyne incognita acrita*, Upland cotton seedlings. *Phytopathology* **50**:673–677.

Canady, C. H. and A. F. Schmitthenner. 1979. The effect of nitrogen on phytophthora root rot of soybeans. *Phytopathology* **69**:539 (abstract)

Clayton, E. E. 1953. Developments in growing tobacco. Pages 540–548 in *USDA Yearbook of Agriculture*. U. S. GPO, Washington, D.C.

Dickson, J. G. 1956. *Diseases of Field Crops*. McGraw-Hill, New York.

Dirks, V. A., T. R. Anderson, and E. F. Bolton. 1980. Effect of fertilizer and drain location on incidence of phytophthora rot in soybeans. *Can. J. Plant Pathol.* **2**:179–183.

Dukes, P. D. and J. L. Apple. 1961. Chemotaxis of zoospores of *Phytophthora parasitica* var. *nicotianae* by plant roots and certain chemical solutions. *Phytopathology* **51**:195–197.

Dukes, P. D. and J. L. Apple. 1962. Relationship of zoospore production potential and zoospore motility with virulence in *Phytophthora parasitica* var. *nicotianae*. *Phytopathology* **52**:191–193.

Elliott, J. A. 1923. Cotton-wilt: a seed-borne disease. *J. Agric. Res.* **23**:387–393.

Fahmy, T. 1927. The fusarium disease (wilt) of cotton and its control. *Phytopathology* **17**:749–767.

Flor, H. H. 1953. Wilt, rust and pasmo of flax. Pages 869–873 in *USDA Yearbook of Agriculture*, U. S. GPO, Washington D.C.

Gaines, J. G. 1960. History of black shank in Georgia flue-cured tobacco including spread of the disease in 1959. *Plant Dis. Reptr.* **44**:155–158.

Gilbert, W. W. 1914. Cotton wilt and root-knot. *USDA Farmers' Bull.* 625.

Herr, L. J. 1957. Factors affecting root rot of soybeans incited by *Phytophthora cactorum*. *Phytopathology* **47**:15–16 (abstract).

Higgins, B. B. 1911. Is *Neo cosmospora vasinfecta* (Atk.) Smith, the perfect stage of the *Fusarium* which causes cowpea wilt? *32nd Ann. Rept., N. C. Exp. Sta.*, 1909. pages 100–116.

Hill, A. F. 1952. *Economic Botany*. McGraw-Hill, New York.

Ibrahim, F. M. 1966. A new race of cotton-wilt *Fusarium* in the Sudan Gezira. *Emp. Cotton Grow. Rev.* **43**:296–299.

Jones, J. E. and W. Birchfield. 1967. Resistance of the experimental cotton variety, Bayou, and related strains to root knot nematode and fusarium wilt. *Phytopathology* **57**:1327–1331.

Kappelman, A. J., Jr. 1971. Inheritance of resistance to Fusarium wilt of cotton. *Crop. Sci.* **11**:672–674.

Kaufmann, M. J. and J. W. Gerdeman. 1958. Root and stem rot of soybean caused by *Phytophthora sojae* n. sp. *Phytopathology* **48**:201-208.

Khadr, A. S., A. A. Salem and B. A. Oteifa. 1972. Varietal susceptibility and significance of the reniform nematode, *Rotycenchus reniformis*, in fusarium wilt of cotton. *Plant Dis. Reptr.* **56**:1040-1042.

Kittle, D. R. and L. E. Gray. 1979. The influence of soil temperature, moisture, porosity, and bulk density on the pathogenicity of *Phytophthora megasperma* var. *sojae. Plant Dis. Reptr.* **63**:231-234.

Klein, H. H. 1959. Etiology of the Phytophthora disease of soybeans. *Phytopathology* **49**:380-383.

Kuan, T. L. and D. C. Erwin. 1980. Formae speciales differentiation of *Phytophthora megasperma* isolates from soybean and alfalfa. *Phytopathology* **70**:333-338.

Lucas, G. B. 1975. *Diseases of Tobacco.* Harold E. Parker and Sons, Printers, Fuguay-Varina, N.C.

Martin, W. J., L. D. Newsom, and J. E. Jones 1956. Relationship of nematodes to the development of Fusarium wilt in cotton. *Phytopathology* **46**:285-289.

McCarter, S. M. 1967. Effect of soil moisture and soil temperature in black shank disease development in tobacco. *Phytopathology* **57**:691-695.

Messiaen, C. M. and R. Cassini. 1981. Taxonomy of *Fusarium* Pages 427-445 in *Fusarium. Diseases, Biology, and Taxonomy.* P. E. Nelson, T. A. Toussoun, and R. J. Cook, eds. Penn. State Univ. Press, University Park.

Morgan, F. L. 1963. Soybean leaf and stem infections by *Phytophthora megasperma* var. *sojae. Plant Dis. Reptr.* **47**:880-882.

Morse, W. J. and J. L. Carter. 1937. Improvement in soybeans. Pages 1154-1189 in *USDA Yearbook of Agriculture.* U. S. GPO, Washington, D.C.

Neal, D. C. 1954. The reniform nematode and its relationship to the incidence of fusarium wilt of cotton at Baton Rouge, Louisiana. *Phytopathology* **44**:447-450.

Nusbaum, C. J. 1952. Host-parasite relations of *Phytophthora parasitica* var. *nicotianae* in roots of resistant and susceptible tobacco varieties. *Phytopathology* **42**:286 (abstr.).

Orton, W. A. 1902. Some diseases of the cowpea. *USDA Bur. Plant Ind. Bull.* 17.

Orton, W. A. 1908. Cotton wilt. *USDA Farmers' Bull.* 333.

Oteifa, B. A. 1952. Potassium nutrition of the host in relation to infection by a root-knot nematode *Meloidogyne incognita. Proc. Helminthol. Soc. Wash.* **19**:99-104.

Powell, N. T. and C. J. Nusbaum. 1960. The black shank root-knot complex in flue-cured tobacco. *Phytopathology* **50**:899-906.

Rosen, H. R. 1928. A consideration of the pathogenicity of the cotton wilt fungus, *Fusarium vasinfectum. Phytopathology* **18**:419-438.

Sasser, J. N., G. B. Lucas and H. R. Powers, Jr. 1955. The relationship of root-knot nematodes to black-shank resistance in tobacco. *Phytopathology* **45**:459-461.

Savage, E. J., C. W. Clayton, J. H. Hunter, J. A. Brenneman, C. Laviola and M. E. Gallegly. 1968. Homothallism, heterothallism, and interspecific hybridization in the genus *Phytophthora. Phytopathology* **58**:1004-1021.

Sinclair, J. B. and M. C. Shurtleff. 1975. *Compendium of Soybean Diseases.* American Phytopathology Society, St. Paul, Minn.

Skotland, C. B. 1955. A Phytophthora damping-off disease of soybean. *Plant Dis. Reptr.* **39**:682–683.

Smith, A. L. 1953 Fusarium and nematodes on cotton. Pages 292–298 in *Plant Diseases. USDA Yearbook of Agric.* U.S. GPO, Washington, D.C.

Snyder, W. C. and H. N. Hansen. 1940. The species concept in *Fusarium. Am. J. Bot.* **27**:64–67.

Suhovecky, A. J. and A. F. Schmitthenner. 1955. Soybeans affected by early root rot. *Ohio Farm Home Res.* **40**:85–86.

Tarr, R. S. and F. M. McMurry. 1904. *A Complete Geography.* Macmillan, New York.

Taubenhaus, J. J. and W. N. Ezekiel. 1939. Seed transmission of cotton wilt. *Science* **76**:61–62.

Tisdale, W. B. 1931. Development of strains of cigar wrapper tobacco resistant to black shank (*Phytophthora nicotianae* Breda de Haan). Fla. *Agric. Exp. Sta. Tech. Bull.* 226.

Todd, L. P. and M. Curti. 1961. *Rise of the American Nation.* Harcourt, Brace & World, New York.

Walker, J. C. and R. E. Foster. 1946. Plant nutrition in relation to disease development. III. Fusarium wilt of tomato. *Am. J. Bot.* **33**:259–264.

Wallace, D. H. and R. E. Wilkinson. 1975. Breeding for resistance in dicotyledonous plants to root rot fungi. Pages 177–184 in *Biology and Control of Soil-Borne Plant Pathogens.* G. W. Bruehl, ed. American Phytopathology Society, St. Paul, Minn.

Walters, H. J. and C. E. Caviness. 1968. Response of Phytophthora resistant and susceptible soybean varieties to 2, 4-DB. *Plant Dis. Reptr.* **52**:355–357.

Ware, J. O. 1936. Plant breeding and the cotton industry. Pages 657–744 in *USDA Yearbook of Agriculture.* U.S. GPO, Washington, D.C.

White, D. M., J. E. Partridge, and J. H. Williams. 1983. Paces of *Phytophthora megasperma* f. sp. *glycinea* on soybeans in eastern Nebraska. *Plant Dis.* **67**:1281–1282.

Wilhelm, S. 1981. Sources and genetics of host resistance in field and fruit crops. Pages 299–376 in *Fungal Wilt Diseases of Plants.* M. E. Mace, A. A. Bell, and C. H. Beckman, eds. Academic Press, New York.

Wollenweber, H. W. 1913. Studies on the Fusarium problem. *Phytopathology* **3**:24–49.

Young, V. H. 1938. Control of cotton wilt and 'rust' or potash hunger by the use of potash controlling fertilizers. *Ark. Agric. Exp. Sta. Bull.* 358.

Chapter 22

Activities of Man

Man accelerates knowledge of many ecological factors by planting a single species in pure stands. Toussoun (1975) quoted from O. A. Reinking's 1934 report of Fusarium wilt in a banana plantation in Central America. The field was planted at the same time with the same cultivar, and all cultural practices were the same. The disease did not radiate out from the diseased area gradually, "but there was a distinct line of demarcation between the two areas in respect to disease that corresponded with the change in soil type. A line could be drawn between severe disease and no or slight disease following the change in the soil. . ."

Production of the same plant over large areas for several years, as in large areas devoted to wheat, aids pathologists in other ways. In 1964, for example, part of a field of winter wheat was severely damaged by *Fusarium culmorum,* and another part of the field was unaffected. The severely diseased part followed a crop of spring oats. Oats are rarely produced in the dryland, summer-fallow region of central Washington, and this was the only field severely affected for miles in any direction. That oats increased the pathogen is undoubted, because the edge of the area previously planted to oats was sharply delineated. Growth of a single cultivar facilitates observation on the effects of cultural practices, including rotation, on soilborne diseases.

TILLAGE

Plants developed long before man tilled the soil and, in early agriculture, tillage must have been minimal. With domestication of draft animals, tillage increased, reaching excessive levels with tractors and cheap fuel. The cycle is now toward reduced tillage, made possible by modern herbicides and encouraged by efforts to reduce fuel consumption and soil erosion.

Russell (1977) reviewed many of the effects of tillage on the soil. Tillage increases the rate of water infiltration, reduces the bulk density of the plowed layer, improves aeration, and reduces the resistance of the soil to penetration by roots. In compact soils roots tend to follow fracture planes or natural fissures within the soil. As soil water content increases, resistance to penetration decreases.

Reduced bulk density following tillage persists for a few months. Plowing increases total pore space, mostly of large pores. A consequence of improved aeration is increased decomposition of residue. Tillage makes surface soils more uniform by frequent mixing with little or no pH gradient within the tilled layer. Soil without residue on the surface is subject to wider and more rapid fluctuations in temperature. In a region with a moderate winter, frost penetrates three times deeper and the frozen layers remain two times longer under conventional tillage than under standing stubble (Almaras et al., Columbia Basin Agriculture Research Center, Pendleton, OR; unpublished). Cultivated soil also dries out more rapidly than minimally tilled soil in the spring.

Long-term effects also result from tillage. Soils cultivated for 50 years have greater bulk densities below tilled layers than uncultivated soils in similar sites (Allmaras et al., 1982). The pH of cultivated soil in a wheat-pea rotation is 5.3 at 0 to 15 cm, 5.2 at 15 to 30 cm, 5.9 at 30 to 45 cm, and 6.1 at 45 to 60 cm deep. This reduction of pH in the surface layers is attributed to the use of NH_4^+ fertilizers. These changes, both in bulk density and pH, have occurred within about 50 years in a dryland agriculture system.

In the central Great Plains of the United States, Doupnik and Boosalis (1980) described the effects of "ecofallow" on stalk rot of sorghum caused by *Fusarium moniliforme*. Sorghum is seeded directly (no-till) into winter wheat refuse from the previous crop. The wheat residue shades the soil surface, reducing the surface soil temperature and conserving soil moisture. Moisture stress is reduced both by lower soil temperature and greater water storage in the soil. Stalk rot of sorghum, which is aggravated by moisture stress, occurs on 39% of plants in conventional tillage and 11% of plants in ecofallow, and the yields under ecofallow are increased 42%.

The same tillage operations done at different times may not have equal effects. Destroying cotton as soon as possible after harvest reduces formation of sclerotia of *Phymatotrichum omnivorum*. Tilling deeply when the soil is dry kills *P. omnivorum* sclerotia by desiccation. Uprooting tobacco plants right after harvest to expose the roots to the air reduces maturation and survival of root-knot nematodes. Tilling wheat stubble right after harvest weakens inoculum of *Gaeumannomyces graminis* by initiating decomposition of refuse as soon as possible. Broad coverage of the effects of tillage on plant pathogens is available in reviews by Cook et al. (1978) and by Sumner et al. (1981).

Burying refuse is a form of sanitation in the case of some pathogens active at or above the soil surface. *Sclerotium rolfsii* is destructive in warm, humid climates on a wide range of hosts (Aycock, 1966) and Boyle (1956) believed that saprophytic colonization of organic debris increased the strength of the attack. He called *S. rolfsii* a necrotrophic fungus, one living on dead tissue. Boyle recommended deep plowing to remove organic matter from the surface, control of leaf spots of peanut so fallen leaflets would not accumulate under the plants, and cultivaton in such a way as to avoid throwing soil (dirting) about the base of the plants to smother small weeds which also smother leaflets and small peanut shoots, weakening them and favoring attack.

Garren and Duke (in Garren, 1959) developed a practical control of *S. rolfsii* by utilizing advanced machinery and dinoseb preemergence herbicide. All refuse was cleanly turned under, leaving no fresh organic matter within the top 10 cm of soil. Planting was in slightly raised beds and dinoseb, surface-active herbicide was sprayed in a band over the row at planting time. The herbicide, applied at planting time, made throwing dirt toward the row unnecessary. Only the centers of the rows were cultivated. Shields were used to prevent soil from being pushed on parts of the peanut plant. Cercospora leaf spot was controlled with fungicides. Clean plowing alone increased yields 388 kg/ha, and cultivating alone so as not to "dirt" plants increased yields by 810 kg/ha. Both measures combined increased yields by 1197 kg/ha, and only 5% of the plants were affected by this disease. Peanuts following peanuts with tillage other than that recommended resulted in 76% diseased plants.

ROTATION

Crop rotation has historically been a major means of disease control in production of annual and biennial crops. Its importance in some cases has declined because of the wide use of fertilizers to maintain soil fertility and improvements in nematicides and other pesticides, advances in plant breeding, and in a few cases in the development of disease suppression following monoculture. Shipton (1977) reviewed monoculture and Curl (1963) reviewed disease control by crop rotation. Both these authors provide extensive literature lists. Many experiment stations no longer maintain rotation experiments, and most biological control projects are aimed at making monoculture successful. In the long-range interest of agriculture, more emphasis of rotation is needed.

Reduced Pathogen Numbers

One of the major benefits of crop rotation is to limit the reproduction of a pathogen by denying it a host on which to increase. This is most successful in cases in which the pathogen survives saprophytically in debris, without a long-lived dormant structure. An example of disease reduction by this means is brown stem rot of soybeans incited by *Phialophora gregata* (Dunleavy and Weber, 1967). When a susceptible soybean was grown for 1 year and corn for 5 years, 6% of the soybean plants were infected; 1 year of soybeans to 4 of corn, 23% infection; 1 year of soybeans to 3 of corn, 45%; 1 year of soybeans to 2 of corn, 83%; 1 year of each 98%; and when soybeans were grown 6 consecutive years, 100% infection. With increased frequency of the suscept, infection increased dramatically. In this experiment propagules of the pathogen per unit of soil was not determined, but I assume increased inoculum accounts for the increased disease.

The root-lesion nematode *Pratylenchus penetrans* causes economic losses on several crops in eastern Canada. Kimpinski and Willis (1980) studied the effects of different hosts on the numbers of this nematode in roots and in the soil. The potato is a relatively poor host, and the experiment followed potatoes so initial populations were low. After three crops, the total population (in roots and in soil) per unit of soil for corn was 71,801; for red clover + timothy, 33,019; for timothy, 12,660; for wheat, 6216; and for potatoes, 1890, direct evidence of the effect of rotation on pathogen populations in the soil.

Nematodes with highly developed host-parasite interactions, such as root-knot nematodes, can be reduced by a resistant cultivar. Second-stage larvae enter them freely, only to starve by failure to establish the giant cells essential to their nutrition. In cases such as these, the "host" can be grown with success equivalent to that of a nonhost. Nematode numbers are reduced with no economic cost to the farmer.

A remarkable survival mechanism for nematodes has been discovered (Ishibashi, 1969). In a warm climate with a compatible host, root-knot nematodes have several generations, reproducing as long as new root tips are available. The eggs hatch with little dormancy and with little or no response to host exudates. Ishibashi noted that old or poorly nourished females of *Meloidogyne incognita* formed brown egg masses containing dormant eggs resistant to stress and stimulated to hatch by root exudates, unlike eggs laid in white egg masses by young or well-nourished females on a favorable host. If a female survives and reproduces on the unfavorable host, dormant, tough eggs responsive to suitable exudates will have survival value. This female, in a poor situation, increases the chances that her progeny will live to attack a more favorable host.

Preventive Value

Rotations have insurance and preventive values. There are many pathogens in soil, several of which could be destructive on a given plant. Changes in farming practice, particularly from rotation to monoculture, often result in new problems. Our ignorance is great, and rotation minimizes known and unknown pathogens. In recent years new lands brought under irrigation in Washington grew their first crops of dry beans. Yields were phenomenal, and some farmers grew several crops without rotation. Within a few years *Fusarium solani* f. sp. *phaseoli* became serious, and yields declined. When a serious problem developed, rotation was adopted. But the population of the pathogen had reached 2000 to 3000 chlamydospores per gram of soil, and when established to this degree, the fungus often maintains inoculum levels capable of economic loss, even with rotation. Once certain pathogens are increased to high population levels, they appear immortal. If a good rotation is followed from the beginning, troubles are usually averted. Mai and Abawi (1980) stressed prevention, that a good rotation should be followed even though the field is not known to be infested. Soil samples can miss a nematode (or other pathogen) present in low numbers. In New York, if a farmer grows cabbage or beets for 1 year, then plants nonhosts of *Heterodera schachtii* for 2 years, no problem develops.

Netscher and Taylor (1979), working in Senegal, identified root-knot nematodes present in rotation plots. *Meloidogyne hapla* was absent, and strawberries should not be galled. They were free of galls for 3 months, but in 6 months some *M. javanica* developed on the strawberry. These workers stressed the value of rotation to prevent increase of special populations. The tobacco cyst-nematode, *Globodera solanacearum,* was detected in Virginia in 1961 in a small field that had produced seven consecutive tobacco crops (Komm et al., 1983). From 1961 to 1978 spread (or increase?) was slow. From 1979 on it has spread rapidly. If the original field had not produced seven consecutive crops of tobacco, this nematode might have remained undetected.

Triantaphyllou (1975) studied the adaptability of races of *Heterodera glycines* to soybeans with different specific resistances. He identified five races on Lee (susceptible), Pickett, P.I. 88788, and P.I. 90763. He then grew several generations on a susceptible and on a resistant soybean and observed the changes in virulence. One population produced 222 cysts on Lee, 49 on Pickett, and 11 on P.I. 88788. After five generations on Pickett, 138 cysts formed on Lee, 101 on Pickett, and 4 on P.I. 88788. The nematode was less fit on Lee and more fit on Pickett. After five generations on P.I.

88788, the population was less fit on Lee (147 cysts) and more fit on P.I. 88788 (58 cysts). Whenever a choice of specific resistances effective against a population is available to a grower, it would be wise to rotate cultivars of the same group with different specific resistances to minimize adaptation of the pathogen to the host. It is expensive, difficult, and time-consuming to develop resistant cultivars, and rotating them could lengthen their useful life. The time required for resistance to be overcome is, however, quite variable.

Young (1984) examined *H. glycines* from 17 commercial fields that had been planted from 3 to 6 years to resistant cultivars without a break. In assays in the greenhouse the nematode reproduced more on the resistant cultivars than formerly, but they were still so resistant that yields of the resistant cultivars in the field were not measurably reduced. For practical purposes, the resistance to *H. glycines* could, in this case, be considered stable(?).

Crop Sequence

In designing rotations it is useful to determine what sequence of crops is best. In actual practice the sequence is influenced by maintenance of soil fertility, reduction of erosion, effect on insects and weeds, and economics, as well as pathogens. In earlier years (1920–1937), when spring wheat and barley were important crops in the Corn Belt, farmers were advised not to plant spring wheat or barley following corn unless the corn refuse was cleanly plowed under or the corn was cut for ensilage. Perithecia of *Gibberella zeae* form in great numbers on corn stalks, and ascosporic inoculum is important. Spring wheat usually suffers less from scab following wheat than corn. This is atypical. Usually a crop is its own worst enemy, harboring all pathogens of itself. But corn is more dangerous to wheat than wheat in this case.

Length of Rotation

Length of an effective rotation depends on the longevity of the means of survival. The life cycle of the wheat gall nematode, *Anguina tritici,* makes it subject to control by a 1-year rotation. The nematode gall disintegrates on moist soil, freeing the second-stage juveniles. The larvae are rich in energy and are able to survive in moist, cool soil up to 7 months, but they die if they do not find a host. The cysts of *Heterodera* spp., in contrast, resist disintegration in soil for long periods, and the eggs within them hatch over an extended period of time, so rotations of 3 or more years are required to reduce their numbers to safe levels.

Another factor of importance in duration of the rotation is the rigor of factors affecting survival. *Cephalosporium gramineum,* causing cephalosporium stripe of wheat, is most destructive in Washington and Idaho following summer fallow or with a winter wheat, fallow, winter wheat succession. The summer is so dry the straw does not decompose during much of the fallow period, making 2 years between winter cereals necessary for control. In Kansas a 1-year rotation is sufficient (W. Bockus, Kansas State Univ., Manhattan; personal communication). In Kansas wheat is harvested in July, and the summer and autumn are warmer and more humid than in Washington. Saprophytic survival is shorter in regions in which the substrate is subjected to warm, moist conditions conducive to rapid decomposition.

Rotations often give variable results. What mixture of species or races of pathogens exists in each field? How good is weed control? Do the pathogens survive or even increase on some of the weeds? Are volunteer (self-sown) plants of the crop present in significant numbers? It was once indicated to me that *Pseudocercosporella herpotrichoides,* a wheat pathogen that dies when its substrate has decomposed, was not eliminated by 5 years of alfalfa. Examination showed the alfalfa field to be filled with weed grasses. The pathogen survived on some of these grasses, not in quantity sufficient to cause serious losses in the first wheat crop, but enough to be easily observed.

Does *Pseudomonas solanacearum* survive 4 years in fallow soil in Sumatra? Or are weeds occasionally present that sustain a small population of the bacteria? Complete weed control is usually impossible, yet rotation remains one of our most useful control measures. Rotation has been, is, and hopefully will continue to be a major means of pathogen control, but scientists must be cautious in interpreting the literature on the longevity of pathogens in soil based on rotations.

Effects on Associated Pathogens

Many years ago cowpeas were used to supply nitrogen and organic matter in light coastal plains soils of the Southeast (Chapter 21). Cowpeas increase root-knot nematodes, aggravating fusarium wilt of cotton. Root-knot reduces the effectiveness of resistance to fusarium wilt of cotton, a serious problem before cottons resistant to both pathogens and effective nematicides were available. A rotation that controls root-knot nematodes reduces Fusarium wilt of some plants, not because it is effective against *F. oxysporum,* but because it is effective against an associated, predisposing pathogen. Similar relationships of historical significance exist between root-knot and *Pseudomonas solanacearum* and root-knot and *Phytophthora parasitica* var. *nicotianae* on tobacco. A rotation may be beneficial for reasons not known at the time.

Clean Seed

The benefits of an effective rotation may be lost if a pathogen is seed-borne and is capable of rapid spread. Pathogens such as *Colletotrichum lin-*

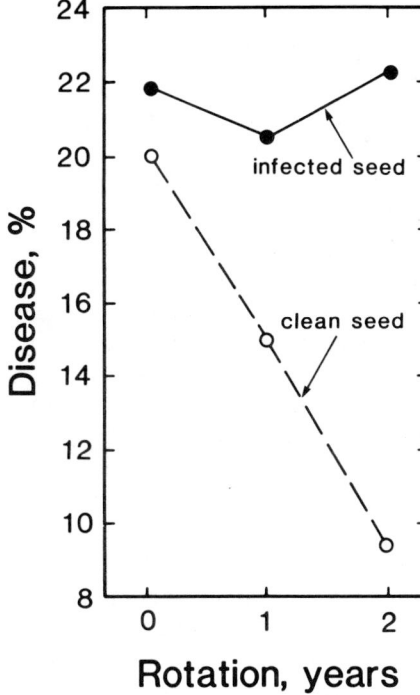

Figure 22.1. *Septoria nodorum* can be controlled on wheat in Florida by rotation plus clean seed, but not by either control measure alone. After Luke et al., 1983.

demuthianum and *Pseudomonas phaseolicola* survive from one season to the next in infested refuse, and short rotations are adequate if pathogen-free seed is used. Glume blotch caused by *Septoria nodorum* was so severe in Florida that yields declined from about 2000 to 1000 kg/ha, and the acreage in wheat declined 60% between 1971 and 1975. Pathologists had not found suitable resistance. Luke et al. (1983) found that *S. nodorum* could be controlled in Florida by growing nonhost crops 2 years between wheat crops and by using clean seed. Clean seed without rotation or rotation without clean seed are ineffective (Fig. 22.1).

PLANTING

An advance, possible in some crops and climates, is to direct seed rather than to transplant seedlings from beds into the field. Tomatoes are seeded directly into the field in the Sacramento Valley, California, with great savings in labor and reduced disease. Many pathogens have every opportunity to spread within plant beds. Pathogens can also be spread during handling of the transplants. Wounds from transplanting are avoided by direct seeding, thus reducing the entry sites and danger from vascular wilt fungi.

Seeding Depth

Seeding depth is governed by the requirements of the plant, with small seeds favored by very shallow seeding and larger seeds favored by somewhat deeper seeding. Farmers cannot afford to alter seeding depths to control disease, but seeding depth affects the development of some diseases. The severities of *Typhula* and *Fusarium* snow molds are increased by seeding winter wheat deeply (Bruehl and Cunfer, 1971). We seeded wheat at 1.2, 2, 4, and 5 cm depth and found that deep seeding delayed emergence, weakened the plants, and reduced resistance to snow mold. Dwarf bunt of winter wheat, caused by *Tilletia controversa,* is a shoot-infecting smut that enters young host tissues near the soil surface. Meiners et al. (1956) found that shallow seeding favored infection. The numbers of bunted wheat heads was 14% in wheat seeded 10 cm deep, 20% when seeded 6 cm deep, 32% when seeded 2.5 cm deep, and 35% when the seed germinated on the soil surface.

Seeding Date

Seeding date is a major factor in determining the percentage of bunted wheat heads. Hoffmann and Purdy (1967) seeded wheat in naturally infested soil augmented with *T. controversa* spores. With natural inoculum and wheat seeded 1 cm deep, 1% of bunted heads developed from 19 August seeding, 25% from 2 September, 53% from 16 September, 56% from 30 September and 11% from 14 October, a very late date. Seeding date alone reduced smut to 1% in the 19 August seeding, compared to 53 to 56% from the mid- to late-September seedings. Dwarf bunt is difficult to control by resistance. Few effective resistance genes exist, and the fungus is highly competent at producing new races. Early seeding produces tillers past the susceptible stage by the time the teliospores germinate, resulting in a nonspecific histological resistance.

Plants susceptible to *Phymatotrichum omnivorum* are grown safely in infested soil in the cool months. Seeding date in temperate climates with a regular sequence of changing temperatures has its effects largely through temperature. These factors are discussed in Chapter 3.

Seeding Rates

Seeding at heavy rates makes some diseases more severe, primarily by increased shading and succulence. Seedlings in plant beds require careful management to escape severe losses from damping-off. Increased seeding rates tend to produce weaker stems. Many yield advances in modern agriculture result from higher numbers of plants per unit area, which would be impossible without advances in lodging and stem-rot resistance. Burke and Nelson (1965) grew beans in rows 17.5, 35, and 55 cm apart with plants, 5,

10, 15, and 20 cm apart in the row, to study the effect of stand density and planting date on Fusarium root rot (*Fusarium solani* f. sp. *phaseoli*). Yields were greatest on both clean and infested soil when seeding was early. The root rot index was lowest in both planting dates at low plant densities. Increasing the stand density, however, even though it increased the root rot index, increased yield (Fig. 22.2). Burke and Nelson proposed high plant densities with close row spacings as a means of increasing bean yields on infested land. Achieving a high plant density by use of narrow row spacings avoids crowding within the row.

In the case of beans attacked by *F. solani* f. sp. *phaseoli*, plant-to-plant spread is not a factor, because the soil contains probably 3000 chlamydospores per gram of soil. In the case of sunflowers attacked by *Sclerotinia sclerotiorum*, a fungus with relatively large sclerotia and few sclerotia per soil volume, plant-to-plant spread is a factor. Hoes and Huang (1985) studied sunflowers in rows of varying width and varying plant spacings within the row. They found that the final percentage of diseased plants was the same regardless of between-row and within-row spacing and that best yields were attained with the same number of plants per hactare in infested soil as in noninfested soil. In close spacings, spread from one plant to its neighbor occurred. In wide spacings the greater root system of larger individual plants

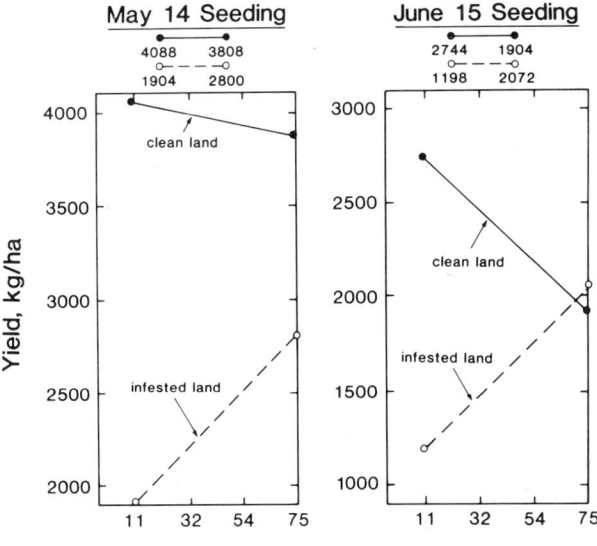

Figure 22.2. On noninfested land, Red Mexican beans yield best at low stand densities. On land heavily infested with *Fusarium solani* f. sp. *phaseoli* yields increase with increased stand density. After Burke and Nelson, 1965.

explored more soil, so that in the end the percentage of diseased plants was the same. The vigorous plants from wider spacings were more resistant, so that disease developed more slowly in them. The net result of the interactions was a balance of responses making the best agronomic plantings the best for yield on clean or infested soil.

Rootstocks

Horticultural plants are often complex, with the roots and stems of grafted plants being from different plant cultivars or species. When important soil-borne pathogens are present, the choice of the rootstock is important. The English walnut is highly susceptible to *Armillaria mellea* in California, and it is grown on resistant native California walnut root stock, with the graft placed well above the soil surface. Collar rot (*Phytophthora cactorum*) of apple became much more serious when Malling and Merton-Malling rootstocks were widely used to produce dwarf trees. Collar rot occurs at the soil surface where the main branch roots join the trunk. Cowling (1978), in describing the increase of collar rot of apples, stated that many seedlings are certified as disease-free, even though they are infected but symptomless at the time of distribution. He also stated that the use of tractor-mounted augers to dig transplant holes increases the danger. The auger exerts pressure on the sides of the hole, forming a water-retaining cylinder in the soil. The soil settles in the hole, forming a depression that favors water retention at the collar region.

IRRIGATION

Diseases favored by water stress, such as charcoal rot of sorghum, can be avoided by irrigation. Proper timing of irrigation can minimize potato scab. Some amelioration of Verticillium wilt in the southwestern United States and northern Mexico is accomplished by growing cotton on ridges and irrigating with reduced frequency, maintaining higher upper tap root and stem temperatures. Irrigation will not be treated here in these respects, because chapters are devoted to host physiology (Chapter 5) and to water (Chapter 2).

Soil Ridges

One of the most interesting papers on early irrigation methods describes irrigating chili (hot) peppers in New Mexico. The chili pepper had been grown for many years in the warmer valleys with irrigation and, so far as is known, the plant had no serious disease problems. According to Garcia (1933) a chili blight became serious after 1908. The cause, *Phytophthora capsici*, was not known when Garcia wrote his paper.

The blight developed severely when land was flood-irrigated in level culture. It was worst on heavy clay soils or in depressions where water stood (Fig 22.3). In the native ridge system ridges 20 to 30 cm high were made. After the ridge was made, water was applied in the ditch by the ridge, and the pepper plants were set at the water line on the ridge after irrigation. In this system water reached the plants with each irrigation. Thus in both level and native ridge culture zoospores (and very wet soil) could be in contact with the lower part of the plants.

In a new system plants were set on level land after irrigation. Then soil was progressively placed around the base of the stem during cultivation as the plants grew. *P. capsici* is not active until high temperatures exist, so that by the time temperatures were favorable the stems were isolated from the irrigation water. A severe disease was controlled before its cause was known.

Pathogens in Water

Collar rot was spreading in apple orchards in British Columbia, and McIntosh (1966) suspected that *Phytophthora cactorum* was introduced into the orchards in irrigation water. McIntosh suspended green pear fruit in perforated tin cans in water in storage dams and in the ditches carrying water to the orchards. *Phytophthora* spp. were recovered from 27 of the 31 irrigation

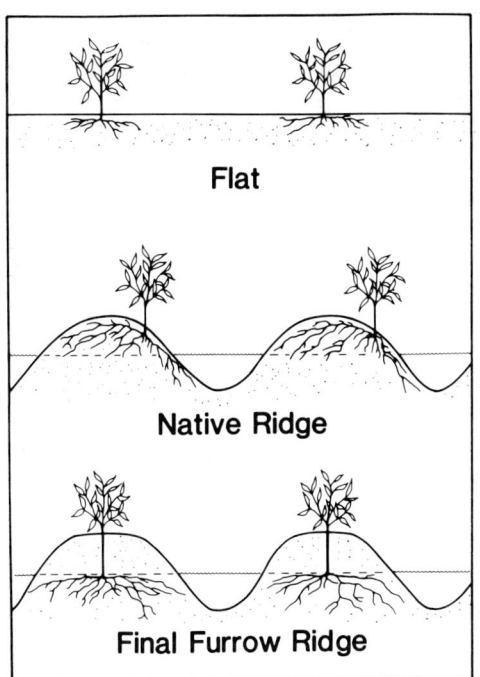

Figure 22.3. Ingenious use of ridging controlled *Phytophthora capsici* on Chile peppers before other control measures were available and before the cause of the disease was known. The straight horizontal line represents the original soil line. In the flood and native ridge systems, water contacts the stems. In the gradual, delayed ridging system (bottom) water does not contact stems when temperatures favor disease. After Garcia, 1933.

sources sampled. *P. cactorum* was obtained from 15, *P. megasperma* from 4, and *P. cambivora* and *P. citricola* from 1 each. The water was not return water but fresh, clear, mountain-stream water that had not been used, so these fungi entered the water from sources other than orchards. In British Columbia collar rot is important only in irrigated orchards, and McIntosh considers water an important means of introducing the pathogen.

The Columbia Basin Irrigation Project in central Washington brought virgin land into cultivation. Pathogens of many types spread with unusual rapidity, some of them unexplainably. Faulkner and Bolander (1966) suspected that nematodes were spreading in the irrigation water. In furrow irrigation, wastewater from the lower end of the field is returned to the main canal from which it will enter a field down stream. Samples of water obtained near the lower end of the canal contained *Ditylenchus, Paratylenchus, Pratylenchus, Tylenchorhyncus,* and *Hemicycliophora.* The water current and nematode buoyancy were such that there was no evidence that the nematodes settled from the main body of moving water. Each irrigation added about 4 to 10 million plant-parasitic nematodes per acre in the lower parts of the system.

Producing vegetable and ornamental transplants for other areas in the United States and Canada is an important industry in southern Georgia. The land has been routinely treated with methyl bromide or methyl bormide + chloropicrin before seeding, but infestation of the treated beds with several pathogens is a problem. Many growers use water from farm ponds to irrigate the beds. Water comes to the ponds by gravity from surrounding slopes and is then static in the pond. Shokes and Carter (1979) studied plant pathogens in pond water and the sediment beneath the water. Most of the pathogens were in the sediment. *Pythium irregulare* and *P. aphanidermatum* were isolated from the water. *Rhizoctonia* spp., *Macrophomina phaseolina, Hoplolaimus* sp. *Tylenchorhynchus* sp., and *Macroposthonia* sp. were found in sediments. The occurrence of *P. irregulare* and *P. aphanideratum* in pond water is particularly dangerous to seedlings. Well water is safe. If pond water is used, the inlet pipe should be well above the bottom so that sediment is not obtained.

Jenkins and Averre (1983) point out that some proponents of hydroponic culture of plants list freedom from soilborne pathogens as one of the advantages, but several hydroponic failures in North Carolina resulted from root and stem diseases. They introduced pathogens of tomato into the recirculating nutrient solution. All plants were diseased 9 days after introduction of *Pythium aphanidermatum,* 10 days for *Pseudomonas solanacearum,* 14 days for *Pythium debaryanum,* 18 days for *P. myriotylum,* and 22 days for *Colletotrichum coccodes.* They also inoculated a single plant with the above pathogens and after 4 days washed all excess inoculum off; the inoculated plant was introduced along with five healthy plants into a system with circulating nutrient solutions. The results were similar to those in which inoculum was added to the nutrient solutions, proving that progagules are produced by and liberated from the inoculated plants.

Root-to-root spread has been proved for several soilborne pathogens, but the impressive thing about these experiments in hydroponics is the ready movement of propagules in water and their ingress into host plants with minimal physical injury to the roots. *Pythium* spp. and *Colletotrichum* should have no need for wounds, but even *Pseudomonas solanacearum* is devastating when an infected plant is introduced among healthy plants. Kelman and Sequeira (1965) had demonstrated root-to-root spread by this bacterium. Jenkins and Averre state that *P. solanacearum* seldom occurs in hydroponic systems, but when it does, 100% loss results.

Constant Wetness

Investors bought cheap, sandy land not suitable for conventional agriculture in parts of south central Washington and adjacent Oregon and planted it to winter wheat. *Gaeumannomyces graminis* var. *tritici*, present in the native vegetation, increased. The soil pH was neutral or above, which is favorable to the pathogen. The soil has little capacity to store water so irrigations were made daily to keep the surface soil moist. Take-all was devastating. Several pathologic problems essentially unknown in this climate of arid summers have resulted from use of circle irrigation systems. When the pivot revolves continuously the central portion of the irrigated area is in an eternal mist.

FERTILIZERS

Infectious diseases usually do not occur every year, but chronic nutrient deficiencies have predictable, reliable consequences, and they should be corrected. Nutrient deficiencies and excesses cause diseases by themselves. Excess nitrogen has harmful effects of its own, beside being a waste and source of pollution of groundwater. Some farmers overfertilized wheat in Washington when anhydrous ammonia was cheap. *Fusarium culmorum* became serious in some fields. The main control was to quit overfertilizing in the dryland region (Cook and Papendick, 1970), a practice which should not have been done anyway. *F. culmorum* becomes virulent when the host is weakened by water stress, and too much nitrogen results in excessive growth and increased water stress.

Nutrient × infectious disease interactions usually first become apparent when the nutrient deficiency is an incipient, clinically undiagnosable, threshold deficiency. A plant with functions impaired by root pathogens expresses the otherwise yet invisible deficiency in an accentuated form. The phosphorus deficient spring wheat in Saskatchewan may have gone unnoticed for several years if it had not been for the interaction with *Pythium* spp. The new semi-dwarf wheats, if not affected by *f. culmorum,* could have withstood overfertilization with little damage, and this uneconomic practice could have continued longer than it did.

In most situations, the farmer should follow best fertility practices, fertilizing for optimum use of cultivars, rain, etc., with little or no consideration of infectious diseases in these decisions. Fertilize to grow a crop, not to control disease, in 90% or more of the situations. Mildew may be worse on vigorous vegetation, but malnutrition is an expensive control at best.

Take-all, caused by *Gaeumannomyces graminis* var. *tritici*, the subject of a book edited by Asher and Shipton (1981), is favored in general by loose, well-aerated neutral or alkaline soils, wet soil, and malnutrition of the host. Take-all was first described and studied in Australia (Garrett, 1970), where it was so destructive on certain soils it received its common name. Rosen and Elliott (1923) found a small field near Fayetteville, Arkansas, severely affected by the "new" disease. They applied farmyard manure to part of the field, limed part, and left part as a control. They came to three conclusions: the disease is favored by malnutrition; lime increases severity of the disease; and the disease is not new to North America.

Why give take-all special treatment under fertilizers? In some cases you should not just fertilize to grow a crop. Garrett, through many observations and experiments over many years, established that low nitrogen during the period not in wheat is beneficial. The take-all fungus survives as a saprophyte, dependent on growth in the low nitrogen, high carbon residue of diseased wheat roots and crowns. It requires nitrogen for active saprophytic growth in high carbon residues. Garrett developed the concept to, "starve" the fungus in the saprophytic phase (low nitrogen), which is particularly easy in wheat-summer fallow systems, and feed the wheat plant. In other words, fertilize just before planting. Lack of adequate nitrogen between crops weakens the inoculum. Fertilizing the crop supports normal host development and promotes production of many roots.

Form of Nitrogen

Huber et al. (1968) fertilized winter wheat under irrigation in southern Idaho with ammonium nitrate and with ammonium sulfate. Ammonium sulfate reduced take-all much more than ammonium nitrate at all rates of application. They concluded that the form of nitrogen is significant, not just the amount. Fewer roots are infected when the sulfate rather than the nitrate form of ammonium fertilizer is used.

Huber (1972) studied spring wheat fertilized in autumn and in spring. Ammonium sulfate applied in the autumn reduces the yield 31%, but when applied in the spring, it increases the yield 51%. Huber attributed the deleterious effect of autumn-applied ammonium sulfate to conversion of ammonium to nitrate in the warm soil. When take-all is not a factor, response to either ammonium- or nitrate-nitrogen is generally equivalent.

The suppression of take-all by ammonium is attributed to lowered pH in the rhizosphere, and the increase of take-all by nitrate to increase of the pH in the rhizosphere (Smiley and Cook, 1973; Smiley, 1974). Plants maintain a balance of cations and anions within the plant with that in the growth medium. When NH_4^+ is absorbed, H^+ is exchanged with the growth medium, increasing acidity at the root surface. A plant nourished by ammonium nitrogen maintains a rhizosphere pH measurably lower than that of the bulk soil pH, by 0.5 to 1.0 pH units. *G. graminis* is favored by alkalinity. NO_3^- nitrogen has the opposite effect. Smiley (1978) found that the numbers of bacteria and streptomycetes in the rhizosphere and rhizoplane are not altered by form of nitrogen, but that ammonium-nitrogen favors antagonists. Huber and Watson (1974) reviewed the literature on the effect of form of nitrogen on plant diseases and found many examples of important effects.

Nitrogen Affects Saprophytic Survival

Fertilizing forest trees increases timber production and is a commercial practice in several regions of the world, but fertilizers can have other effects. Forest soils are usually low in nitrogen, and increased nitrogen could speed decomposition of dead roots within soil. Nelson (1970) buried wood cubes of Douglas fir naturally infested with *Poria weirii*. When the nitrogen sources were mixed uniformly within soil and the blocks incubated at 15°C for 6 months in moist soil, the results were striking. At 6 months no block of 30 under test with NH_4Cl had zone lines, and no *P. weirii* was recovered. In the controls, 15 blocks had zone lines and *P. weirii* was isolated from 14 blocks. *P. weiri* is less severe where red alder, a nitrogen-fixing tree, forms part of the forest composition. It is possible that nitrogen, released deep in the soil around alder roots, plays a role in nature in reducing the survival of *P. weirii* in dead roots.

CHEMICALS

Herbicides, nematicides, insecticides and fungicides used to control pests are increasingly powerful and selective. In spite of increasing selectivity, these biologically active materials have effects on nontarget organisms. These relationships have been reviewed by Katan and Eshel (1973), Altman and Campbell (1977), Papavizas and Lewis (1979), Rodriguez-Kabana and Curl (1980), and Griffiths (1981). These reviews are recommended for more extensive treatment of this important subject.

Pentachloronitrobenzene (PCNB) is widely used to control *Rhizoctonia solani* and *Sclerotium* spp., and even though a fungicide, it reduces reproduction of *Longidorus elongatus* and *Xiphinema diversicaudatum*. Murant and Taylor (1965) found PCNB useful in reducing the spread of nematode-transmitted viruses in strawberry and raspberry.

Griffiths (1981) lists changes in host structure and composition from hormone type compounds, increased leakage of metabolites by roots and hypocotyls following soil-applied herbicides, and reduction of wound-healing in potato tubers treated wtih the sprout-inhibitor, napthalene acetic acid (Cunningham, 1953). The possible interactions brought about by effects on the plant, the pathogen, other pathogens, and antagonists are extensive. The pathologist should be aware of the chemicals used on a crop and be alert for unexpected results.

Beneficial Side Effects

Aphanomyces Root Rot Hopen observed that trifluralin herbicide reduced pea root rot (*Aphanomyces euteiches*) in a growth chamber, and Carlson and Hopen (1971) observed disease reduction in the field. Trifluralin is incorporated preplant into the surface soil, and 1 kg/ha is equivalent to about 0.5 μg/g of soil. The only practical limit to Aphanomyces root rot of green peas in Wisconsin is to avoid heavily infested fields; rotation, chemicals, and breeding have failed to adequately control the disease (Harvey et al., 1975). Trifluralin and dinoseb are among the common herbicides used in processing peas, and 60% of the pea acreage was treated with herbicides in 1972. If these herbicides had a disease-controlling effect it would constitute a "free" disease control. In 1972 and 1973 trifluralin was incorporated in a heavily infested silt loam about 6 days prior to seeding. By comparing the yields with those obtained with other herbicides giving equal weed control, they concluded that trifluralin gave a 29% increase in yield not attributable to weed control and that this increase resulted from root rot control.

Dinitramine and trifluralin both interfere with formation of motile zoospores by *Aphanomyces euteiches* (Grau, 1977). The motility of zoospores produced in the absence of these herbicides is reduced when the zoospores are placed in their presence (Grau, 1977). Zoospores produced in the absence of these chemicals germinate on media containing them and germ tube elongation is normal, so the chemicals are not lethal to the zoospores. Plants inoculated with zoospores in the presence of herbicide are less diseased than in the absence of the herbicide. Trifluralin and dinoseb reduce the growth of *A. euteiches* in culture. Trifluralin reduces motility of zoospores, but it stimulates the production of oospores in vitro. Dinoseb does not stimulate oospore formation in culture, and it decreases the production of primary zoospores, but it does not prevent their subsequent motility, so the action of the two herbicides is not identical. Dinoseb is recommended where grassy annual weeds are anticipated.

In contrast to the beneficial effect of this herbicide on pea root rot, trifluralin increases the severity of *Phytophthora megasperma* f. sp *glycinae* on susceptible soybeans (Duncan and Paxton, 1981). Trifluralin is used on

60% of the soybean acreage in Illinois and the authors emphasize the importance of resistant cultivars under these conditions.

Growth Regulator Effects Treatment of soybeans with TIBA (2,3,5-triiodobenzoic acid) reduces charcoal rot (*Macrophomina phaseolina*) (Oswald and Wyllie, 1973). Kroll and Moore (1981) confirmed this observation and related the reduction to altered host anatomy. If properly applied, the growth regulator reduces lodging and plant height. It is most effective when applied before 10% of the flowers have emerged. Two cultivars were sprayed, and the early cultivar, Wayne, was past the stage of optimum response, in that 60% of the flowers had emerged. The vessel pattern in Amsoy 71 soybean stems was altered, in that new vessels were not formed during mid-season. Vessels formed early and late. In untreated Amsoy 71, vessels were formed continuously. The vessel pattern in Wayne, which was treated too late, was unaltered. Microsclerotia were produced from the soil surface up 33% of the stem in untreated Amsoy 71 and up 18% of the stem length of treated Amsoy 71. In Wayne there was no significant difference in the length of the colonized portion of the stem, treated or untreated. In this example a growth regulator changes host anatomy in such a way as to restrict development of the pathogen with no indication of complex interactions. The authors believe that in the normal vessel pattern the fungus is relatively unimpeded within the xylem, but that in the altered patterns its growth within the stem is restricted. The chemical has little toxicity toward the fungus.

Harmful Side Effects

Rhizoctonia solani Aldicarb, a systemic insecticide, increases damping-off of sugarbeet seedlings caused by *Rhizoctonia solani* in Colorado (Tisserat et al., 1977). Aldicarb at 4 μg/g of soil has no apparent effect on seedling survival in the absence of *R. solani*, but when the pathogen is present, it reduces survival. Aldicarb at 8 μg per gram of soil increases colonization of bean hypocotyls buried in soil from 35% in the checks to 62% with aldicarb. At 16 μg/g of soil, aldicarb is somewhat phytotoxic. This concentration can be reached in banded treatments in commercial practice.

The green peach aphid, *Myzus persicae*, and the Colorado potato beetle, *Leptinotarsa decemlineata*, can be economically controlled by systemic insecticides applied at planting. Maine potato growers began to change from disulfoton to aldicarb in 1975 because aldicarb was more effective against aphids. Following this change, *Rhizoctonia solani* became serious in some fields. Leach and Frank (1982) applied aldicarb and carbofuran, effective against some nematodes and insects, and disulfoton, an insecticide, as a granular in-row preplant or banded with fertilizer at seeding. Aldicarb did not reduce yield, but it reduced the percent of marketable potatoes from

76% in the control to 35% with aldicarb. Leach and Frank were unable to determine the cause of the disease increase. They had no evidence for phytotoxicity or of any direct effect on the pathogen or on the soil microflora.

In the coastal plain region of Georgia, snap beans can be grown from spring, summer, and autumn plantings. Nematicides and insecticides are commonly used in combinations on this crop. Sumner (1974) observed that these treatments influenced root diseases in the field, and he therefore studied the use of these chemicals alone and in combination in greenhouse trials. Ethoprop nematicide and DCPA (dimethyl tetrachloroteterephalate) herbicide alone caused slight reductions in early bean development. Root rot caused by *Pythium myriotylum* was not influenced by applications of pesticides. DCPA increased the severity of root rots caused by a mixture of pathogens and by *R. solani* alone. Trifluralin and dinoseb plus ethoprop in combination (T + D + E) led to increased severity of disease from *F. roseum*. Maximum use of chemicals can obviously cause situations that cannot be explained by a field pathologist.

A Perfect Canopy The narrative (under tillage) on cultural control of *Sclerotium rolfsii* in peanuts stresses prevention of leaf spots that cause blighted, abscissed leaflets to accumulate about the base of the peanut plant which could serve as food for *S. rolfsii*. Fungicides now give essentially perfect control of leaf spots (*Cercospora arachidicola* and *Cercosporidium personatum*) (Backman et al. 1975). *S. rolfsii* did not respond to leaf spot control as predicted. All fungicides resulted in small increases in dead plants. The prevention of leaf fall did not decrease white mold of peanuts. Even though the fungicide trials were made during three growing seasons, all three seasons were drier than normal. Backman et al. attributed most of the increase of *S. rolfsii* by leaf spot control to the perfect canopy that shaded the soil surface and increased humidity over that in the unsprayed controls in these dry seasons. The paper did not report cultural practices, but they must have been excellent, because the highest incidence of plants killed by white rot in any treatment in 3 years was 6%. Results in a wet season could be quite different. In tillage studies reported by Garren (1959) 76% of the plants in the controls were diseased.

COMMENTS

In the introduction I stressed that the cultivation of a single plant over large areas facilitates observations on the effects of environment on disease development. This may not be true, however, if a supersusceptible (or super-resistant) cultivar occupies large acreages. For many years flag smut of wheat (*Urocystis agropyri*) was confined to a single county in Washington, held there by unknown environmental factors. Then, after a succession of highly susceptible wheats, flag smut appeared essentially everywhere in eastern Washington. For many years Cephalosporium stripe had a rather definite distribution in Washington. It was favored by wet soils with poor internal drainage. Then, after Stephens, a very susceptible cultivar, became widely grown, Cephalosporium stripe became serious in a wider geographic area. Either extreme susceptibility or resistance can obscure environmental relationships.

Monoculture seems to be increasing at an alarming rate. Reduction of disease with monoculture (Chapter 18) is an amazing biological phenomenon, but how many diseases will respond in this way? Experiment stations should return to long-term rotation experiments. They would serve not only for determining the effects of rotation on diseases, but also to aid in the discovery of more disease-suppressing agents.

The discussion of irrigation in this chapter is sobering. Apparently only well water is free of plant pathogens.

Epidemiology is complicated by the diverse, unpredictable effects of pesticides on nontarget organisms. New farm chemicals are being constantly developed, and old ones find new applications. It is too much to expect experiments to reveal in advance, before wide farm use, all the effects these chemicals may have. Pathologists will have to remain alert for unusual consequences.

At one time I advised farmers to fertilize to grow a crop, to add so much nitrogen, according to total need, not knowing that the form of nitrogen can modify disease reactions. Modern, technical agriculture demands much of the applied pathologist.

KEY REFERENCES

Doupnik, B., Jr. and M. G. Boosalis. 1980. Ecofallow—a reduced tillage system and plant diseases. *Plant Dis.* **1**:31–35.

Faulkner, L. R. and W. J. Bolander. 1966. Occurrence of large nematode populations in irrigation canals of south central Washington. *Nematologica* **12**:591–600.

Grau, C. R. 1977. Effect of dinitramine and trifluralin on growth, reproduction, and infectivity of *Aphanomyces euteiches*. *Phytopathology* **67**:551–556.

Huber, D. M. 1972. Spring versus fall nitrogen fertilization and take-all of spring wheat. *Phytopathology* **62**:434–436.

Kroll, T. K. and L. D. Moore. 1981. Effect of 2,35-triiodobenzoic acid on the susceptibility of soybeans to *Macrophomina phaseolina*. *Plant Dis.* **65**:483–485.

Smiley, R. W. 1978. Colonization of wheat roots by *Gaeumannomyces graminis* inhibited by specific soil microorganisms and ammonium-nitrogen. *Soil Biol. Biochem.* **10**:175–179.

Triantaphyllou, A. C. 1975. Genetic structure of races of *Heterodera glycines* and inheritance and ability to reproduce on resistant soybeans. *J. Nematology* **7**:356–364.

REFERENCES

Allmaras, R. R., K. Ward, C. L. Douglas, Jr., and L. G. Ekin. 1982. Long-term cultivation effects on hydraulic properties of a Walla Walla silt loam. *Soil Tillage Res.* **2**:265–279.
Altman, J. and C. L. Campbell. 1977. Effect of herbicides on plant diseases. *Ann. Rev. Phytopathol.* **15**:361–385.
Asher, M. J. C. and and P. J. Shipton. 1981. *Biology and Control of Take-all.* Academic Press, New York.
Aycock, R. 1966. Stem rot and other diseases caused by *Sclerotium rolfsii*. *N. C. Agric. Exp. Sta. Tech. Bull.* 174.
Backman, P. A., R. Rodriguez-Kabana, and J. C. Williams. 1975. The effect of peanut leaf spot fungicides on the nontarget pathogen, *Sclerotium rolfsii*. *Phytopathology* **65**:773–776.
Boyle, L. W. 1956,. Fundamental concepts in the development of control measures for southern blight and root rot of peanuts. *Plant Dis. Reptr.* **40**:661–665.
Bruehl, G. W. and B. Cunfer. 1971. Physiologic and environmental factors that affect the severity of snow mold of wheat. *Phytopahology* **61**:792–799.
Burke, D. W. and C. E. Nelson. 1965. Effects of row and plant spacings on yields of dry beans in *Fusarium*-infested and noninfested fields. *Wash. Agric. Expt. Sta. Bull.* 664.
Carlson, W. C. and H. J. Hopen. 1971. The effects of six herbicides on root rots in seedling vegetable legumes. *Proc. North Cent. Weed Control Conf.* **25**:79.
Cook, R. J., M. G. Bousalis, and B. Doupnik. 1978. Influence of crop residues on plant diseases. Pages 147–163 in *Crop Residue Management Systems.* Am. Soc. Agron. Spec. Publ. 31.
Cook, R. J. and R. I. Papendick. 1970. Soil water potential as a factor in the ecology of *Fusarium roseum* f sp. *cerealis* 'Culmorum'. *Plant Soil* **32**:131–145.
Cowling, E. B. 1978. Agricultural and forest practices that favor epidemics. Vol. II. Pages 361–381 in *Plant Disease: An Advanced Treatise.* J. G. Horsfall and E. B. Cowling, eds. Academic Press, New York.
Cunningham, H. S. 1953. A histological study of the influence of sprout inhibitors on *Fusarium* infection of potato tubers. *Phytopathology* **43**:95–98.
Curl, E. A. 1963. Control of plant diseases by crop rotation. *Bot. Rev.* **29**:413–479.
Duncan, D. R. and J. D. Paxton. 1981. Trifluralin enhancement of Phytophthora root rot of soybean. *Plant Dis.* **65**:435–546.
Dunleavy, J. M. and C. R. Weber. 1967. Control of brown stem rot of soybeans with corn-soybean rotations. *Phytopathology* **57**:114–117.
Garcia, F. 1933. Reduction of chile wilt by cultural methods. *N. Mex. Agric. Exp. Sta. Bull.* 216.
Garren, K. H. 1959. The stem rot of peanuts and its control. *Agric. Exp. Sta. Bull.* 144.
Garrett, S. D. 1970. Pathogenic Root-Infecting Fungi. Cambridge Univ. Press, London, UK
Griffiths, E. 1981. Iatrogenic plant diseases. *Ann. Rev. Phytopathol.* **19**:69–82.
Harvey, R. G., D. J. Hagedorn, and R. L. De Loughery. 1975. Influence of herbicides on root rot in processing peas. *Crop Sci.* **15**:67–71.
Hoes, J. A. and H. C. Huang. 1985. Effect of between-row and within-row spacings on

development of sclerotinia wilt and yield of sunflower. *Can. J. Plant Pathol.* **7**:98–102.
Hoffmann, J. A. and L. H. Purdy. 1967. Effect of stage of development of winter wheat on infection by *Tilletia controversa*. *Phytopathology* **57**:410–413.
Huber, D. M., C. G. Painter, H. C. McKay, and D. L. Peterson. 1968. Effect of nitrogen fertilization on take-all of winter wheat. *Phytopathology* **58**:1470–1472.
Huber, D. M. and R. D. Watson. 1970. Effect of organic amendment on soilborne plant pathogens. *Phytopathology* **60**:22–26.
Ishibashi, N. 1969. Studies on the propagation of the root-knot nematode, *Meloidogyne incognita* (Kofoid and White) Chitwood, 1949. *Rev. Plant Protec. Res.* **2**:125–129.
Jenkins, S. F. Jr. and C. W. Averre. 1983. Root diseases of vegetables in hydroponic culture systems in North Carolina greenhouses. *Plant Dis.* **67**:968–970.
Katan, J. and Y. Eshel. 1973. Interactions between herbicides and plant pathogens. *Residue Rev.* **45**:145–177.
Kelman, A. and L. Sequeira. 1965. Root-to-root spread of *Pseudomonas solanacearum*. *Phytopathology* **55**:304–309.
Kimpinski, J. and C. B. Willis. 1980. Influence of crops in the field on numbers of root lesion and stunt nematodes. *Can. J. Plant Pathol.* **2**:33–36.
Komm, D. A., J. J. Reilly, and A. P. Elliott. 1983. Epidemiology of a tobacco cyst nematode (*Globodera solanacearum*) in Virginia. *Plant Dis.* **67**:1249–1251.
Leach, S. S. and J. A. Frank. 1982. Influence of three systemic insecticides on Verticillium wilt and Rhizoctonia disease complex of potato. *Plant Dis.* **66**:1180–1182.
Luke, H. H., P. L. Pfahler and R. D. Barnett. 1983. Control of *Septoria nodorum* on wheat with crop rotations and seed treatments. *Plant Dis.* **67**:949–951.
Mai, W. F. and G. S. Abawi. 1980. Influence of crop rotation on spread and density of *Heterodera schachtii* on a commercial vegetable farm in New York. *Plant Dis.* **64**:302–305.
McIntosh, D. L. 1966. The occurrence of *Phytophthora* spp. in irrigation systems in British Columbia. *Can. J. Bot.* **44**:1591–1596.
Meiners, J. P., E. L. Kendrick, and C. S. Holton. 1956. Depth of seeding as a factor in the incidence of dwarf bunt and its possible relationship to spore germination on or near the soil surface. *Plant Dis. Reptr.* **40**:242–243.
Murant, A. F. and C. E. Taylor. 1965. Treatment of soil with chemicals to prevent transmission of tomato black ring and raspberry ringspot viruses of *Longidorus elongatus* (de Man). *Ann. Appl. Biol.* **55**:227–237.
Nelson, E. 1970. Effects of nitrogen fertilizer on survival of *Poria weirii* and populations of soil fungi and aerobic actinomycetes. *Northwest Sci.* **44**:102–106.
Netscher, C. and D. P. Taylor. 1979. Physiologic variation within the genus *Meloidogyne* and its implications on integrated control. Pages 269–293 in *Root-Knot Nematodes*. F. Lamberti and C. E. Taylor, eds. Academic Press, New York.
Oswald, T. H. and T. D. Wyllie. 1973. Effects of growth regulator treatments on severity of charcoal rot diseases of soybean. *Plant Dis. Reptr.* **57**:789–792.
Papavizas, G. C. and J. A. Lewis. 1979. Side-effects of pesticides on soilborne plant pathogens. Pages 483–505 in *Soil-borne Plant Pathogens*. B. Schippers and W. Gams, eds. Academic Press, London.

Roderiguez-Kabana, R. and E. A. Curl. 1980. Nontarget effects of pesticides on soilborne pathogens and disease. *Ann. Rev. Phytopathol.* **18**:311–332.

Rosen, H. R. and J. A. Elliott. 1923. Pathogenicity of *Ophiobolus cariceti* in its relationship to weakened plants. *J. Agric. Res.* **25**:351–358.

Russell, R. S. 1977. *Plant Root Systems: Their Function and Interaction with the Soil.* McGraw-Hill, London.

Shipton, P. J. 1977. Monoculture and soilborne plant pathogens. *Ann. Rev. Phytopathol.* **15**:387–407.

Shokes, F. M. and S. M. Carter. 1979. Occurrence, dissemination, and survival of plant pathogens in surface irrigation ponds in southern Georgia. *Phytopathology* **69**:510–516.

Smiley, R. W. 1974. Take-all of wheat as influenced by organic amendments and nitrogen fertilizers. *Phytopathology* **64**:822–825.

Smiley, R. W. and R. J. Cook. 1973. Relationship between take-all of wheat and rhizosphere pH in soils fertilized with ammonium vs. nitrate-nitrogen. *Phytopathology* **63**:882–890.

Summer, D. R. 1974. Interaction of herbicides and nematicides with root diseases of snap bean and Southern pea. *Phytopathology* **64**:1353–1358.

Summer, D. R., B. Dupnik, Jr., and M. G. Boosalis. 1981. Effects of reduced tillage and multiple cropping on plant diseases. *Ann. Rev. Phytopathol.* **19**:167–187.

Tisserat, N., J. Altman and C. L. Campbell. 1977. Pesticide plant disease interactions: the influence of aldicarb on growth of *Rhizoctonia solani* and damping-off of sugar beet seedlings. *Phytopathology* **67**:791–793.

Toussoun, T. A. 1975. Fusarium-suppressive soils. Pages 145–151 in *Biology and Control of Soilborne Plant Pathogens.* American Phytopathology Society, St. Paul, Minn.

Young, L. D. 1984. Effects of continuous culture of resistant soybean cultivars on soybean cyst nematode reproduction. *Plant Dis.* **68**:237–239.

Chapter 23

Numbers

Yarwood (1956) presented a diagram comparing the sizes and rates of reproduction by life forms. The conclusion was obvious. Viruses multiply fast, bacteria less fast, etc., past cats, dogs, cows, to whales and trees. In general, the larger and more complex the organism, the lower the population density per unit area of the earth and the slower the reproduction rate. A large organism requires more food, and thus a greater area in which to forage. Chuang and Ko (1981) tested this generality with a large assemblage of propagules of fungi and found that, using the size of the propagule, it was possible to predict the maximum population to be expected in a given volume of soil. *Aspergillus* and *Penicillium* spp., with small conidia, are often found in the range of many thousands per gram of soil. *Fusarium* chlamydospores are often present in thousands, microsclerotia of *Macrophomina phaseolina* and *Verticillium dahliae,* usually hundreds or less, and true sclerotia of *Sclerotinia* and *Sclerotium* spp. from several to fewer than one per gram of soil.

Allowing for variation and a degree of error, the findings of Chuang and Ko imply that a wide range of pathogens are essentially equal in obtaining food from host tissues and in the proportion of the food that is stored to provide for the future. The enzymes produced by a wide range of organisms attack different substrates, yet the end result has remarkable similarity. Their findings have other implications as well. Assuming the general validity of size versus numbers, the fact that different propagules contain different stored foods, such as lipids or glycogen, must not be of great significance.

Garrett (1956) addressed the size versus numbers relationship from a different perspective when he defined inoculum potential. "Inoculum potential may be defined as the energy of growth of a parasite available for infection of a host at the surface of the host organ to be infected. Inoculum potential may be increased in either or both of two ways: (a) by increase in the number of infecting units or propagules of the fungus per unit area of root surface... (b) by increase in the nutritional status of such units."

Bruehl (1976) discussed the adaptation of propagules in terms of energy requirements with the target. Microconidia of the vascular fusaria require little endogenous nutrients, because the xylem fluids are a suitable medium for growth (Kessler, 1966). Basidiospores and conidia of *Fomes annosus* can infect a freshly cut tree stump, but not unwounded tree trunks or large roots with bark. Infecting the latter requires a great energy reserve, as from an infected woody root. A weakened host may be invaded by a weaker than normal propagule, and the environment influences the success of the pathogen. A precise measure of inoculum potential may not be possible.

Some organisms produce different infective units, often in mixtures within the soil. How does this affect the inoculum potential? Ko and Chan (1974) tested the ability of chlamydospores, sporangia, and zoospores of *Phytophthora palmivora* to produce disease on papaya seedlings in natural soil. With equal numbers of propagules (24,000/ml), sporangia produced the most disease, chlamydospores second, and zoospores the least. A sporangium can produce 16 zoospores, so 16 zoospores minus the energy required for differentiation should be roughly equivalent to one sporangium. Chlamydospores have the mass of about 10 zoospores. Sporangia produced 98% seedling mortality, chlamydospores produced 49%, and motile zoospores produced 16% seedling blight. Sporangia (5000/ml) equivalent to the zoospores (83,000/ml) produced much more seedling blight. This lesser effectiveness of zoospores on a mass basis may be offset by their distribution and searching ability. It is not likely that evolution preserved functional motility if it were not advantageous to the species.

Baker (1978), in his review of inoculum potential stated, "The inoculum density of a given soil-borne pathogen usually is no greater than that needed for reasonably efficient positioning in the three-dimensional space to cover the available infection courts." This straightforward statement is questionable. Living forms, unless restrained, reproduce in excess.

CAUTIONS

Enumeration of propagules by dilution plate counts can not reveal endogenously dormant propagules or distinguish avirulent from virulent isolates of a species. If microsclerotia of *Verticillium dahliae* in soil bear external conidia, unrealistic counts of "microsclerotia" may be obtained by dilution plating. Menzies (1963) reviewed methods of direct assay of plant pathogen populations in soil, and his paper is recommended. Menzies also cautioned that propagules in soil may not be accurately counted if many of them are still contained within undecomposed host fragments.

Biologically competent means being capable of infecting if located near enough to the root surface (Grogan et al., 1980). Being able to germinate and grow on nutrient media is not necessarily equivalent to competency. Newly formed sclerotia of *Sclerotinia minor* can germinate myceliogenically by few hyphae in soil, but they are incapable of infecting a host without an exogenous source of nutrition, such as a dead leaf. After aging and drying, they can germinate eruptively (many hyphae emerge simultaneously from a portion of the sclerotium), and after germinating in this manner, in which stored food reserves are exhaustively used, exogenous food is unnecessary. Sclerotia of the same species and of the same size are therefore not equal. The younger (and undried) sclerotia have at least as much endogenous energy as the older, conditioned sclerotia that germinate eruptively, but physiologic changes within them make them unequal.

Grogan et al. continued the competence theme to include competence distance—the maximum distance a propagule can lie from a root surface and still cause infection in the presence of the soil microflora and without exogenous food. They give 1 cm as the competence distance of sclerotia of *Sclerotinia minor*, 3 cm for *Sclerotium rolfsii*. If hyphae grow equally in all directions within soil from these sclerotia, the soil volume within which sclerotia of *S. minor* could function is 4.2 cm^3, and that of *S. rolfsii* is 113 cm^3. On a theoretical basis, 238 competent sclerotia of *S. minor* would ensure infection of any susceptible structure within 1000 cm^3 of soil, and 8.8 sclerotia of *S. rolfsii* would be adequate. Thus competence distances are magnified when transformed to soil volume by the formula $4/3\pi r^3$.

COUNTING

Use of an Infection Index

Before differential media were available for identifying *Verticillium dahliae* propagules in soil, Wilhelm (1951) grew tomato plants in infested soil for 3 weeks and expressed soil infectivity on the basis of Verticillium wilt in tomato plants. Wilhelm mixed a naturally infested soil with chloropicrin-treated compost mix in varying proportions and determined the relation between inoculum density and disease. When the proportion of naturally infested soil to chloropicrin-treated soil were 1:0, 1:5, 1:10, 1:25, 1:50, 1:100, 1:500, 1:1000, and 0:1, the disease indexes were 82, 59, 45, 22, 5, 2, 1, 0, and 0, respectively. This method does not reveal how many propagules are present in the soil, but it provides relative relationships based on disease production and is not confused by propagules incapable of inciting disease but capable of growing on laboratory media.

Enumerating Rhizoctonia solani

Sneh et al. (1966) compared baiting with plant segments, soil immersion tubes, isolating from infested debris from soil, and infection of seedlings for reliability in estimating the population of *Rhizoctonia solani* in naturally infested soil. The agreement between assessment with plant segments as bait, soil immersion tubes, or the number of infested debris particles in the soil was good ($r = 0.96$). Use of seedlings was less precise. They used stem pieces of 6-week-old bean and cotton plants, mature stems of wheat, and several other baits. Bean stems were most reliable. They screened debris from the soil as done by Boosalis and Scharen (1959), except they washed the debris very gently. Boosalis and Scharen and Davey and Papavizas (1962) obtained low recoveries from debris (under 10%), but Sneh et al, reached 50%. They suggest that very gentle washing avoids loss of some sclerotia.

Wet Sieving of Sclerotia

Sclerotia of *Sclerotium rolfsii* were quantitatively recovered from sugar beet fields by washing soil on a series of screens of graded pore size (Leach and Davey, 1938). About 80% of the sclerotia occurred in the top 15 cm, but a few were found 45 to 60 cm deep. Sclerotia were found as far as 50 cm from a diseased beet (most of the sclerotia were found within 10 to 13 cm of the beet). By not forming all the sclerotia on the diseased host, distribution within the soil was enhanced. In spite of this characteristic, sclerotium concentrations within the fields were usually erratic and nonuniform. Leach and Davey (1938) used 20 cm as the sampling depth. When 50% of the beets were infected in an area, up to 5000 sclerotia/ft^2 (.09 m^2) were found. Viability was determined by placing sclerotia without surface disinfection on natural, moist peat soil in Petri dishes and incubating them 5 days at 30°C. This simple method gave good results and was superior to the more laborious methods tried. More than 200 viable sclerotia/0.09m^2 resulted in 15% infection of sugar beets, fewer than 100 sclerotia/0.09 m^2 had infection less than 10%. Predictions based on sclerotial numbers are sufficiently accurate for use in determining that certain fields should not be planted to sugar beets until the infestation is reduced.

Direct Assay of Sclerotia

Previously, researchers wet-screened sclerotia of *S. rolfsii* from soil or used flotation plus screening, followed by tests of germinability. Rodriguez-Kabana et al. (1980) developed a one-step procedure. They air-dired the soil, sieved it through a 2-mm screen, and spread it uniformly over the surface of filter paper covering the bottom of a sieve. The sieve with the soil on it was placed in a dish with sufficient water or stimulating solution (1.33% v/v methanol in water) to attain 60% of field capacity. Germination in soil wet by water alone was 65%; when wet with 1.33% methanol, 100%. After the soil was moistened, the sieve with the soil was placed inside a desiccator and incubated 4 days at 20°C. A Petri dish with 5 g of BaO_2 and 10 ml of water in the bottom of the desiccator supplied O_2 and removed CO_2. The germinated sclerotia were easily seen and counted directly in or on the soil.

The thinnest soil layers yielded the highest percentage of germination. Germination with 25 g on the sieve did not differ significantly from 50 g of soil, so they used 50 g. The depth of soil at this rate was about 2 mm. Sclerotia of *S. rolfsii* germinate best on or near the soil surface, a fact usually attributed to its highly aerobic nature. Results of this one-step procedure correlated highly with the more labor-intensive two-step methods.

NUMBERS VERSUS DISEASE

Disease Incidence

Mitchell (1975) added oospores of *Pythium myriotylum* to autoclaved Arredondo fine sand to provide varying inoculum levels. Surface-disinfected rye seeds were pregerminated and planted in a band of sterile soil above the soil containing oospores. Every 48 hr the containers were flooded and allowed to drain. Half the containers contained rye seedlings, half did not. The number of viable propagules in the controls that lacked rye plants remained constant with no oospore germination, evidence that autoclaving the soil did not release enough food to break dormancy. The number of propagules in the containers with rye seedlings did not increase, evidence that germination was by germ tube only, with no zoospores being produced. Infection of rye seedlings with 1 oospore/g was 2%; 10/g, 34%; 25/g, 56%; 50/g, 71%; 100/g, 81%, and 150/g or higher, 100%. Mitchell calculated that 90 to 100% infection occurred when oospores were 1 mm apart in the soil, with about 30% infection when oospores were 5 mm apart, and near zero infected seedlings with oospores 1 cm apart. There was no evidence of synergistic action between propagules, meaning that individual spores incite infections alone (Figure 23.1).

Hani (1981) inoculated surface-disinfected wheat seed with logarithmic spore concentrations of conidia of *Fusarium nivale* per seed. Seedling blight increased in a linear fashion as the spore load increased logarithmetrically, to 10^4 spores per seed. Beyond that level increased inoculum accomplished little, because mortality approached 100%. Disease increased linearly with logarithmic inoculum level whether the seeds were incubated on paper or in soil. This implies a minimal effect for the soil microflora when the pathogen was present on the seed.

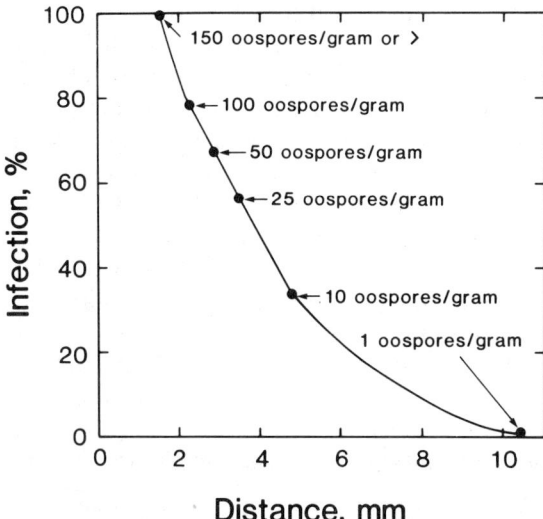

Figure 23.1. Relationship between number of diseased rye seedlings and calculated distance in millimeters between oospores of *Pythium myriotylum* in soil when zoospores are not produced. Disease not directly proportional to inoculum density. After Mitchell, 1975.

Inoculum Not the Limiting Factor

Pythium aphanidermatum is a vigorous high-temperature species. In Arizona it attacks sugar beets months after planting, not because seedlings or young plants are resistant, but because soil temperatures must reach 27°C or higher for the pathogen to become virulent (Stanghellini et al., 1983). When a large beet is infected, it rots rapidly. Beets must be dug at the proper time to count individual lesions on the swollen root. Stanghellini et al. removed roots from field soil and collected soil from the root-soil interface when the beets were 8 to 9 months of age. From one to five lesions were found per root with a surface area of 500 cm^2.

The population of oospores in the 1-mm layer of soil at the root-soil interface (scraped from the root imprint in the soil) was determined. The 1-mm depth into the root imprint was chosen because infectivity studies indicate that oospores within that distance from a target are effective. The beet during growth displaces much soil, yet inoculum concentration near its surface was the same as in bulk soil. Soil displacement apparently moves the soil mass nondifferentially, so that propagules per unit of soil remain the same.

The number of oospores within 1 mm of the beet-soil interface was enough to result in 170 lesions per sugar beet tap root. Only 1.6 lesions per root were observed. Inoculum was not the limiting factor, as about 1% of the lesions possible from available inoculum actually developed. These soil samples were also tested on fresh potato tuber slices. To approach 100%

inoculum efficiency required 48 hr at 27 to 34°C under water-saturated conditions. Maximum disease develops at 0 to −0.01 bar water potential. In irrigated beets these soil conditions were not maintained for sustained periods. The authors concluded that water, not inoculum, was the limiting factor.

Plant-to-Plant Spread

Scott (1956a, b) reported plant-to-plant spread of *Sclerotium cepivorum* and that hyphae from a colonized food base could grow a few centimeters through soil. When Crowe and Hall (1980) buried sclerotia 30 cm deep, roots of garlic plants were colonized and the fungus grew upward within and upon the surface of the roots. Symptoms appeared above ground about 12 weeks after emergence of the seedlings. Plant-to-plant spread was accomplished when roots of diseased and healthy plants intermingled. Greatest root-to-root spread occurred 2 to 4 cm below the stem plate, the level of greatest root density. Plant-to-plant spread was observable in soils with few sclerotia.

Radial growth of the fungus is significant because clusters of up to 20 infected plants develop from single infections by the end of the growing season. Because the disease loci develop over periods of time and tend to merge as they expand, the relationship between sclerotial numbers and numbers of diseased plants is lessened.

Effect of Host Response on Numbers

If the pathogen and the host both thrive, the host is highly tolerant. If the pathogen thrives but the host is severely damaged, the host is susceptible. If the host does well and the pathogen reproduces little or slowly, the host is resistant. If the pathogen and host both do poorly, the host is intolerant, being severely damaged by little pathogen so that neither thrives. If the pathogen cannot reproduce at all the host is immune. Rohde (1972) has diagrammed these relationships.

Unless the organism is acting as a vector or the pathogen becomes systemic, it is logical to assume a relationship between numbers of the organism in the soil and its effect on the host. Jones (1959) illustrated a general relationship between numbers of parasitic nematodes and damage to the host. A plant tolerates a certain population with no detectable economic loss. At some threshold density losses begin, and losses are proportional to increased numbers of the pathogen. Then, at some point the host no longer suffers in proportion to the increase in numbers of the pathogen. At some further point the host, if still alive, will not support further increase of the parasite, unless the parasite is capable of saprophytic growth on the dead plant.

Many types of disease resistance are relative, not absolute, and to differentiate useful levels of resistance requires that the test be neither too severe nor too light. This means the breeder must establish proper levels of inoculum. Only a few (1 to 5) *Ditylenchus dipsaci*/500 g of soil can cause damage to onions. From 1 to 5 *Pratylenchus penetrans*/500 g of soil can damage daffodils, but more than 500/500 g are required to damage potato (from table in Barker and Nusbaum, 1971). The threshold levels for measurable loss varies greatly with the host-pathogen combination. Barker and Imbriana (1984) give a table of numbers of several nematodes versus several crops required to reach damaging population thresholds.

Davis et al. (1983) assayed the degree of soil infestation in field soils and found a straight line increase in Verticillium wilt with pseudosclerotia per gram of dry soil from 0 to 100 ppg. In addition, inoculum production within stems of susceptible potatoes increased in a straight line, beginning 19 July and proceeding for 5 weeks. Susceptibility of potatoes to Verticillium wilt, judged by symptoms and yield, was proportional to the number of propagules of *V. dahliae* per gram of host tissue produced above the soil surface. The potatoes were resistant rather than tolerant because the amount of inoculum produced in them was proportional to susceptibility. Use of resistant cultivars should reduce the pathogen population in the soil. Butte (susceptible) produced more inoculum than Russet Burbank, and two numbered resistant lines produced little inoculum. When Davis et al. (1985) grew the potato lines on the same soil for 5 successive years, colony-forming units were 315/g of soil after Russet Burbank, 81/g after a highly resistant selection. The beneficial result of resistance on propagules per gram of soil was less than I would have anticipated. Some saprophytic development probably occurred.

Flowers and Hendrix (1972) estimated propagule numbers of *Phytophthora parasitica* var. *nicotianae* in the rhizosphere of tobacco. Propagules increased rapidly in the rhizospheres of susceptible plants, reaching a maximum when the plants died. In contrast, the fungus increased gradually under resistant cultivars, reaching a maximum at the end of the growing season. When a highly susceptible plant died at an early age, less inoculum was produced than on a moderately resistant plant that survived until it was killed by frost.

Multiplication was rapid on susceptible tobacco (cultivar A), relatively slow on a moderately resistant cultivar (cultivar B), but the final population was about the same. If counts were taken only near the end of the growing season, the conclusion could be that cultivar B was tolerant; it grew relatively well and sustained the same numbers of the pathogen that killed the susceptible tobacco, yet cultivar B is really resistant, not tolerant. The rate of reproduction of the pathogen was fast on A (susceptible) and slow on B (resistant).

Propagules were concentrated near the diseased plant, near the tap root and stem. Soil samples vary greatly in pathogen counts from spot to spot after production of plants spaced rather far apart.

Nematodes after Fumigation

Nematodes, being obligate parasites, flourish when the host flourishes. Minton et al. (1962) studied the population of root-knot larvae in highly infested land at a cotton breeding station. They determined the number of larvae before fumigation (7 April 1960), fumigated the soil with nemagon (DBCP) or ethylene dibromide, and planted three cotton cultivars 18 April. Fumigants reduced the larval population to very low levels. Nematode numbers in the soil declined in all treatments during April and May because of larvae entering the roots and to death of those free in the soil that did not establish within the host. A rapid increase in nematode population occurred in unfumigated soil planted to susceptible cotton, reaching over 300 larvae per pint of soil by 27 June, after which little increase was observed. In fumigated soil, 300 larvae per pint of soil was not reached until maturity of the crop, between 10 October and 7 November. Crop loss, however, was slight after fumigation, because the cotton on fumigated land was vigorous throughout the growing season. The large, vigorous plants produced many nematodes late in host development with little damage. They tolerated the late infections. The final population was equal with or without fumigation. Fumigation would be required the following year if a susceptible plant were to be grown (Fig. 23.2).

When nematicides are injected just beneath the row of susceptible annual crops, great increases in yield result, even though many nematodes survive some distance from the seedlings. These results indicate that protection during the seedling stage is adequate to minimize losses from root-knot nematodes. Bergeson (1968) inoculated tomatoes at different stages of development with different numbers of second-stage juveniles. Inoculation of

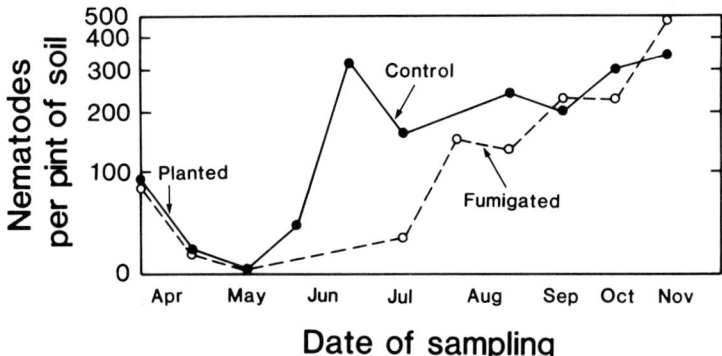

Figure 23.2. Soil fumigation prevents economic loss from root-knot nematodes in cotton but by the end of the growing season survivors on the vigorous cotton have increased to numbers equal to or greater than on the diseased, unprotected cotton. After Minton et al., 1962.

tomatoes in the cotyledon stage with 1000 larvae per pot reduced growth 81%. In the three-to-four-petiole stage, 1000 larvae per pot did not affect growth, but 50,000 larvae reduced growth 81%, and at the six-to-seven-petiole stage growth this population caused a 33% reduction. Thus a given number of larvae per volume of soil has different effects on plants of different sizes. Protection during the seedling stage is the reason planting cool-weather crops such as potatoes or sugar beets early, when soil temperatures are low, gives some reduction in loss from cyst-nematodes. The plant develops more rapidly at low temperature than the nematodes, enabling them to escape severe early damage.

EFFECT OF CROPPING ON POPULATIONS

A very sandy, formerly forested area in Central Wisconsin produces beans economically with irrigation. Reeleder and Hagedorn (1981) could not detect *Pythium* spp. with selective media in 10 of 11 forest sites. When brought into cultivation, after one bean or corn crop, one to eight propagules per gram, mainly of *P. irregulare,* were detected. After two to three bean crops, the propagule count increased to 80 to 282 propagules per gram, and *P. ultimum* had become the dominant species. *P. ultimum* in this environment is a vigorous colonist, reaching dangerous levels quickly in response to opportunity.

EFFECT OF GERMINATION STIMULATORS

Sclerotia of *Sclerotium cepivorum* germinate in soil in response to specific volatile stimulants from host roots. Sclerotia present at planting time decline during the growing season as roots penetrate the soil, stimulating germination of sclerotia with exhaustion of the intial sclerotia (Crowe et al., 1980). Sclerotia at the end of the growing season are new sclerotia formed on the diseased plants of that season. At a very high initial inoculum level (7.4 sclerotia per gram of soil) resulting in severe early infection, host development is restricted. Host plant volatiles stimulate germination of preplant sclerotia, with an ultimate reduction of total sclerotia present after harvest. The increase per gram of soil is inversely related to the initial population (Table 23.1).

In contrast to the results of Crowe et al. (1980) in California, Adams (1981) noted no decrease in sclerotial numbers in sandy loam and loamy sand soils of New Jersey 5 weeks after *Allium* spp. were planted.

COMMENTS

Numbers are used in all our studies: growth rates, reproduction, longevity, survival. We use numbers to determine the effects of temperature, water, etc., on either the pathogen or the disease. Standardizing numbers of

Table 23.1. Increase of sclerotia of *Sclerotium cepivorum* on garlic at Davis, California as influenced by varying initial sclerotia per gram of soil

	Number of sclerotia per g of soil				
Preplant	0	0.0001	0.0005	0.0018	0.0072
Post harvest	0	0.41	0.63	1.44	6.15
Gain or loss	0	X4100	X1260	X800	X851
Preplant	0.029	0.12	0.46	1.9	7.4
Post-harvest	9.22	3.41	2.73	0.93	0.87
Gain or loss	X318	X41	X1.3	X0.5	X0.1

Source: Crowe et al. (1980).

pathogen propagules is important in determining the resistance of cultivars. The relative increase or decrease of a pathogen on a host determines whether the host is intolerant, tolerant, susceptible, or resistant. Mitchell (1975), by using a graded inoculum series, determined that, under the conditions of his study, *Pythium myriotylum* did not produce zoospores and that an oospore had to be within 1 mm of the host tissue to have much chance of being effective. These results were achieved through use of mathematics, not by direct observation.

An example of the use of numbers to determine the effect of environmental factors on pathogen ecology is the study of Fidler and Bevan (1963). They read the work of Wallace (see Chapter 2) on the relationship of water-filled soil pores to the movement of juveniles of *Heterodera avenae,* the cereal-cyst nematode. When pores smaller than 30 μm in diameter are emptied of water, water films are too thin to permit much movement. When pores greater than 100 μm in diameter are filled with water, oxygen is deficient. Nematode activity in soil is greatest when pores 30 to 60 μm in diameter are just filled with water, extending to but not exceeding 100 μm pores. Fidler and Beven sampled spots within fields that had vigorous or diseased oats for cysts after the crop. They determined the water-holding and pore characteristics of the soil spots and obtained a high correlation between the available pore space (30 to 100 μm) and number of cysts. Nematode numbers were highly correlated with pore space. By using numbers of cysts they confirmed the work of Wallace, that distribution of water-filled pores within the soil is the most important ecological factor in oat fields in England as far as *H. avenae* is concerned.

This chapter will disappoint many, because no modeling is presented. While statistical studies of populations are essential to progress, I have doubts as to how much precision is possible or practicable in relation to most soilborne plant pathogens and to the relation of their numbers to disease severity. For those interested in this subject, see Pfender (1982). I have used the term numbers rather than populations as the title of this chapter. Populations is a broader term. It may include genetic differences among members of a species.

KEY REFERENCES

Bergeson, G. B. 1968. Evaluation of factors contributing to the pathogenicity of *Meloidogyne incognita*. *Phytopathology* **58**:49-53.
Crowe, F. J., D. H. Hall, A. S. Greathead, and K. G. Baghott. 1980. Inoculum density of *Sclerotium cepivorum* and the incidence of white rot of onion and garlic. *Phytopathology* **70**:64-69.
Grogan, R. G., M. A. Sall, and Z. K. Punja. 1980. Concepts for modeling root infection by soilborne fungi. *Phytopathology* **70**:361-363.
Mitchell, D. J. 1975. Density of *Pythium myriotylum* oospores in soil in relation to infection of rye. *Phytopathology* **65**:570-575.
Stanghellini, M. E., L. J. Stowell, W. C. Kronland, and P. von Bretzel. 1983. Distribution of *Pythium aphanidermatum* in rhizosphere soil and factors affecting expression of the absolute inoculum level. *Phytopathology* **73**:1463-1466.

REFERENCES

Adams, P. B. 1981. Forecasting onion white rot disease. *Phytopathology* **71**:1178-1181.
Baker, R. 1978. Inoculum potential. Pages 137-157 in *Plant Disease—An Advanced Treatise*. Vol. 2. J. G. Horsfall and E. B. Cowling, eds. Academic Press, New York.
Barker, K. R. and J. L. Imbriani. 1984. Nematode advisory programs—status and prospects. *Plant Dis.* **68**:735-741.
Barker, K. R. and C. J. Nusbaum. 1971. Diagnostic and advisory programs. Pages 281-301 in *Plant Parasitic Nematodes*. Vol. 1. B. M. Zuckerman, W. F. Mai, and R. A. Rodhe, eds. Academic Press, New York.
Boosalis, M. G. and A. L. Scharen. 1959. Methods for microscopic detection of *Aphanomyces euteiches* and *Rhizoctonia solani* and for isolation of *Rhizoctonia solani* associated with plant debris. *Phytopathology* **49**:192-198.
Bruehl, G. W. 1976. Management of food resources by fungal colonists of cultivated soils. *Ann. Rev. Phytopathol.* **14**:247-264.
Chuang, T. Y. and W. H. Ko. 1981. Propagule size: Its relation to population density of microorganisms in soil. *Soil Biol. Biochem.* **13**:185-190.
Crowe, F. J. and D. H. Hall. 1980. Vertical distribution of sclerotia of *Sclerotium capivorum* and host root systems relative to white rot of onion and garlic. *Phytopathology* **70**:70-73.
Davey, C. B. and G. C. Papavizas. 1962. Comparison of methods for isolating *Rhizoctonia* from soil. *Can. J. Microbiol.* **8**:847-853.
Davis, J. R., J. J. Pavek, and D. L. Corsini. 1983. A sensitive method for quantifying

Verticillium dahliae colonization in plant tissue and evaluating resistance among potato genotypes. *Phytopathology* **73**:1009-1014.

Davis, J. R., J. J. Pavek, D. L. Corsini, L. H. Sorenson, and S. L. Hafez. 1984. Evaluation of field resistance to *Verticillium dahliae* among potato clones and relationships of potato clones to the soil environment. Pages 327-328. *Abstracts of Conference Papers,* EAPR, Interlaken, Switzerland, 1-6 July 1984.

Fidler, J. H. and W. J. Bevan. 1963. Some soil factors influencing the density of cereal root eelworm (*Heterodera avenae* Woll.) populations and their damage to the oat crop. *Nematologica* **9**:412-420.

Flowers, R. A. and J. W. Hendrix. 1972. Population density of *Phytophthora parasitica* var. *nicotianae* in relation to pathogenesis and season. *Phytopathology* **62**:474-477.

Garrett, S. D. 1956. *Biology of Root-Infecting Fungi.* Cambridge Univ. Press. London.

Hani, F. 1981. Zur Biologie und Bekampfung von Fusariosen bei Weizen und Roggen. *Phytopathol. Z.* **100**:44-87.

Jones, F. G. W. 1959. Ecological relationships of nematodes. Pages 395-411 in *Plant Pathology, Problems and Progress.* C. S. Holton, G. W. Fischer, R. W. Fulton, H. Hart, and S. A. E. McCallan, eds. Univ. Wis. Press, Madison.

Kessler, K. J., Jr. 1966. Xylem sap as a growth medium for four tree wilt fungi. *Phytopathology* **56**:1165-1169.

Ko, W. H. and M. J. Chan. 1974. Infection and colonization potential of sporangia, zoospores, and chlamydospores of *Phytopathora palmivora* in soil. *Phytopathology* **64**:1307-1309.

Leach, L. D. and A. E. Davey. 1938. Determining the sclerotial population of *Sclerotium rolfsii* by soil analysis and predicting losses of sugar beets on the basis of these analyses. *J. Agric. Res.* **56**:619-632.

Menzies, J. D. 1963. The direct assay of plant pathogen populations in soil. *Ann. Rev. Phytopathol.* **1**:127-142.

Minton, N. A., E. J. Cairns, and A. L. Smith. 1962. Resistant cottons, nematicides, and fallow vs. nematodes. *Highlights of Agric. Res.* **9**. Auburn Univ., Agric. Exp. Sta., Auburn, AL.

Pfender, W. F. 1982. Monocyclic and polycyclic root diseases: distinguishing between the nature of the disease cycle and the shape of the disease progress curve. *Phytopathology* **72**:31-32.

Reeleder, R. D. and D. J. Hagedorn. 1981. *Pythium* populations in Wisconsin bean fields. *Can. J. Plant Pathol.* **3**:90-93.

Rodriguez-Kabana, R., M. K. Beute, and P. A. Backman. 1980. A method for estimating numbers of viable sclerotia of *Sclerotium rolfsii* in soil. *Phytopathology* **70**:917-919.

Rohde, R. A. 1972. Expression of resistance in plants to nematodes. *Ann. Rev. Phytopathol.* **12**:233-252.

Ross, J. P. 1972. Influence of *Endogone* mycorrhiza on Phytophthora rot of soybean. *Phytopathology* **62**:896-897.

Scott, M. R. 1956a. Studies of the biology of *Sclerotium cepivorum* Berk. I. Growth of the mycelium in soil. *Ann. Appl. Biol.* **44**:576-583.

Scott, M. R. 1956b. Studies of the biology of *Sclerotium cepivorum* Berk. II. The spread of white rot from plant to plant. *Ann. Appl. Biol.* **44**:584-589.

Sneh, B., J. Katan, Y. Henis, and I. Whal. 1966. Methods for evaluating inoculum density of *Rhizoctonia* in naturally infested soil. *Phytopathology* **56**:74–78.

Wilhelm, S. 1951. Effects of various soil amendments on the inoculum potential of the Verticillium wilt fungus. *Phytopathology* **41**:684–690.

Yarwood, C. E. 1956. Generation time and the biological nature of viruses. *Am. Naturalist* **90**:97–102.

Chapter 24

Evolution and Taxonomy

Soil microbiologists stress the complexity and rigor of the soil environment. Is this overemphasized? How different are soil bacteria from bacteria in other places? Soilborne fungal pathogens with true aquatic relatives have undergone essentially no structural changes. Plant parasitic nematodes have stylets or spears with which to pierce plant cells or fungal hyphae, but these are adaptations to parasitize plants, not for life in soil. Nematodes originated in an aquatic environment, and they are regarded as aquatic, even though the ones of concern to us occur in soil. The differences between soilborne pathogens (bacteria, fungi) and their close relatives are primarily physiologic. If adapting to life in the soil required no obvious morphologic changes, maybe the ecology of soilborne plant pathogens is not so different from ecology of other plant pathogens. The complexity of the soil microbiota has endowed it with a mystique that may be unjustified.

ANTIQUITY OF SOILBORNE PLANT PATHOGENS

If all the bacteria, fungi, and nematodes on earth were known, it is probable that most of them would be in the soil. It is also probable that of this vast assemblage, most of the important plant pathogens are already known. Pathogens constitute a remarkably small part of the total. Cooke (1979) gives the origin of bacteria at about 3.25 billion years ago; of eukaryotes about 1 billion years ago, including fungi; of *Stereum* about 150 million years ago; and of *Fomes* about 20 million years ago. Stanghellini (1974) estimates that *Pythium* existed 400 million years ago. It makes little difference how accurate these estimates are. The important thing is that time has developed the general architecture and life cycles of organisms, that the size of oospores, eggs, or chlamydospores was determined long before agriculture. Agriculture did not shape these organisms—it selected from the vast background of living forms those best adapted to conditions afforded by production of plants in agriculture or forestry.

The antiquity of plant pathogens or their progenitors assisted in the distribution, many of which existed in native vegetation prior to agriculture. E. Forbes in 1846 (in Wulff, 1943) concluded that most of the plants of the British Isles were present before the islands separated from continental Europe. Wulff noted endemic species of *Nicotiana* in North and South America and in Austrlia, and he believed they existed before the continents separated. Willis (1922) calculated that a species that originated in one location at one time and spread 1 m per year would occupy 1,000,000 square miles in a million years. Again, regardless of the precision of these statements, even if bacteria, fungi, or nematodes had no means of spread other than their own activities, many could be widely distributed today.

Seasonal changes have forced seasonal development of higher plants in much of the world, and many pathogens have developed seasonal rhythms. Basidiocarps of *Typhula idahoensis* form just prior to the normal time of snow fall. This psychrophile competes only under snow. Its sclerotia are dormant all summer, but they sporulate in autumn. The perithecia of the cereal fusaria develop at a time when ascospores have a chance of infecting inflorescences. Cyst-nematodes adapted to winter cereals often require a cool treatment to hatch. In contrast, tropical root-knot nematodes have no marked seasonal rhythms, but develop continuously so long as young root tips are available. The development of seasonal rhythms has required many physiologic adaptations acquired over long periods of time.

Cephalosporium gramineum, a fungus with no known persistent spore stage, illustrates another type of "timing." *C. gramineum* infects winter wheat from November on through winter, but leaf symptoms are not apparent until growth has resumed the next spring. The fungus becomes systemic in stems. Development of the wheat stem is essential to its saprophytic survival. If the host is killed quickly, it can not persist long in the leafy remains of juvenile plants. To be successful, it must develop slowly, not killing before the culm is well developed. A smut fungus that reproduces in the inflorescence must not kill the host prematurely. Developing a system of survival surely involves much time.

EVOLUTION AND TAXONOMY

POLYMORPHISM AND PHENOTYPIC PLASTICITY

It was not until the studies of the Tulasne brothers, DeBary and Leveille, and others that polymorphism, the existence of different spore forms within a species, became established in mycology. Kendrick (1979) remarked that a single genotype with several phenotypes should not seem strange. He mentioned that caterpillars become butterflies, that tadpoles become frogs. The various states of fungi have evolved to survive hazards in the environment or to more efficiently exploit opportunities. The contents of macroconidia of *Fusarium culmorum* can form chlamydospores. The chlamydospores are long-lived in soil; the conidia are not. *Pythium graminicola* takes several forms, including oospores, sporangia, vesicles, zoospores, hyphae, each of which performs a function in the life cycle, including assimilation, reproduction, dissemination, and long-time survival. This fractionation of function among forms, of polymorphism and phenotypic plasticity, serves the organism.

Weresub and Pirozynski (1977) emphasized the use of anamorph for the asexual state, teleomorph for the perfect or sexual state, and holomorph for the entire or whole fungus. *Fusarium graminearum* is the anamorph (asexual state) and *Gibberella zeae* the teleomorph (sexual or perfect state). The holomorph of this fungus is *Fusarium graminearum* plus *Gibberella zeae*, normally referred to as *G. zeae*. Pathologists are aware of these relationships, but many prefer to use the name of the asexual state when the fungus is seldom encountered in the sexual state. The teleomorph *Glomerella graminicola* is known, but almost all workers report working with *Colletotrichum graminicola,* the anamorph, because the teleomorph is seldom seen. Most mycologists state that the anamorph is usually a relatively simple structure and is produced quickly, adapting it to rapid propagation and spread . The teleomorph is usually more complex, is less quickly formed, and serves not only as a source of propagules but also for genetic recombination. The teleomorph also often functions for long-term survival and is often linked with a particular season of the year. In general, propagules of teleomorphs are timed to disperse the fungus after a critical period of adverse conditions (Kendrick, 1979).

Mather (1965) offered two possibilities for the preservation of the sexual state among soilborne fungi. Rare beneficial sexual recombinations may have sufficient merit to warrant preservation of the sexual state, or sexual reproduction may produce the only resting spore available within the life cycle. Many important soilborne fungi have no known sexual state (*Cephalosporium gramineum, Fusarium culmorum, F. oxysporum, Macrophomina phaseolina, Periconia circinata, Phialophora gregata, Phymatotrichum omnivorum, Pseudocercosporella herpotrichoides, Thielaviopsis basicola, Verticillium dahliae*). Some widespread soilborne pathogens have sexual states of little or no known importance in nature (*Cochliobolus sa-*

tivus, Phytophthora cinnamomi in most of its range, *Fusarium solani,* and *Sclerotium rolfsii*). Most soilborne fungi dependent on a sexual state are homothallic (*Aphanomyces cochlioides, A. euteiches, A. raphani, Phytophthora cactorum, P. fragariae, P. megasperma, Pythium aphanidermatum, P. arrhenomanes, P. ultimum*).

Burnett (1965) observed that single sexual types (A^1 or A^2) of heterothallic fungi are repeatedly found in soil (as in *Phytophthora cinnamomi*), that zygotes are usually abundant only among homothallic species, and that sexual reproduction in the sense of outbreeding is not common. If fitness, with reduced variability, is more important to soilborne pathogens than variability, it implies a relatively uniform environment for life, and the most uniform feature in the environment is the host. *Fusarium solani* f. sp. *phaseoli* is adapted primarily to beans, whether in Washington, Wisconsin, or Australia. The sexual state, when important, is important for reasons other than as a source of variation (Bruehl, 1976).

Mating types in heterothallic fungi give clues as to the origin of some species. *Phytophthora infestans,* the late blight of potato pathogen, was introduced to Europe with the common potato from the Western Hemisphere. From Europe it spread to many parts of the world, including the United States and Canada. The sexual state did not develop, and it was not until Gallegly and Gallindo (1958) found A^1 and A^2 isolates in essentially a 1:1 ratio in Central Mexico that the explanation became apparent. Isolates from Canada, the United States, Western Europe, South Africa, and the West Indies, 105 in all, were all A^1. When 95 isolates from Central Mexico were studied the 1:1 ratio, $A^1:A^2$, was found. Oospores were produced in abundance in diseased potato tissues in the Toluca Valley of Mexico. *P. infestans* must have originated in Mexico, and by chance only one mating type of this heterothallic species was disseminated.

The same phenomenon is illustrated by some other *Phytophthora* spp. *P. infestans* is not soilborne to any extent, but *P. cinnamomi* is. In *P. cinnamomi,* A^2 is generally distributed, with both mating types present in New Guinea, Celebes (Indonesia), and Taiwan (Ko et al., 1978). This soilborne fungus survives in most of its geographic range as asexual chlamydospores rather than as oospores.

COLONISTS OF CULTIVATED SOILS

Because the number of important pathogenic species of soilborne fungi is small, it is useful to consider the attributes of those that are established extensively in cultivated soils. Why have so few become so widely important? What characteristics have made them successful? Success to them is reproduction, survival, and establishment in wide expanses. Pathogenicity only brings them to our attention and subjects them to our studies (see Bruehl, 1976, for discussion of pioneer colonists).

Some were present in the soil when it was brought under cultivation.

Some started somewhere in the world and by means of effective dispersal mechanisms, usually by being seed-borne to some extent, followed the host relentlessly as the host was brought to new land. *Rhizoctonia solani* and *Pythium* spp. are examples of widely disseminated pathogens present in soils before cultivation. The relatively specialized forms of *Fusarium oxysporum* and *F. solani* are examples of pathogens that relentlessly follow the host as the host is disseminated by man.

It is advantageous to a pathogen not to depend on a single host for survival. A host can nourish and sustain a population without necessarily sustaining loss; a suscept is subject to real damage by the pathogen. The most successful colonists are physiologically "primitive," able to reproduce in several hosts. They can invade several species without eliciting pronounced resistance reactions. In early stages of infection they are subtle, establishing themselves and occupying as much tissue as possible before meaningful host response. It is illogical to consider these pathogens primitive, while those with highly specific host-pathogen relationships, such as rusts, are considered advanced.

Soilborne pathogens are spatially restricted. They do not spread rapidly enough to render cropping histories meaningless. They must, therefore, depend on adequate means of surviving periods of adversity. In cultivated soils a good crop rotation subjects the organisms to unlimited food, followed by a period of little or no food. The tilled layers of soil may be too dry, too cold, or too hot. Pathogens adapted to cultivated soils must have a means of surviving changes in plant species (rotation) and in the physical environment, particularly in temperate climates (oospores, chlamydospores, sclerotia, eggs, etc.).

Dead leaves, stems, and roots of most herbaceous plants decompose rapidly, so the ideal colonist does not depend on active saprophytic survival in transient debris; it relies on dormant structures that remain viable after host debris has disintegrated. The dormant stage should not germinate in response to physical conditions alone, but in response to exudates from a potential host or substrate.

After germination (hatching), the germ tube should respond chemotropically by growing toward the potential substrate, or it should respond chemotactically (swim or move) toward the substrate. If an infection is not established, the organism should salvage its protoplasm in a replacement chlamydospore or encysted zoospore; it should have the capability of recycling, even if at a reduced energy level.

Quickness of response to the presence of a potential host may be critical. If a fallow, cultivated field is seeded all at once with soybean seeds, a seed-rotter or seedling-blighter must become active immediately in order to establish in juvenile tissues before resistance develops. Another type of quickness is also important. Most herbaceous cultivated plants occupy the land for a few months, and the period of susceptibility may involve only part of this time. The ideal pathogen should have an uncomplicated, short life cycle, especially if more than one infection and reproduction cycle is possible on the host.

Ideally, each propagule should be independent, capable of establishing a new colony by itself. A single oospore, chlamydospore, or nematode should have enough energy (inoculum potential) to establish an infection. If sexual reproduction is essential, homothallism is advantageous. Most successful colonists, with the exception of most nematodes, do not require a mate. The motility and energy level of nematodes and response to pheromones or sex attractants facilitate mating, even in relatively low populations.

The propagules should be as small as possible, yet should contain sufficient energy for infection. Garrett saw the importance of matching energy level of the propagule with the energy required to infect the target, expressed in his concept of inoculum potential.

The ideal colonist of herbaceous residue will invade the substrate as thoroughly as possible, digest all it can, store it in suitable propagules, and abandon the spent residue completely. It should not leave stable antibiotics within the residue. The tissues containing the dormant propagules should disintegrate to free the propagules for maximum dispersal when soil is cultivated.

If the pathogen lacks a true dormant state and must survive by continued activity within dead host debris, it should digest the substrate slowly but maintain the necessary threshold of metabolism to maintain possession. It should not respond to favorable physical conditions by excessive activity. The ideal saprophytic survivor (active possessor of substrate) prolongs the usefulness of the substrate as long as possible. It should not produce excess extracellular enzymes to produce surplus soluble foods that would alert dormant competitors to the presence of the debris. It should produce staling products or antibiotics in quantities only sufficient to retard would-be invaders.

The ideal active possessor of herbaceous residues, able to maintain exclusive possession of substrate over fluctuating environmental conditions and changing chemical nature of the substrate, will obviously never exist (Bruehl, 1975). The scheme of nature is to recycle nutrients, to break residues down, and to employ a succession of organisms capable of digesting different chemicals. The wonder is that fungi like *Gaeumannomyces graminis*, *Cephalosporium gramineum* and *Verticillium albo-atrum* do as well as they do.

NEMATODES

Groupings

Nematodes have been grouped according to host-parasite relations and according to ecological relations. Nematologists use groupings such as migratory ectoparasites, migratory endoparasites, and sedentary endoparasites, based on host-parasite feeding relationships. They usually consider sedentary endoparasites, such as root knot nematodes that feed on giant cells, as the most advanced groups, stressing the most delicate host-parasite relationships as the most advanced. Wallace and Doncaster (in Wallace, 1973) divided nematodes into three groups, according to motility in water. One group can swim in deep water, true aquatics. One group can swim in thick water films, which enables them to escape from the soil and move up a plant when the plant is wet. *Ditylenchus* and *Aphelenchus* spp. belong to this group. The vast majority of plant pathogens belong to the third group, able to move well only in thin water films. These are confined to soil.

Triantaphyllou and Hirschmann (1980) discussed categories of nematodes based on cytology. Diploid and tetraploid races of *Ditylenchus dipsaci* and *Globodera rostochiensis* are called cytological races. Twenty-six species of cyst-nematodes have been studied; twenty are diploid and reproduce sexually (amphimictally); six reproduce asexually, some by mitotic parthenogenesis and some by meiotic parthenogenesis. Among *Meloidogyne* spp., at least three species reproduce sexually; at least five species can reproduce sexually and by meiotic parthenogenesis when males are absent; at least five species reproduce only by mitotic parthenogenesis. Some races of *M. arenaria* and *M. incognita* are 3n, and some are 2n. Within *Meloidogyne* are species wholly dependent on sexual reproduction (mating of males and females), some that reproduce sexually and asexually, and some that reproduce only asexually. Triantaphyllou and Hirschmann, after reviewing reproduction among nematodes parasitic on plants, concluded that sexual reproduction is favored. No major taxonomic group has become asexual. They attribute the success of some parthenogenetic groups to interaction with the unnatural environments of "continuous monoculture of susceptible crop plants."

I have stressed the advantages of independent propagules (no need for a mating partner as well as adequate energy) for pioneer colonists of cultivated lands (Bruehl, 1976). It seems significant that *Meloidogyne arenaria, M. incognita,* and *M. javanica* reproduce parthenogenetically, that race A of *M. hapla* can reproduce sexually or asexually, and that race B of *M. hapla* is apomictic. It is not without significance that the above four species, mostly asexual, are the most important root-knot nematodes in cultivated soils in the world.

Sexual reproduction is favored among cyst-nematodes, but cysts persist in soil and cysts confine the larvae to a small spot. On liberation from the cyst both males and females should be relatively close together so that mating can be easy, especially since the female protrudes from the root (facilitating mating?) and is not within a gall, as may be true with root-knot females. In addition, nematodes are capable of significant movement, so there is little selection against the need of a partner for reproduction.

Origin

Plant parasitic nematodes evolved from fungal feeders with delicate stylets adapted to feeding on hyphae (Maggenti, 1981). Nematodes with very long stylets would be at a disadvantage trying to feed on hyphae. One line of evolution progressed from fungal feeders to plant parasites with slender stylets, many of which feed on above-ground plant parts. Some plant parasitic nematodes retained the ability to feed on both fungi and higher plants. The other evolutionary line developed many nematodes with large, stout spears.

Norton (1978) stated that nematodes parasitic on higher plants originated from algal or fungal feeders or from predators. Norton's book contains an illustration of wavy tracks in Eocene rock in Utah. These tracks are like those made by contemporary nematodes on agar, and Norton states that no other known form of life can make such tracks. Fungi are known from Precambrian strata, and land plants are not known in fossils of the Cambrian era, so plant feeders may be derived from fungal feeders. Filamentous fungi preceded roots. Norton presented evidence that many nematodes were distributed by continental drift to the present continents.

Another discussion by Norton stresses what can be learned from the ice age. Ice covered much of present North America, northern Europe and Asia for thousands of years; the ice disappeared 8000 to 10,000 years ago in the United States and part of Canada. The soils formed by the glaciers were colonized by a wide variety of bacteria, fungi, and nematodes prior to agriculture. These organisms, many with very limited ability to self-disseminate, colonized these vast areas rather quickly.

The problems of nematologists in both classification and evolution are discussed by Stone (1977). *Heterodera schachtii* was described in 1871. Between 1871 and 1900 about two species were known; by 1940, 10; by 1960, 20; by 1970, 50, and the list is growing. Stone listed nine genera under Heteroderidae. One genus was described in 1871 (*Heterodera*), one in 1956 (*Meloidodera*), one in 1959 (*Globodera*), one in 1966 (*Cryphodera*), and five in the 1970s, to the time of Stone's article. Nematology is advancing at such speed that in some ways it is still in an exploratory stage of development! Tests of relatedness by hybridization of species among cyst-nematodes have been made, and Stone lists several. These efforts revealed relatedness, but Stone emphasized that the results should not be misinterpreted. Few viable offspring resulted, and those few may not have been hybrids. In general, hybridization experiments supported the species as erected by taxonomists on the basis of morphology.

RADIATE EVOLUTION

Evidence is strong that many complexes of pathogens arose from common progenitors. The many *formae speciales* of *Fusarium oxysporum*, pathogenically specialized toward many plants, are examples of an organism becoming adapted physiologically to different suscepts, yet retaining the basic physiology and morphology of the ancestral form. The many anastomosis groups of *Rhizoctonia solani* and the various breeding groups of *Armillaria mellea* probably arose from a common ancestor rather than by any type of convergent evolution. Bacteriologists found so much in common among certain former *Pseudomonas* species that they reduced them to pathovars of *P. syringae* (*tabaci, lachrymans, syringae, pisi, tomato, savastanoi, coronafaciens, glycinea, phaseolicola*, etc.). The hosts isolate these entities, and in time they often diverge sufficiently to warrant being recognized as species.

VARIABILITY VERSUS FITNESS

General biologists view reduced variation as likely to lead to extinction, to an inability to survive significant alterations in the environment. What is the role of sexual reproduction? When selection reduces sexual species to asexual species, and this has occurred many times, there are reasons. Sexual reproduction in these cases was of no survival value. Many find this hard to accept, even in the face of much evidence. Many plant pathologists spend their lives working with asexual or selfing (considered asexual in the context of reducing variability) organisms or organisms that seldom reproduce sexually. Viruses, bacteria, some of the most important nematodes, and many soilborne fungi are essentially asexual. Stebbins (in Lemke, 1973) stated that organisms with short life cycles favor fitness over variability. Transient conditions favorable for growth and reproduction are most readily exploited by fit, closely selected, rapidly reproducing organisms.

In terms of geologic time, or in eventually providing new forms of life, the emphasis on the advantages of sexual reproduction is justified. Plant pathologists and farmers, however, need not be concerned about organisms in terms of geologic time. *Fusarium culmorum* may be extinct 10,000,000 years from now, but it will survive for a sufficient period of time to concern us. We must view organisms from the short-range perspective.

Van der Plank (1968) stressed homeostasis, or maintenance of the original constitution, in his treatment of responses to host-specific plant pathogens. Flor (1953) reported that the races of flax rust (*Melampsora lini*) that predominated in North Dakota were those with the fewest possible genes for virulence. Even though all *M. lini* in North Dakota is initiated in the spring by sexual spores, and annual recombination provides many races, those with excess virulence genes are at a disadvantage. In *Phytophthora infestans* (in Van der Plank) and *Tilletia caries*, the same selection against excess virulence operates. Van der Plank considered the tendency of the pathogen to maintain the original, minimal virulence as a form of homeostasis. *Melampsora lini* is basically adapted to the flax plant. *Phytophthora infestans* is basically adapted to the potato. *Tilletia caries* is basically adapted to the wheat plant. When these obligate parasites encounter specific resistance, specific resistance may be viewed as environmental factors not normally encountered.

Dobzhansky (1951) stressed that modifications beneficial against conditions normally encountered within the environment are adaptive, that mutations to rare conditions seldom are. Adaptive modifications maintain the normal equilibrium of physiologic processes (homeostasis, according to Dobzhansky), as well as the harmony between the organism and the external world. Genetic responses to environmental factors seldom encountered in the environment are seldom beneficial. He gave as an example a bacterium resistant to a bacteriophage. This bacterium requires a substance not required by the wild-type bacterium. In the absence of the phage the susceptible bacterium dominates. The resistant mutants save the species in the presence of the phage. The tendency for races of *Melampsora lini* and *Phytophthora infestans* with the fewest possible virulence genes to predominate also fits the adaptive process. Specific resistance genes have not been an important part of the environment encountered by the pathogen during its long period of adaptation to the host species, and in their absence the unneeded virulence genes may have had a negative effect.

EXAMPLES OF EVOLUTION

Fusarium graminearum

Even though *Fusarium graminearum* is one of the most studied fungi in the world, important new things about it are still being learned. Purss (1969) in Queensland, Australia, found that isolates from the Atherton Tablelands from maize did not produce typical crown rot of wheat as did isolates from wheat in the Darling Downs area of southern Queensland, where wheat is important. In addition, maize isolates readily produced the sexual state (*Gibberella zeae*), and those from the wheat area did not. Purss (1971) found that wheat crown rot isolates produced both head blight of wheat and stalk rot of maize, so they had not lost that capacity. In contrast, only the wheat crown rot isolates produced crown rot of wheat. A stalk rot of maize isolate from the United States produces less than 1% crown rot of wheat, an Australian stalk rot of maize isolate 6%, an isolate from head-blighted (scabbed) wheat, 2%, and a crown rot of wheat isolate, 100%.

Wheat in Australia is planted in the autumn and harvested in early summer. The winters are mild, and rainfall is greatest in the winter months in the southern portions of the wheat belt, where summer rains are too light to allow maize or sorghum production without irrigation. Burgess et al. (1975) surveyed the *F. graminearum* population in the eastern Australia wheat belt. They classified isolates of *F. graminearum* that produced crown rot of wheat and no or few perithecia as group I and isolates that did not produce crown rot of wheat but did produce perithecia and were important in head blight (scab) of wheat and as pathogens of maize as group II.

Francis and Burgess (1977) reported the following: group I isolates grow more slowly, only two matings produced perithecia, and they soon lost this ability. They are either heterothallic, rarely fertile, or infertile. Group I isolates predominate in areas with an arid summer. Group II isolates grow faster, are homothallic, and predominate in areas of maize culture and humid mid-summer weather. It is primarily aerial, attacking organs above the soil (wheat heads and maize stalks and ears). They concluded that microevolution produced two ecotypes within *F. graminearum,* each with a different niche.

Sung and Cook (1981) studied *F. graminearum* group I isolates from dryland, south central Washington and group II isolates from Pennsylvania on media adjusted to different water potentials. Group I isolates produce the greatest number of macroconidia at about -15 bars. Group II isolates produce the greatest number of conidia at -1.4 to -3.0 bars (the basal media without added osmotica). Group I isolates from Washington produce no perithecia; group II isolates from Pennsylvania produce perithecia. It is logical to assume that the whole fungus, the sexual state plus the asexual state, is the ancestral form and that adaptation to dry summers, where the sexual state would seldom function, results in alterations in water relations and the loss of the sexual state.

Pseudomonas solanacearum

Pathologists frequently observe changes in virulence of pathogens to cultivars of a host species, but well-documented examples of host range changes are rare. *Pseudomonas solanacearum,* the southern bacterial wilt pathogen, is endemic in many soils about the warm (45°N to 45°S latitude) part of the world. As a species, it attacks many plants in several families, especially Solanaceae. Race 1 has the widest host range, particularly on solanaceous plants, and I will refer to it as the tomato race. Race 2 occurs mainly on banana and *Heliconia* spp. The most interesting aspect of *P. solanacearum* is its documented adaptation to cultivated bananas.

The banana is a native of the Eastern Hemisphere, with wild bananas (2n) in southeast Asia and the eastern Pacific Islands (Simmonds, 1976). Most cultivated bananas are sterile triploids (AAA, AAB, ABB genomes). Even though bananas have been grown for centuries in southeast Asia, bacterial wilt of cultivated banana, or Moko disease, is unknown in that region. Buddenhagen and Kelman (1964) suggested that the progenitor strains of *P. solanacearum* present in the soil in the Orient differs from those in the Western Hemisphere in some way that makes them unable to adapt to triploid bananas.

The first reliable description of bacterial wilt of banana was by Rorer in

1911 from Trinidad. The disease on banana was studied by Buddenhagen (1960) in Costa Rica. The tomato strain is avirulent on cultivated banana. *Heliconia latispatha,* an endigenous weed, is diseased in abandoned banana plantations. *Heliconia* spp. are the only members of the Musaceae (banana family) native in the Western Hemisphere. Sequeira and Averre (1961) surveyed 20,000 acres of virgin woodland in the Coto Valley of Costa Rica and found three species of *Heliconia* infected. The bacteria from *Heliconia* spp. produces a distortion syndrome on cultivated banana, not the rapid destructive wilt. *P. solanacearum* produces indole acetic acid, and this causes the curvature of the banana pseudostem (Sequeira and Kelman, 1962). The strain from *Heliconia* became known as the distortion or D-strain.

The D-strain is common in Central America on *Heliconia* spp., but the tomato strain occurs in a large number of native plants. It is logical to assume the D-strain originated from the tomato strain. After several passages through cultivated banana, the D-strain becomes virulent on banana. It is assumed the B-strain arose from the D-strain. Figuring backward from the establishment of clean propagating materials on cleared jungle sites, it is estimated that 30 years elapsed before the B-strain was established. The B-strain is virulent on *Heliconia*.

The next observed change in virulence was the development of the SFR-strain, the "small, fluidal, round" strain (Buddenhagen and Elsasser, 1962). This strain is highly virulent on banana. The SFR-strain is especially dangerous because it produces bacterial exudate on male flowers of the banana, and this exudate is attractive to insects. The SFR-strain thus can be spread by insects, making it more dangerous than the B strain. SFR is avirulent on *Heliconia* spp. Stover (1972) stated that fallowing for 12 months eliminates the B-strain and for 6 months eliminates the SFR strain.

Thus, as emphasized by Buddenhagen (1965), we have record of change within this species. If all the assumptions are true, the D-strain originated on native *Heliconia* spp. from the tomato strain. When cultivated *Musa* triploids were introduced, the D-strain infected them, and within 30 years it became the virulent B-strain, but only here and there, not everywhere. Then SFR developed from B. This progression was accompanied by reduced longevity in soil and reduced host range, and some advance was made toward becoming airborne (ooze from flowers that can be dispersed by rain or insects). The D-strain is avirulent on banana. The B-strain is virulent on *Heliconia* and banana. The SFR-strain is not virulent on *Heliconia*. This sequence constitutes an example of the bridging-host theory postulated by Ward (1903).

Wheat Soilborne Mosaic Virus

Wheat mosaic virus has a split-genome, each genome within particles of different length. The longer rigid rod is nearly 300 nm in length and about 20 nm in width, with a hollow center. Measurements of the shorter rod vary from 110 to 160 nm in length, depending on the isolate (Tsuchizaki et al. 1973). The short rods are noninfectious (Gumpf, 1971). Tsuchizaki et al. (1975) added short rods to long rods in varying proportions. Infectivity increased as short rods were added, proving that both particles are required for infectivity. They mixed long and short rods of different strains. The short particles governed the length of the short particles, the serotype, and the type of inclusion bodies. The long rods governed infectivity on tobacco and virus concentration in spinach leaves. Both particles influenced symptoms on rye and on *Tetragonia expansa,* but the long particles exerted the greater effect. American and Japanese isolates had much in common, but were distinct enough to be classified as serotypes I (Japanese isolates) and II (American isolates).

Powell (1976) noted that tobacco mosaic virus and wheat mosaic virus both have rod-shaped particles 300 nm long (ignoring the short rod of WMV) and that potato mop top virus was serologically related to TMV (Cooper and Harrison, 1972). Powell found that WMV is as close serologically to TMV as some strains of TMV are to each other. The molecular weights of the coat proteins are similar. The proteins of the long and short rods of WMV are similar. Powell concluded, "WMV is the second soilborne virus with a fungal vector shown to be related to TMV, potato mop top virus being the first" (Kassanis et al., 1972). This is significant, because TMV is soilborne to a limited extent, although it has no known vector.

TMV and WMV were isolated from winter wheat in Kansas (Paulsen et al., 1975). S. A. Tolin (personal communication to Hamilton and Nichols, 1977) stated that TMV can be recovered from a wheat cover crop following TMV-infected flue-cured tobacco in Virginia. WMV has an intimate relationship with its vector, *Polymyxa graminis,* TMV for the most part lacks a vector, other than man. It appears logical to assume that TMV arose from WMV rather than the other way around. It is easier to lose a vector (*Polymyxa graminis*) than to gain one. It would be interesting to try to develop TMV from WMV by repeated passage of WMV through tobacco. The results of Tsuchizaki et al. (1975) would indicate promise.

BIOLOGIC SPECIES, CRYPTIC SPECIES

Classic taxonomy, with reliance primarily on morphology, is inadequate for pathologists and ecologists within certain groups of pathogens (viruses and bacteria, certainly, but these are excluded from this treatment). Dobzhansky (1972) noted that borderline taxonomic units are rare, that they annoy classifiers, but that they are precious to evolutionists. He warned that species may look alike and yet to be different. Organisms of sexual species should recognize members of the same species, and within the restraint of incompatibility factors, should exchange germ plasm, or reproduce in fertile sexual matings.

Rhizoctonia solani

Baker (1970) stated that, "*Rhizoctonia solani*, as presently understood, probably causes more different types of diseases to a wider variety of plants, over a larger part of the world, and under more diverse environmental conditions, than any other plant-pathogenic species." Baker summarized the wide environments and groups of these diseases. Some examples include seed decay, damping-off of seedlings, stem rot of plants more advanced in development, root rots, endophytic mycorrhizal association with orchid seedlings, collar rot of mature tea shrubs, stem rot of aquatic plants, lettuce and cabbage head rot, soft rots of fruits and vegetables in contact with soil, black scurf on potato tubers, and aerial web, leaf, and thread blights above ground in warm, humid climates. This rough summation of types of diseases certainly justifies Baker's assessment. Any single species capable of such variation in type of attack, host range, and adaptation to a wide variety of environments is a wonder species. *Pythium* as a genus, and *Fusarium* as a genus compete with or exceed the above listing, but no species within those genera are as diverse as described above (or could any virus, bacterium, nematode, or higher plant be?). Baker, who knows these diseases and the fungus intimately, states that this wide adaptability results from its many strains, strains that differ in host range, virulence on a given host, type of attack, temperature relations, survival in soil, response to CO_2, etc. He concluded that *R. solani* has to be a very ancient fungus to become so widespread and genetically stablized into many strains to occupy such diverse ecological niches.

Anastomosis Groups Anderson (1982) provided additional insight into the truly remarkable versatility of *Rhizoctonia solani*. The perfect state, *Thanatephorus cucumeris,* is considered to be a "collective species." The asexual state is fragmented into several "anastomosis groups" determined by observing the reaction of hyphae of different isolates when they come in contact. Isolates belonging to different anastomosis groups do not anastomose, so genetic exchange does not occur between them. Anderson stated that the anastomosis groups are somewhat similar morphologically but isolated genetically. The complexity within this species accounts for many failures in the past to develop resistant varieties or rotations.

The anastomosis group concept began in Germany in 1937 when it was suggested by Schultz. Anderson cited Richter and Schneider in Germany, Watanabe and Matsuda in Japan, and Parmeter, Sherwood, and Platt in the United States as its main architects. Anastomosis group I (AG-1) is distributed worldwide, and causes seed and hypocotyl rots and aerial blights of many plants in diverse climatic zones. Ogoshi divided AG-1 into two subgroups, one including what was formerly known as *Corticium microsclerotia* and one based on the former *Corticium sasaki.*

AG-2 causes cankers on root crops and root diseases of crucifers.

Subgroups exist within AG-2, but they have not completely diverged. In Minnesota, Anderson found AG-2 type 1 pathogenic to radish and AG-2 type 2 pathogenic to radishes and carrots. In Minnesota radishes, carrots, and potatoes are often grown on the same truck farm. AG-2 type 1 and AG-2 type 2 do not attack potatoes, but AG-3 does. In Minnesota they recommend rotating potatoes (susceptible to AG-3) followed by radish (susceptible to AG-2 type 1 that does not attack carrots) followed by carrots. In this sequence no crop materially increases the inoculum for the next crop.

AG-4 attacks seed and hypocotyls of a very wide range of plants, and it is apparently distributed worldwide. It represents the type described as *Corticium practicola* by Kotila. It has no subgroups. Anderson (1982) believed the AG-4 group constitutes a biologic species and that legumes are important hosts on a worldwide basis. He concluded his paper by stating that progress in obtaining disease resistance is possible if pathologists work within particular AG groups. The above, abridged treatment of variation within *Rhizoctonia solani* illustrates the value of precise taxonomy and also the difficulty of evaluating literature predating the use of anastomosis groupings in reporting research results relating to "*R. solani.*"

Sclerotia of AG-1 have a central region of dense cells, surrounded by an outer layer of empty cells bordered by a darkly pigmented mucilaginous surface layer (Naiki and Ui, 1978). Sclerotia of AG-2 type 1 have two well-defined layers. The outer, darkly pigmented mucilaginous layer is absent. Sclerotia of all other anastomosis groups consist of loosely arranged brown cells without well-defined zones. Sclerotia of AG-1 are quite spherical; those of AG-2 type 2, AG-3, AG-4, and AG-5 are quite irregular. Those without internal differentiation are irregular, apparently more primitive. AG-1 sclerotia, the most differentiated, survive longest in soil. Naiki and Ui concluded that the use of AG group classification would reduce conflicting statements as to sclerotial morphology and longevity within *R. solani.* Storck and Alexopoulos, cited in Kuninaga and Yokosawa (1980), had concluded that higher guanine-cytosine content is associated with evolutionary advancement. Kuninaga and Yokosawa found AG-1 isolates highest in GC, corroborating the generality.

Sherwood (1969) studied 27 isolates of AG-1, 14 of AG-2, 9 of AG-3, and 25 of AG-4 in culture. Most AG-1 and AG-4 isolates were recognizable in culture, but AG-2 and AG-3 isolates were similar in appearance. Considerable variation among isolates occurred in all characters studied, so anastomosis phenomena were the only reliable means of identification.

Use of Genetic Markers in Epidemiology

Armillaria mellea produces tan-yellowish mushrooms about the base of dead or diseased trees or shrubs. The mushrooms appear in September, October, and November in north temperate climates. *A. mellea* is an assemblage of biologic species, rather than a single, interbreeding entity. Anderson and Ullrich (1979) obtained haploid tester lines from 97 fruit bodies in North America and found 10 distinct, intersterile groups. Group I, from 36 fruit bodies, was interfertile when testers with proper incompatibility factors were mated. It was tetrapolar, with alleles at A and B loci. Group I testers were incompatible with testers of other compatibility groups, regardless of their incompatibility factors.

Adams (1974) paired cultures of *A. mellea* obtained near Lapine, Oregon. His cultures were mycelial (diploid), not basidiospore-cultures. When cultures from the same source met, hyphae intermingled with no visible reaction. When cultures of different origin met, a dark line of demarcation developed where the cultures met. Adams obtained cultures from three infection centers and found three distinct mycelia, as judged by demarcation lines. Each center apparently resulted from a single infection that spread in all directions. Spread of this type is by rhizomorphs bridging from root to root.

Following this discovery, Shaw and Roth (1976) found that three infections in Ponderosa pine in Klickitat County, Washington, consisted of three clones. One infection center covered 600 ha. Shaw and Roth estimated that it took from 460 to 1000 years for the fungus to invade this area, if it spread a maximum of 1 m per year from a single infection. Sporocarps were obtained from the three infections, and Anderson et al. (1979) obtained basidiospore cultures from them. Matings with tester lines proved that each infection site was occupied by a different clone and that each site was a pure culture of a single clone. Among long-lived hosts, mycelial spread (rhizomorphs) over periods of many years is significant, and even a rarely functional spore stage is significant.

The study areas in central Oregon and Washington are in the rain shadow of the Cascade Mountains and summers are dry. L. F. Roth of Oregon State University observed an area in central Oregon for 30 years without finding a single sporophore of *A. mellea*. In 1973, the wettest autumn on record, sporophores appeared. In this dry region the sexual state functions rarely. Anderson et al. (1979) by using incompatibility alleles, found much smaller clones in the humid Northeast of the United States. The sexual state functions more frequently there.

Typhula spp. have tetrapolar, bifactorial incompatibility governed by alleles at the A and B loci. Everyone who has studied these species has found many combinations of A and B alleles in small areas, evidence that the sexual state gives rise to new combinations maintained vegetatively in sclerotia from season to season.

Agrobacterium tumefaciens as a species has one of the widest host ranges known, but Anderson and Moore (1979) demonstrated that individual isolates differ in host range and that the reported wide host range is the collective total of entities within the species. Loper and Kado (1979) transformed a strain virulent only on grape to one with a host range of 26 species

EVOLUTION AND TAXONOMY

by transferring a plasmid from a strain with 26 hosts to the limited host range bacterium, indicating that plasmids govern not only pathogenicity but host range as well. Knauf et al. (1982) found that the host bacterium (the recipient of the plasmid) also influences plant host range, that the regular bacterial genome as well as the plasmid has influence. In 20 of 22 cases in which plasmids were transferred, the host range of the recipient became that of the donor of the plasmid. In two cases the host range of the transformed recipient differed from that of the donor of the plasmid, proving that the chromosome of the recipient cell affected host-pathogen relationships. These studies prove the probable origin of *A. tumefaciens* from a soil bacterium, *Agrobacterium radiobacter*. *A. radiobacter* is a nonpathogenic soil bacterium without the Ti plasmid.

MATING POPULATIONS NOT CONSIDERED BIOLOGIC SPECIES

The future taxonomy of plant pathogens fills with challenge as new knowledge accumulates. Within the species *Fusarium solani*, as modified by Snyder and Hansen (1940), are several formae speciales distinguished on the basis of host range. Matuo and Snyder (1973) identified seven populations within *F. solani* based on mating reactions. All pathogenic isolates that were fertile were heterothallic (all homothallic isolates they studied were nonpathogenic). The sexual state of all seven mating groups was the same, *Nectria haematococca*. Isolates were fertile within groups but not between groups. *F. solani* is a natural group with the same sexual state, yet within it are distinct mating populations.

Van Etten (1978) made an intensive study of mating group VI (f. sp. *pisi*). He found both+ and −isolates common in the United States, making it strange that no one has found the sexual state in nature on peas. Unfortunately, Van Etten found members of mating group VI that were pathogenic on other plants but not on peas, so the correlation of mating group and host reactions are not adequate for identification, at least not for f. sp. *pisi*. Even though Van Etten considers the mating groups to be biologic species, he favors retention of the present f. sp. system within *F. solani*.

Hsieh et al. (1977) identified mating group A of *Fusarium moniliforme*, mostly from maize, mating group B from sugarcane in Taiwan, and mating group C from rice in Taiwan. Their study was of limited extent, but it is highly significant that isolates from sugarcane from Taiwan were fertile among themselves but not with isolates from rice. That three intra- but not interfertile groups of isolates were found on an island the size of Taiwan makes one wonder what would be the result of a truly extensive study? *F. moniliforme* (*Gibberella fujikuroi*) is heterothallic.

Verticillium dahliae has no sexual state, but Puhalla and Hummel (1983) studied vegetative compatibility, the ability of one thallus to exchange nuclei with another to form heterokaryons. Among 96 isolates from 38 host species from 15 countries they identified 16 vegetative compatibility groups. Nuclear exchange could occur within groups but not between groups.

Fungi develop means of limiting genetic exchange. The result in most cases is increased fitness on a narrower host range. These sterility groups are evidence of microevolution, differences earlier students of the species did not even know existed. At present, their discovery certainly does not justify changing names of the pathogens involved, but researches such as these can add to our understanding of evolution and epidemiology.

SPLITTING VERSUS LUMPING

Taxonomy is essential to sustained progress. We know that Tillet and Prevost worked with *Tilletia caries*, and what they discovered is useful today. Tyler (1933b) determined that a root-knot nematode produced 12 generations without mating. She (1933a) also learned the number of days it took for this nematode to reproduce at various temperatures. Her classic studies have diminished value, because we will never know the species with which she worked. She studied a root-knot nematode on tomato, and more than one species attacks tomatoes in California. I studied the root-knot nematode, "*Heterodera marioni*," as an undergraduate in my first plant pathology course. The professor was up-to-date (1939). It was not until 1949 that Chitwood separated the cyst-nematodes from the root-knot nematodes and described several species within the genus *Meloidogyne*. There are now about 50 species of cyst nematodes (*Heterodera*) and about 37 species of root-knot nematodes. Can you imagine the confusion and lack of progress there would be today if nematologists did not distinguish the species of these nematodes? What about geographic distribution, rotations, and breeding for resistance?

Within my lifetime some pathologists referred to some fusaria pathogenic to cereals as *Fusarium roseum*, or as *F. roseum* f. sp. *cerealis*, thinking they had identified a pathogen. Until the cultivars Avenaceum, Culmorum, and Graminearum were added, their designations were worthless. Works using *F. avenaceum*, *F. culmorum* and *F. graminearum* are unambiguous.

In Europe most workers recognized *Verticillium dahliae* and *Verticillium albo-atrum*. For years most American workers referred to *V. albo-atrum* only. If they did not designate *V. albo-atrum*, microsclerotial form or dark, mycelial form, if they worked where both species exist and on a host susceptible to both species, we will never know with what they worked.

Taxonomists are called "lumpers" if they tend to recognize larger groups and "splitters" if they tend to recognize smaller differences. When justified, as in the wilt forms of *Fusarium oxysporum*, lumping represents progress and is an aid to science. When lumping is unjustified and is followed, lost information results. If in doubt, don't "lump." Everyone welcomes lumping when justified, because it simplifies our work and reduces the clutter of literature, but caution should precede lumping.

COMMENTS

The antiquity of living organisms, especially widespread soilborne plant pathogens, is evidence that they have evolved means of reproducing and surviving over many years. The structures and processes were developed prior to agriculture, and organisms of greatest significance to us were selected, in large part, by their ability to benefit from the activities of man. They have much in common with weeds. The most common weeds of cultivated soils are annuals, plants with short life cycles. Many of the annual weeds are self-pollinated, comparable to homothallic fungi. Weeds of uncultivated lands tend to be perennials, and outcrossing is more common among them, comparable to the higher basidiomycetes so important as pathogens of trees.

Evidence is increasing that substantiates the observation of Stebbins (in Lemke, 1973), that organisms that develop in transient environments tend to reduce genetic variation. Rapid reproduction (viruses, bacteria, many fungi, some important nematodes) is favored by asexual reproduction (Kendrick, 1979b). Examples familiar to plant pathologists are the reproduction of cereal rusts by urediospores and of aphids by parthenogenetic females. The sexual states are of greatest importance as means of surviving periods of adversity (teliospores of *Puccinia graminis*, eggs of aphids) or in producing forcibly ejected propagules (ascospores, basidiospores) important in aerial dissemination.

Mycologists and plant pathologists, in my opinion, have exaggerated the significance of genetic recombination resulting from the sexual state. Leaf rust of wheat (*Puccinia recondita*) and stripe rust (*P. striiformis*) develop new races in abundance, the former with little use of the sexual state, the latter with none. Mutation can be easily underestimated. Bruehl (1976) hypothesized that if a mutation at one locus occurred once in every 1×10^6 spores, and soil contained 1000 spores/g, a mutant at that locus would occur in every kilogram of infested soil. In haploid forms, maximum fitness could be maintained in the bulk population at little cost to the species. The vast majority of propagules could be at a high state of fitness. The mutants, most of which would be inferior, would die. If, on the other hand, the one per million was superior, with one per kilogram of soil, within a hectare of land at least some of them would infect a plant to survive and increase. Regular recombination in contrast would result in many unfit individuals in situations placing a premium upon fitness.

Most plant pathologists are so engrossed with achieving control that they spend a minimum of effort on taxonomy, but as exact identification of organisms as is possible and practicable aids in accumulation of knowledge. Taxonomists deserve greater support and appreciation.

KEY REFERENCES

Anderson, J. B. and R. C. Ullrich. 1979. Biological species of *Armillaria mellea* in North America. *Mycologia* **71**:402–414.
Anderson, N. A. 1982. The genetics and pathology of *Rhizoctonia solani*. *Ann. Rev. Phytopathol.* **20**:329–347.
Bruehl, G. W. 1976. Management of food resources by fungal colonists of cultivated soils. *Ann. Rev. Phytopathol.* **14**:247–264.
Buddenhagen, I. W. 1965. The relation of plant-pathogenic bacteria to the soil. Pages 269–282 in *Ecology of Soilborne Plant Pathogens*. K. F. Baker and W. C. Snyder, eds. Univ. of California Press., Berkeley.
Francis, R. G. and L. W. Burgess. 1977. Characteristics of two populations of *Fusarium roseum* 'Graminearum' in eastern Australia. *Trans. Brit. Mycol. Soc.* **68**:421–427.

REFERENCES

Adams, D. H. 1974. Identification of clones of *Armillaria mellea* in young-growth ponderosa pine. *Northwest Sci.* **48**:21–28.
Anderson, A. R. and L. W. Moore. 1979. Host specificity in the genus *Agrobacterium*. *Phytopathology* **69**:320–323.
Anderson, J. B., R. C. Ullrich, L. F. Roth, and G. M. Filip. 1979. Genetic identification of clones of *Armillaria mellea* in coniferous forests in Washington. *Phytopathology* **69**:1109–1111.
Baker, K. F. 1970. Types of Rhizoctonia diseases and their occurrence. Pages 125–148 in *Rhizoctonia solani, Biology and Pathology*. J. R. Parmeter, Jr., ed. Univ. of California Press, Berkeley.
Bruehl, G. W. 1975. Systems and mechanisms of residue possession by pioneer fungal colonists. Pages 77–83 in *Biology and Control of Soilborne Plant Pathogens*. G. W. Bruehl, ed. American Phytopathology Society, St. Paul, Minn.
Buddenhagen, I. W. 1960. Strains of *Pseudomonas solanacearum* in indigenous hosts in banana plantations of Costa Rica, and their relationship to bacterial wilt of bananas. *Phytopathology* **50**:660–664.
Buddenhagen, I. W. and T. A. Elsasser. 1962. An insect-spread bacterial wilt epiphytotic of Bluggoe banana. *Nature* **194**:164–165.
Buddenhagen, I. W. and A. Kelman. 1964. Biological and physiologic aspects of bacterial wilt caused by *Pseudomonas solanacearum*. *Ann. Rev. Phytopathol.* **2**:203–230.
Burgess, L. W., A. H. Wearing, and T. A. Toussoun. 1975. Surveys of Fusaria associated with crown rot of wheat in eastern Australia. *Aust. J. Agric. Res.* **26**:791–799.
Burnett, J. H. 1965. The natural history of recombination systems. Pages 98–113 in *Incompatibility in Fungi*. K. Esser and J. R. Raper, eds. Springer-Verlag, New York.
Chitwood, B. G. 1949. "Root-knot nematodes"—Part I. A revision of the genus *Meloidogyne* Goeldi, 1887. *Proc. Helminthol. Soc. Wash. D.C.* **16**:90–104.
Cooke, W. B. 1979. *The Ecology of Fungi*. CRC Press, Boca Raton, Fla.

Cooper, J. I. and B. D. Harrison. 1972. Potato mop-top virus. *Rep. Scott. Hort. Res. Inst.* **1971**:63.
Dobzhansky, T. 1951. *Genetics and the Origin of Species.* Columbia Univ. Press, New York.
Dobzhansky, T. 1972. Species of *Drosophila.* New excitement in an old field. *Science* **177**:664–669.
Flor, H. H. 1953. Epidemiology of flax rust in the north central states. *Phytopathology* **43**:624–628.
Gallegly, M. E. and J. Gallindo. 1958. Mating types and oospores of *Phytophthora infestans* in nature in Mexico. *Phytopathology* **48**:274–277.
Gumpf, D. J. 1971. Purification and properties of soilborne wheat mosaic virus. *Virology* **43**:588–596.
Hsieh, W. H., S. N. Smith, and W. C. Snyder. 1977. Mating groups in *Fusarium moniliforme. Phytopathology* **67**:1041–1043.
Kassanis, B., R. D. Woods, and R. F. White. 1972. Some properties of potato mop-top virus and its serological relationship to tobacco mosaic virus. *J. Gen. Virol.* **14**:123–132.
Kendrick, B. 1979. Introduction. Pages 11–15 in *The Whole Fungus,* Vol. 1. Proc. Second Intern. Mycol. Conf., Kananaskis, Alberta, Canada. B. Kendrick, ed. National Museum of Natural Sciences, Ottawa.
Knauf, V. C., C. G. Panagopoulos, and E. W. Nester. 1982. Genetic factors controlling the host range of *Agrobacterium tumefaciens. Phytopathology* **72**:1545–1549.
Ko, W. H., H. S. Chang, and H. J. Su. 1978. Isolates of *Phytophthora cinnamomi* from Taiwan as evidence for an Asian origin of the species. *Trans. Br. Mycol. Soc.* **71**:496–499.
Kuninaga, S. and R. Yokosawa. 1980. A comparison of DNA base compositions among anastomosis groups in *Rhizoctonia solani* Kuhn. *Ann. Phytopathol. Soc. Jap.* **46**:150–158.
Lemke, P. A. 1973. Isolating mechanisms in fungi - prezygotic, postzygotic and azygotic. *Persoonia* **7**:249–260.
Loper, J. E. and C. Kado. 1979. Host range conferred by the virulence-specific plasmid of *Agrobacterium tumefaciens. J. Bacteriol.* **139**:591–596.
Maggenti, A. 1981. *General Nematology.* Springer-Verlag, New York.
Mather, K. 1965. The genetic interest of incompatibility in fungi. Pages 113–117 in *Incompatibility in Fungi.* K. Esser and J. R. Raper, eds. Springer-Verlag, New York.
Matuo, T. and W. C. Snyder. 1973. Use of morphology and mating populations in the identification of formae speciales in *Fusarium solani. Phytopathology* **63**:562–565
Naiki, T. and T. Ui. 1978. Ecological and morphological characteristics of the sclerotia of *Rhizoctonia solani* Kuhn produced in soil. *Soil Biol. Biochem.* **10**:471–478.
Norton, D. C. 1978. *Ecology of Plant Parasitic Nematodes.* Wiley, New York.
Paulsen, A., C. L. Niblett, and W. G. Willis. 1975. Natural occurrence of tobacco mosaic virus in wheat. *Plant Dis. Reptr.* **59**:747–750.
Powell, C. A. 1976. The relationship between soilborne wheat mosaic virus and tobacco mosaic virus. *Virology* **71**:453–462.

Puhalla, J. E. and M. Hummel. 1983. Vegetative compatibility groups within *Verticillium dahliae*. *Phytopathology* **73**:1305–1308.

Purss, G. S. 1969. The relationships between strains of *Fusarium graminearum* Schwabe causing crown rot of various gramineous hosts and stalk rot of maize in Queensland. *Aust. J. Agric. Res.* **20**:257–264.

Purss, G. S. 1971. Pathogenic specialization in *Fusarium graminearum*. *Aust. J. Agric. Res.* **22**:553–561.

Sequeira, L. and C. W. Averre III. 1961. Distribution and pathogenicity of strains of *Pseudomonas solanacearum* from virgin soils in Costa Rica. *Plant Dis. Reprtr.* **45**:435–440.

Sequeira, L. and A. Kelman. 1962. The accumulation of growth substances in plants infected by *Pseudomonas solanacearum*. *Phytopathology* **52**:439–448.

Shaw, C. G., III and L. F. Roth. 1976. Persistence and distribution of a clone of *Armillaria mellea* in a ponderosa pine forest. *Phytopathology* **66**:1210–1213.

Sherwood, R. T. 1969. Morphology and physiology of four anastomosis groups of *Thanatephorus cucumeris*. *Phytopathology* **59**:1924–1929.

Simmonds, N. W. 1976. *Evolution of Crop Plants*. Longman, London.

Snyder, W. C. and H. N. Hansen. 1940. The species concept in *Fusarium*. *Amer. J. Bot.* **27**:64–67.

Stanghellini, M. E. 1974. Spore germination, growth and survival of *Pythium* in soil. *Proc. Am. Phytopathol. Soc.* **1**:211–214.

Stone, A. R. 1977. Recent developments and some problems in the taxonomy of cyst-nematodes, with a classification of the Heteroderoidea. *Nematologica* **23**:273–288.

Stover, R. H. 1972. *Banana, Plantain and Abaca Diseases*. Commonwealth Mycological Institute, Kew, Surrey, England.

Sung, J. M. and R. J. Cook. 1981. Effect of water potential on reproduction and spore germination of *Fusarium roseum* 'Graminearum', 'Culmorum' and 'Avenaceum'. *Phytopathology* **71**:499–504.

Triantaphyllou, A. C. and H. Hirschmann. 1980. Cytogenetics and morphology in relation to evolution and speciation of plant-parasitic nematodes. *Ann. Rev. Phytopathol.* **18**:333–359.

Tsuchizaki, T., H. Hibino and Y. Saito. 1973. Comparisons of soilborne wheat mosaic virus isolates from Japan and the United States. *Phytopathology* **63**:634–639.

Tsuchizaki, T., H. Hibino, and Y. Saito. 1975. The biological functions of short and long particles of soilborne wheat mosaic virus. *Phytopathology* **65**:523–532.

Tyler, J. 1933a. Development of the root-knot nematode as affected by temperature. *Hilgardia* **7**:391–415.

Tyler, J. 1933b. Reproduction without males in aseptic root cultures of the root-knot nematode. *Hilgardia* **7**:373–388.

Van der Plank, J. E. 1968. *Disease Resistance in Plants*. Academic Press, New York.

Van Etten, H. D. 1978. Identification of additional habitats of *Nectria haematococca* mating population VI. *Phytopathology* **68**:1552–1556.

Wallace, H. R. 1973. Nematode Ecology and Plant Disease. Arnold, London.

Ward, H. M. 1903. Further observations on the brown rust of the bromes, *Puccinia dispersa* (Erikss.), and its adaptive parasitism. *Ann. Mycol.* **1**:132–151.

Weresub, L. K. and K. A. Pirozynski. 1977. Pleomorphism of fungi as treated in the history of mycology and nomenclature. *Mycotaxon* **6**:207–211.

Willis, J. C. 1922. *Age and Area. A Study of Geographic Distribution and Origin of Species*. Cambridge Univ. Press, Cambridge. UK.

Wulff, E. V. 1943. *An Introduction to Historical Plant Geography*. Chronica Botanica, Waltham, Mass.

Index

(Boldface indicates Figures and Tables)

Actinostasis, 94
Active possession of substrate (*see* Saprophytic survival)
Aeration, 58–69 (*see* Oxygen)
  carbon dioxide, effects of (*see* Carbon dioxide)
  role in soil formation, 58
Agrobacterium tumefaciens, 140–141, 185, 350–351
Ammonia, 126
Amoebae as predators, 227–230
Anastomosis groups, 348
Anguina tritici, 303
Antagonism:
  concept, 196
  in potato lenticels, 30–31
  sensitivity and tolerance to, 30
  in spermosphere, 30, 117
Antibiotics, 135–143
  in bulk soil vs. microsites, 135–136
  adsorption by soil, 135–136
  aid saprophytic survival, 146, 147–151
  effect of water potential on, 136–138
  in fungistasis, 139–140
  leakage from sclerotia, 175

Leucopaxillus cerealis vs. *Phythophthora cinnamomi,* 250
  vs. pathogenic bacteria, 140
  production of, 149–151
Aphanomyces euteiches, 29, 314
Aphelenchoides besseyi, 226
Aphelenchus avenae, 9–10
Armillaria mellea:
  a "sugar" fungus, 155–156
  attacked by *Trichoderma viridis,* 221–222
  genetic markers, 349–350
  rhizomorph development, effect of oxygen upon, 60–61
  root stocks, 308
  size of clones, 350
  stimulated by tannic acid, 155
Azotobacter chroococcum, 107, 210

Bacillus cereus, 23, 144, 185
B. spp., 107, 210, 218
Background microflora, 38, 197–199
Bacteria:
  adsorption in soil, 23
  competition in lenticels, 30

Bacteria: (*continued*)
 entrapment in host, 85
 entrapment in soil, 23
 germling survival affected by, 30
 movement by water, 23, 185
 rapid stress of, 191–192
 relation to food in soil solution, 184–185
 in the rhizosphere, 105–107, 111, 185, 188, 189, 190
 role in fungistasis, 94–97, 127, 132
 role of exudates, 191
 survival: in phyllosphere, 188
  in soil, 184–185
  in water, 185
Bacterial soft rot of potato (*see Erwinia carotovora*)
Bacteriocins, 140–141
Bacteriostasis, 94
Bacterization, 209
 (*see also* Microbialization)
Bean root rot (*see Fusarium solani* and *F. solani* f. sp. *phaseoli*)
Big-vein of lettuce, 244–245, 269–270
Biologic balance (*see* Microflora, general)
Biologic control, 210
Biologic species, 347
Bipolaris sorokiniana:
 attacked by amoebae, **227**
 natural microbialization, 211
 relation to soil and rainfall, 4, 14
 response in different soils, 97, 98
 response to fungistasis, 92, 95, 101–102
 rhizosphere role in resistance, 110
 survival as spores, 162–163, 165
B. victoriae, 95, 96, 211–212
Black shank of tobacco, 285–289
Bulk density of soil, 5, 7–9

Carbon dioxide:
 colonization of substrate by *Rhizoctonia solani*, 67–68
 concentration in soil, 59
 differential tolerance in *Rhizoctonia solani*, 66
 Gaeumannomyces graminis, 66
 morphogenetic effects on *Phymatotrichum omnivorum*, 67
 saprophytic relationships influenced by, 67–68
 soil pH, effect of, 66
 tolerance in fungi to, 65–66, 68–69
Carbon: nitrogen (C:N) ratios of substrate, 130
Carpogenic germination of sclerotia, 168–169, 178
Cation exchange capacity, 11
Cellulolysis adequacy index, 152–153
Cephalosporium gramineum:
 antibiotics aid survival of, 147–149, 156

 composition of residue changed by, 149–150
 fire destroys inoculum, 54
 frozen roots predisposed, 108
 production of antibiotics by, 137
 rotation, 303–304
 soil pH affects survival, 149
Chaetomium sp. on seed, 212
Charcoal rot, 50–51, 81–82
Chelation, 214–215, 218
Chemotaxis, 105, 339
Chemotropism, **99**, 105, 339
Clay, 8–11, 23
Closteridium spp., 63–64, 107, 189
Club root of cabbage, 12–13, 24–25
Colonists of cultivated soils, 338–340
Colonization of substrate, 144–147, 184, 190–191, 196–197, 201–205
Competency of propagules, 177, 178, 323
Competition, 196–207
 factors that reduce competition, 197, 200–201
 for food, 196–197
 between interformae speciales, 200–201
 for iron, 95–96, 108
 saprophytic (*see* Competitive saprophytic ability), 201
Competitive saprophytic ability, 201–205
Coniothyrium minitans, 223–224
Cortical death, 75, 110–111
Cortical rots (*see* Predisposition)
Cotton wilt, 279–284
Criconemella xenoplax, **232**
Cylindrocladium crotalariae, 83

Delia platura, 63
Diplodia zeae, 74
Direct germination of sporangia, 20
Disease curve, 37–41
Disease decline, 238–246
 of oat cyst-nematode, 241–243
 of potato scab, 239–240
 of *Rhizoctonia solani*, 243–244
 specific disease decline concept, 238–239, 244–245
 of take-all, 240–241
Disease suppressive soils, proper use of, 244–245
 (*see also* Disease decline)
Dormancy, endogenous, 160–161
Dormant survival, 145
Drought stress, 27, 41, 81–82

Endogone sp., 251 (*see* Mycorrhizae)
Epiphytic survival of bacteria, 188, 192
Eruptive germination of sclerotia (*see* Sclerotia)

INDEX

Erwinia carotovora:
  control by microbialization, 215
  nature of parasitism, 144
  oxygen deficiency, 63–64
  survival in soil, 188–190
Ethylene, 130
Evolution:
  adjusted death of host, 336
  antiquity of pathogens, 335
  colonists of cultivated soils, 338
  Fusarium graminearum, 344
  independence of propagules, 340
  mating types, 351
  origin of plant parasitic nematodes, 342
  phenotypic plasticity, 337
  polymorphism, 337
  Pseudomonas solanacearum, 345
  quickness of response, 339
  radiate evolution, 200–201, 342
  relation to antibiotics, 340
  role of sexual state, 338, 353
  seasonal rhythms, 336
  sexual nematodes, 341
  sexual states, 337
  size of propagules, 340
  slow digestion of substrate, 340
  variability vs. fitness, 343
  wheat soilborne mosaic virus, 346
Exchangeable bases, 11
Exudates, 107–113
  bacterial, 191
  fungal, 191
  in fungistasis, 82, 92–93
  oxygen as an exudate, 64
  from roots, 107–113
  from seeds, 117, 121
  stimulate germination of sclerotia, 82
  toxic, 113

Fertilizers, 311–313
  deficiencies, 78–79, 82, 111, 312
  excesses, 27, 311
  form of nitrogen, 111, 312–313
  nitrogen affects saprophytic survival, 313
Fire and smoke, 53–54
Flax wilt, 39–40
Forest soils, 3–4
Fruiting stress, 74, 81–82
Fungistasis, 91–102
  abiotic factors, 96
  as a timing mechanism, 94
  bacterial siderophores, 95–96
  carbon- or energy-deficiency, 96–97
  concept, 92–93
  effect of speed of germination, 98
  importance of, 98–99
  induction of, 94–97
  iron deficiency, 95–96
  pine litter, 98–100
  propagules differ in response to, 97–98
  release from, 98
  role of bacteria, 94–95
  soils differ, 97
  volatile factors, 96, 127, 132
Fusarium avenaceum, 213
F. culmorum:
  effect of water stress on the host, 27–28, 311
  fire does not eliminate from stubble, 54
  germling survival, 30
  saprophytic colonist, 202–204
F. graminearum:
  evolution within, 344–345
  predisposition by temperature, 38, 45–46
  rotation, 303
  saprophytic survival, 151–152
  tissue senescence, 74
F. moniliforme:
  effect of water on sporulation, 26
  mating groups, 351
  no-till vs. sorghum stalk rot, 299
F. oxysporum:
  death under pine litter, 98–99
  fungistasis in, 95–96, 97–98
  radiate evolution, 200–201
  trapped microconidia, 84–85
F. oxysporum f. sp. *conglutinans,* 40
F. oxysproum f. sp. *cubense,* 84
F. oxysporum f. sp. *lini,* 39–40, 277, 278–279
F. oxysporum f. sp. *melonis,* 41
F. oxysporum f. sp. *pisi,* 40
F. oxysporum f. sp. *vasinfectum:*
  discovery, 270, 280–281
  nematode interactions, 283–284
  potassium deficiency, 284
  races, 281–282
  resistance, 282–283
  response to temperature, 40
  spread, 281
Fusarium root rot of bean (*see F. solani* f. sp. *phaseoli*)
Fusarium seedling blight of maize, 38, 45–46
Fusarium solani, 201, 351
F. solani f. sp. *phaseoli:*
  bulk density, pressure pan, 7–8
  not adapted to spermosphere, 117, 121
  predisposition by oxygen deficiency, 62–63
  response to exudates, 109
  seeding rate, 306–307
  use of volatiles, 128
Fusarium spp., chlamydospores vs. fungistasis, 98–100
Fusarium wilts, 277–278
  cotton, 40, 279–284
  flax, 39–40, 278–279
  temperature effects on, 38–41

Gaeumannomyces graminis var. *tritici:*
   aeration, 66, 69
   decline, 240−241
   fertilization, 312
   host lignification, 73−74
   host nutrition, 12, 152−153, 156
   irrigation, 311
   microbialization, 215−216
   response to temperature, 38
   rhizosphere pH, 11
   tillage, 300
General soil microflora, 197−200
Geocarposphere, 105
Geographic distribution, effect of
   temperature on, 37, 42, 46−49
Germination:
   inhibitors, 92−93, 95, 96, 161
   stimulants, 92−93, 98−100
Germling survival, 30, 117, 121
Gibberella zeae (see *Fusarium graminearum*)
Gliocladium virens, 222
Globodera rostochiensis, 112
G. solanacearum, 302
G. spp., temperature, 48
Glomus fasiculatus, 252
G. macrocarpum, 252
Grapevine fan-leaf virus, 256
Growth regulators, 315

Hatching factors, 111−113
Hemicycliophora arenaria, oxygen
   requirements, 62
Herbicides, 292, 314, 315
Heterodera avenae, 51, 241−243
H. glycines, 49, 52, 82
H. schachtii, 111, 302−303
Hirschmaniella oryzae, 64−65
Hirsutella rhossiliensis, attacks nematodes, **232**
Humidity, effect on hyphal growth, 21−22, 25−26
Humus, 9−10
Hydrogen sulfide, effect on nematodes, 64−65
Hydrophilic and hydrophobic propagules, 22−23
Hyperparasites:
   Aphelenchoides besseyi, 226
   Aphelenchus avenae, 226, 227
   Arachnula impatiens, 227−228
   Arthrobotrys, 231−232
   Bacillus penetrans, 233−234
   baiting them from soil, 225−226
   Coniothyrium minitans, 223
   fungi attacking nematodes, 231−233
   Gliocladium virens, 222
   Hirsutella rhossiliensis, 232−233
   mycoviruses, 230−231
   Nematophthora gynophila, 241−243
   Pythium spp., 225−226
   Sporodesmium sclerotivorum, 224−225
   Theratromyxa weberi, 227
   Trichoderma spp., 222
   viruses, 230−231
Hyphae as survival structures, 159, 163−164, 165

Infection index, 323
Inoculum potential, 322−325
Insecticides affect *Rhizoctonia solani,* 315−316
Iron deficiency, 95−96, 108
Iron pan, 6−7
Iron toxicity, 64, **65**
Irrigation:
   constant wetness, 30−31
   in ditches or flat, 44−45
   frequency, 44
   pathogens in irrigation water, 309−311

Jarrah (*Eucalyptus marginata*) dieback, 6−7
Juvenile susceptibility, 72−74

Kaolinite, 9−10

Laimosphere, 105, 117
Lectins, 85
Leucopxillus cerealis vs. *Phytophthora cinnamomi,* 250
Longidorus spp.:
   distribution in soil, 259, 260
   importance of weeds, 262−263
   longevity, 258, 262
   loss of virus from, 258−262
   toxicity of raspberry exudates, 113
   vector of virus, 257
   vector specificity, 259

Macrophomina phaseolina:
   complex host stresses, 81−82
   diurnal temperature fluctuation, 50
   effects of growth regulators on susceptibility, 315
   nutrient loss by microsclerotia, 96
Macrophysiology, 72
Macroposthonia ornata, interaction in peanut black rot, 83
Man, 298−317
   clean seed, 304−305
   fertilizers, 311−313
   growth regulators, 315

harmful side effects of chemicals, 315–316
herbicides, 313–315
irrigation, 308–311
planting, 305–308
root stocks, 308
rotation, 300–305
tillage, 299–300
Meloidogyne graminis, sex ratios, 51–52
M. incognita, 83, 233–234, 301
M. javanica, water and oxygen on hatching, 31–32
M. spp., 9, 37, 113
Microbialization, 209–218
  broad spectrum, 216–218
  concept, 209–210
  damping-off, 213
  early attempts, 212
  natural, 211–212
  specific, 140–141
Microbiostasis, 93–94
Microconidia, 84–85, 97
Microflora, general, 197–199
Microsclerotia:
  aeration affects formation of, 68–69
  death by soil solarization, 52–53
  nutrient loss from, 96
  release of microsclerotia from host residue, 164
  response to fungistasis, 81–82
  as survival structures, 163–164
  volatiles stimulate germination, 128
Montmorillonite, 10–11
Mucigel, 109
Mycorrhizae, 248–252
  attacked by *Aphelenchus avenae*, 227
  ectotrophic, 248–250
  effect of volatiles upon, 127
  Glomus spp., 252
  vesicular-arbuscular, 250–252
Mycostasis, 92
Mycoviruses, 230–231

Nematodes:
  aerobic, 61–62, 64–65
  attacked by amoebae, 227–229
  carrying capacity of the host, 83
  depth in soil, 51
  effect of a second disease upon, 83
  effect of plant size on damage, 329–330
  hatching, 31–32, 111–113
  host defoliation, 83
  host nutrition, 82–83
  interactions of, 83, 283–284, 287–288
  in irrigation water, 310
  movement in soil, 9, 10
  number of generations, 52
  oat cyst, 51, 241–243
  recovery after fumigation, 329
  sex ratios, 51
  sexual reproduction, 341
  taxonomy, 340–342
  toxic residues, 113
  as vectors of viruses, 257–263
Nematophthora gynophila, 241–243
Niche, 197
Nitrogen:
  effect of form N on rhizosphere pH, 111
  effect upon saprophytic survival, 152–154
  recylcing by wood fungi, 154–155
Numbers, 81
  baits, 324
  cautions, 322
  direct assay of sclerotia, 324–325
  disease incidence as a measure of, 325
  after fumigation, 329
  host response to, 327
  infection index as a measure of, 323
  not the limiting factor, 326
  plant size vs. numbers, 329–330
  plant-to-plant spread, 327
  relation to propagule size, 321–322
  response to germination stimulators, 330
  rotations, 330

Oat cyst nematode, 51, 241–243
Obligate parasites, 144
Olpidium brassicae:
  adsorption of virus, 255, 270
  big-vein of lettuce virus, 269–270
  tobacco necrosis virus, 270–271
  vector of virus, 254, 255
  water relations, 244–245
Osmotica, 19
Oxygen:
  anaerobic microsites, 59–60
  bean root rot, 62–63
  decomposition of residues, 60
  diffusion, 24, 59–60
  effect on distribution of roots, 58
  escape from rice roots, 64–65
  in germinating seeds, **28**
  gradients, 60
  hatching nematode eggs, 31–32
  importance of soil pore space, 8, 9, 24, 31–32
  macerating enzymes affected by, 63–64
  predisposition by deficiency of, 28, 62–64, 121
  quantity required, 58–59, 111
  rhizomorph growth, 60–61
  role of water films, 24, 59–60
  sensitivity in nematodes, 61–62
  wound healing in potatoes, 63–64

Pans, 5–8

Paratrichodorus anemones, vector of virus, 257
Paratylenchus dianthus response to exudate, 112–113
Passive possession of substrate and dormant survival, 145
Pathogen suppressive soils, 238–239, 244
Penicillium oxalicum, microbialization, 213, 216–218
P. spp., 66, 94
Periconia circinata, 144
pH, 11–14, 111, 149, 215, 244
Phenotypic plasticity, 337
Phialphora gregatum, 150, 301
Phloem transport of viruses, 85–86
Phosphorus deficiency, 78–79
Phyllosphere, 188, 192
Phymatotrichum omnivorum:
  biological control, 199–200
  carbon dioxide, 67
  cortical sloughing, 75
  geographic distribution, 46–47
  microbialization, 212
  sclerotia killed by desiccation, 169
  seeding date, 49, 306
  temperature, 46–47, 49
  tillage, 299–300
Phytophthora cactorum, root stocks, 308
P. capsici, spread by irrigation, 308–309
P. cinnamomi:
  iron pan increases Jarrah dieback, 6–7
  protection from by mycorrhizae, 250
  response to matric and osmotic water potential, 2, 20
  water affects type of germination, 28–29
P. infestans, sporangia in soil, 22–23
Phytophthora megasperma f. sp. *glycinea,* 289–293
  age of host, 292
  herbicide effects, 292, 314
  interaction with mycorrhizae, 251–252
  pathogenic adaptability, 292–293
  soil, 291–292
  spread, 291
P. palmivora, inoculum potential, 322
P. parasitica var. *nicotianae,* 285–289
  inoculum, 288–289, 328
  nematode interactions, 287–288
  pH, 13–14
  resistance, 286–289
  spread, 285–286
  temperature, 47–48, 289
  water, 289
Phytophthora root rot of soybean (*see P. megasperma* f. sp. *glycinea*)
Phytophthora spp.:
  in irrigation water, 309–310
  temperature, 47–48
  water relations, 20, 22, 24, 28–29

Planting:
  clean seed, 304–305
  date, 49, 306
  depth, 306
  rate, 306–307
Plant-to-plant spread, 327, 350
Plasmids, 141, 350–351
Plasmodiophora brassicae, 12–13, 24–25
Polymyxa graminis, 254, 265
Polysaccharide slimes, 85, 191
Poria weirii, survival affected by nitrogen, 313
Possession of substrate (*see* Saprophytic survival)
Potassium deficiency, 64–65
Potato corky ringspot virus, 260
Potato scab, 30–31, 239–240
Pratylenchus penetrans, rotation, 301
P. zeae, soil texture on movement, 9–10
Predisposition:
  cotton wilt, 283–284
  juvenile tissue, 72–74
  by nematodes, 283–284, 287–288
  nutrient imbalance, 78–79, 82–83
  sclerotia, 177
  (*see also* Stress)
Pressure, effect on sclerotia, 15–16
Pressure pan, 5, 7–8
Prior colonization of substrate (*see* Saprophytic survival)
Pseudobactin, 214
Pseudocercosporella herpotrichoides, 60, 146–147, 304
Pseudomonas fluorescens:
  in bacterization, 209, 210, 212, 214, 215–216
  Gaeumannomyces graminis, 215–216
  in rhizosphere, 107
  siderophores, 95–96, 214
P. herbicola in bacterization, 209
P. putida, 191–192, 210
P. solanacearum:
  agglutination by lectins, 85
  evolution, 345
  in hydroponics, 310–311
  rotation, 304
  survival in soil, 186–187
  temperature, 37, 46, 54
P. spp. production of antibiotics by, 139
P. syringae survival, 187–188
P. tabaci, survival, 188
Puccinia spp., temperature affects sequence of organisms, 49, 54
Pythium aphanidermatum, 326
P. arrhenomanes:
  vs. antibiotics, 138–139
  attacked by *Aphelenchus avenae,* 226
  climatic adaptation, 79–80
  effect upon nutrient uptake, 79
  nutrient imbalance, 78–79

INDEX

predisposition to, 78–80, 125–126
toxic volatiles, 125–126
P. iwayami, effect of water quality, 34
P. myriotylum, propagule number vs. disease, 325
Pythium root rot (*see P. arrhenomanes*)
  reduces nutrient utilization, 79
  in spring wheat and barley, 78–80
  in sugarcane, 78–79
Pythium spp.:
  adapted to spermosphere, 117
  attacked by *Pythium* spp., 225–226
  fungistasis, 94, 98, 99, 101–102
  injured seeds, 118–120
  juvenile susceptibility, 72–73
  sensitivity to ammonia, 126
  temperature, 49, 75–76
  tissue maturation, 73
  water, 28
P. ultimum:
  adaptation to spermosphere, 30, 120
  attacked by *Gliocladium virens*, 222
  dormant oospores, 160, 161
  response to seed anoxia, 120–121
  response to seed leachates, 28, 120–121
  sensitivity to antibiotics, 137–138

Radiate evolution, 342
Radopholus similis, temperature influences depth in soil, 51
Raspberry ringspot virus, 258–259
Resistance:
  climatic adaptation, 79–80
  cortical sloughing, 75
  gels and tyloses in vessels, 84–85
  juvenile tissue, 72–74
  localization of bacteria, 85
  senescence, 74–75
  tissue maturation, 73–74
  (*see also* Predisposition; Stress)
Resting structures:
  adaptations, 159
  release from dormancy, 160–161
  release from refuse, 164
  toughness, 162
Rhizobacteria, 210
Rhizoctonia solani:
  anastomosis groups, 348–349
  attacked: by *Gliocladium virens*, 222
    by virus, 230–231
  buoyant sclerotia, 169, 173, 175
  carbon dioxide ecotypes, 66
  carbon dioxide on saprophytic ability, 67–68
  C:N ratio of substrate, 130, 204
  counting propagules, 324
  decline, 243–244
  effect of volatiles on, 130
  evolution, 348

nutrients from soil, 205
responds to insecticide, 315–316
saprophytic ability, 171–172
size of sclerotia, 170, 178–179
survival, 164
temperature, 37–38
R. tuliparum, antibiotic in sclerotia, 175
Rhizomorphs, 60–61
Rhizosphere, 105–123
  adaptation of organisms to, 106–107
  antibiotic producers in, 106
  changes with host development, 109–110
  changes with host stress, 111
  concept, 105–107
  importance in bacterial survival, 186–191, 192
  oxygen consumption, 111
  pH, 111
  role in host resistance, 110
  special habitat, 106
  stability, 106
Rhizoplane, 106, 108–109
Root exudates (*see* Rhizosphere)
  beneficial role, 107–108
  freezing roots, effect of, 108
  gradients, 108–109
  hatching factors, 111–113
  host genetics, 110
  iron absorption, effects on, 108
  qualitative differences, 109
  quantity, 107–108
  root age, effect of, 109–110
  toxic, 113
Root systems, asymmetry, **58**
Rootstocks, 308
Rotation:
  associated pathogens, effect on, 304
  crop sequence, 303
  duration, 303–304
  importance of clean seed, 190, **304–305**
  length of, 303–304
  numbers, effect on, 301
  preventive value, 302
Rotylenchus reniformis, 283–284

Salicylic aldehyde, 125
Salinity, 20–21
Saprophytic survival, 144–166
  active, 60, 67–68, 146–147
  aided by antibiotics, 147–149
  effect of commensals upon, 155
  heat and drought, 150–152
  nitrogen, 152–154
  prior colonization of substrate, 144–147
  requirements for, 146–147
  soil pH, 149
  value of prior colonization, 144
  water, 150–152

Saprophytic survival: (*continued*)
  in wood, 154–156
Sclerotia, 168–183
  carpogenic germination, 168–169
  density, 169, 173–175, 176–177
  effect of CO_2, 67
  eruptive germination, 128, 132, 168–169, 177
  exudate from, 173
  formation, 170–171
  leakage by, 174–175
  longevity of, 128
  myceliogenic germination, 168–169
  regulation of water by, 176
  rind, 168–169, 176
  size and function, 170
  sporogenic germination, 168–169
  translocation into, 171–172
  type of germination, 168–196
  weakened sclerotia, 177
Sclerotinia borealis, 22
S. minor, 224–225, 323
S. sclerotiorum, **223**, 306
Sclerotium cepivorum, 128–129, 327, **330**
S. oryzae, buoyant sclerotia, 173
S. rolfsii:
  canopy, effect of upon, 316
  direct assay of sclerotia, 324–325
  effect of pressure upon sclerotia, 16
  eruptive germination, 132, 177, 178, 180
  leakage from sclerotia, 172, 174
  production of sclerotia, 67, 171
  resistance to desiccation, 176
  stimulation of sclerotia by volatiles, 128, 131, 132
  survival of sclerotia, 128, 176
  tillage, 303
  wet sieving of sclerotia, 324
Seed:
  age vs. vigor, 118–119
  anaerobiosis in, 121
  antagonism near seeds, 30, 117
  bruising vs. cracking, 119–120
  clean seed, 190, 304–305
  date of seeding, 49
  in dry soil, 26–27
  effect of temperature, 37–38, 49
  effect of wounds, 119, 120
  exudates, 28, 120–121
  histological complexities, 118
  maturity vs. seed rot, 118
  oxygen response, 28, 121–122
  quantity of leakage, 120–121
  rate of imbibition, 38–39
  rot and damping-off, 117–123
  seeding rate, 306–308
  structure of, 117–118
Sexual reproduction, 337, 341
Siderophores, 108, 214–215, 218
Snow mold, 76–78

Soil:
  aeration, 58–70
  bulk density, 8–9
  cation exchange capacity, 11
  colloids, 9–11
  horizon, 5
  pans, 5–8
  pH, 11–14
  pores, 8, 24–28
  pressure, 16
  profile, 5
  salinity, 20–21
  solarization, 52–53, 69
  structure, 8
  temperature, effect of slope, 4
  texture, 5, 8, 9
  water, 18–20
Soilborne concept, 90
Soils:
  azonal, 2
  pathogen-suppressive, 238–246
  zonal, 1–4
Solarization, 52–53, 69
Solutes vs. volatiles, 125
Southern bacterial wilt (*see Pseudomonas solanacearum*)
Spermosphere, 105, 117–123
  adaptations of seed rotters to the, 30, 117
  age of seed, 118–119
  injured seed, 119–120
  maturity of seed, 118
  oxygen, 121
  special nature of, 30
  temperature, 121–122
Spores, 159–166
  dormant survival, 159
  germinability, 160–161
  germination inhibitors, 161
  requirements for germination, 12–13, 161 (*see* Fungistasis)
Sporidesmium sclerotivorum attacks *Sclerotinia* spp., 224–225
Staling products, 95, 135, 139–140
Stalk rot, 27–28, 74–75, 81–82
Storage fungi, 26–27
Streptomyces spp., 127, 132, 137–139
S. scabies:
  decline, 239–240
  no increase in soil, 94
  water, control by, 30–31
Stress:
  carbohydrate starvation, 76–78
  carrying capacity, 83
  climatic adaptation, 79–81
  complex stress, 81–82
  defoliation affects nematodes, 83
  drought, 27, 41
  fruiting, 74, 80, 81–82
  oxygen, 28, 62–64, 121
  rapid, 191–192

INDEX

temperature, 38, 39, 45–46, 50–51, 75–76
wounds, 38–39
(*see also* Predisposition; Resistance)
Striga sp., 130, 165
Strigol, 130
Substrate:
   active possession of, 67–68, 144–166
   carbon:nitrogen ratio, 204
   condition of, 202–204
   passive possession of, 145
   saprophytic colonization of, 201–205
Super oganisms, 206
Suppressive soils, 238–246
   oat-cyst nematode, 241–243
   potato scab, 239–240
   Rhizoctonia solani, 243–244
   take-all, 240–241
Survival:
   active (saprophytic), 144–166
   of bacteria: in the phyllosphere, 188
      in the rhizosphere, 186, 188
   dormant vs. active, 145
   physiologic adaptations in the life cycle, 159
   role of antibiotics in, 146–151, 221–222
   in soil of: *Erwinia carotovora*, 188–189
      Pseudomonas solancearum, 186–187
      Pseudomonas syringae, 187–188
      Xanthomonas spp., 190
Susceptibility (*see* Resistance)

Take-all (*see Gaeumannomyces graminis*)
Take-all decline, 240–241
Tannic acid, 155
Taxonomy:
   anastomosis groups, 348
   biological species, 347
   cytologic races, 341
   genetic markers, 349
   incompatibility factors, 349–350
   mating groups, 351
   nematodes: groupings of, 340
      problems in, 342
   plasmids, 350–351
   splitting vs. lumping, 352
Teleomorph, 337
Temperature, 37–55
   cortical rots, **41**, 45–46
   disease and growth curves, 38–40
   diurnal fluctuations, 50
   effect of irrigation on soil, 44
   effect of seeding dates, 49
   escape from climatic rigors, 48
   Fusarium wilts, 39–41
   generations of nematodes, 52
   geographic distribution, 37, 46–49
   hatching nematode eggs, 51
   influence of altitude, 48
   predisposition, 38, 45–46
   recovery from Verticillium wilt, 45
   and sequence of pathogens, 49, 54
   sex ratios and nematodes affected by, 51–52
   soil and air temperatures, 43
   and soil microflora, 38
   and soil moisture, 41
   in tests for resistance, 40, 45
   transition zones, 47–48
   vertical distribution of nematodes, 51
   Verticillium wilts, 41–45
Tillage, 299–300
Tilletia caries, 90–92
T. controversa, 91, 92, 306
T. spp., temperature, 49
Tissue maturation, 72–74
Tissue senescence, 74
Tobacco mosaic virus in soil, 255
Tobacco necrosis virus, 255, 270–271
Topography, 4–5
Translocation, 85–86, 171–172
Trapping propagules by host, 84–85
Trichoderma harzianum, 243–244
Trichoderma spp., 177
T. viride:
   antibiotics, 136, 140
   attacks *Armillaria mellea*, 221–222
   carbon dioxide, 66
Trichodorus spp., 257–259, 260
Tylenchulus semipenetrans, 61–62, 198–199
Typhula spp., 76–78, 350

Urocystis agropyri temperature, 49
U. cepulae, temperature, 46
Ustilago maydis 91–92, 230

Variability vs. fitness, 343
Vascular occlusions, 84–85
Vertical distribution:
   of fungal spores, 6–7, 22–23
   of nematodes, 51
Verticillium albo-atrum, temperature, 42
V. dahliae, 41–45
   carbon dioxide, 68–69
   effect of stimulatory volatiles upon, 128
   hyphal growth, 21–22
   general microflora limits, 197–198
   infection index, 323
   mating groups, 351–352
   release of microsclerotia from host debris, 164
   soil solarization, 52–53
   survival structures, 163–164
   temperature, 21–22, 42–45
   volatiles, 128
Verticillum wilts, 41–45

Vertisol, 11
Vesicular-arbuscular mycorrhizae, 250–252
Viruses, 254–272
　big-vein of lettuce, 269–270
　corky ringspot of potato, 260
　free in soil, 255
　fungal vectors, 263–271
　importance of weeds, 262–263
　in fungi, 263–271
　loss of infectivity in nematodes, 262
　nematode-transmitted, 257–263
　raspberry ringspot, 259
　rate of spread, 259–260
　in roots, 255–256
　study of vectors, 256–257
　tobacco necrosis, 270–271
　translocation of, 85–86
　transmitted by fungi, 263–271
Volatiles, 125–132
　C:N ratios of residues, 130
　in fungistasis, 127, 132
　inhibitory, 126–127
　predisposition by, 125
　produced by *Streptomyces* spp., 127
　regulators of mycorrhizae, 127
　stimulants, 127–132
　toxic, 125–126

Water, 18–34
　activity, 19, 21
　bars, 18–19
　capillary, 23–24
　constant wetness, 311
　effects of matric and osmotic water potential, 20
　effects of temperature upon, 21
　expression of soil wetness, 18–19
　gravitational, 22–23
　grouping organisms by water requirements, 25–26
　hatching nematode eggs, 31–32
　hydrophobic and hydrophilic propagules, 22–23
　hygroscopic, 25
　hyphal growth, 21
　interacts with oxygen, 28, 31–32
　matric, 19
　morphogenetic effects, 26, 28–29
　movement in the plant, 34
　movement of propagules by, 22–23, 309–311
　osmotic forces, 19
　oxygen, diffusion in, 24
　potential, 18–20
　quality, 34
　regulation by sclerotia, 176
　sporangium formation and zoospore release, 28–29
　stress within the plant, 27–28
　zoospore movement, 24–25
Wheat root rot, common (*see Bipolaris sorokiniana*)
Wheat soilborne mosaic virus, 85–86, 264–269
White rot (*see Sclerotium rolfsii*)
Witchweed, 130
Wounds differ from stress, 38–39

Xanthamonas campestris survival in soil, 191–192
X. malvacearum, survival in soil, 190
X. musicola, limited by general soil microflora, 198
X. translucens, 190
Xiphinema americanum, 261, 263
X. diversicaudatum, 259–260
X. index, 254, 256, 258
X. vuittenezi, vertical distribution in soil, **260**

Yield compensation, 292

Zonal soils, 1–4
Zygorhynchus vuillemini, carbon dioxide, 66

DAVIDSON COLLEGE